STERLING
Test Prep

SAT

BIOLOGY E/M

Practice Questions

7th edition

www.Sterling-Prep.com

7 6 5 4 3 2 1

ISBN-13: 978-1-9475561-4-0

Sterling Test Prep products are available at special quantity discounts for sales, promotions, premed counseling offices and other educational purposes.

For more information contact our Sales Department at:

Sterling Test Prep
6 Liberty Square #11
Boston, MA 02109

info@sterling-prep.com

Congratulations on choosing this book as part of your SAT Biology preparation!

Scoring well on the SAT Biology is important for admission into college. To achieve a high score, you need to develop skills to properly apply the knowledge you have and quickly choose the correct answer. You must solve numerous practice questions that represent the style and content of the SAT questions. Understanding key science concepts is more valuable than memorizing terms.

This book provides 1,567 biology practice questions that test your knowledge of all SAT Biology topics, both E and M formats. In the second part of the book, you will find answer keys and detailed explanations to questions, except those that are self explanatory. These explanations discuss why the answer is correct and – more importantly – why another answer that may have seemed correct is the wrong choice. The explanations include the foundations and details of important science topics needed to answer related questions on the SAT. By reading these explanations carefully and understanding how they apply to solving the question, you will learn important biology concepts and the relationships between them. This will prepare you for the SAT Biology E/M test and will significantly improve your score.

All the questions are prepared by our science editors who possess extensive credentials and are educated in top colleges and universities. Our editors are experts on teaching sciences, preparing students for standardized science tests and have coached thousands of undergraduate and graduate school applicants on admission strategies.

We wish you great success in your future academic achievements and look forward to being an important part of your successful preparation for SAT Biology!

Sterling Test Prep Team

180526gdx

Visit www.Sterling-Prep.com for SAT online practice tests

Our advanced online testing platform allows you to practice these and other SAT questions on the computer and generate Diagnostic Reports for each test.

By using our online SAT tests and Diagnostic Reports, you will be able to:

- Assess your knowledge of different topics tested on your SAT subject test

- Identify your areas of strength and weakness

- Learn important scientific topics and concepts

- Improve your test taking skills by solving numerous practice questions

To access the online tests at a special pricing
go to page 565 for web address

This book should be supplemented by "SAT Biology Review" book or online practice material at www.Sterling-Prep.com

Table of Contents

Table of Contents (*Continued*)

Table of Contents (*Continued*)

Our Commitment to the Environment

Sterling Test Prep is committed to protecting our planet's resources by supporting environmental organizations with proven track records of conservation, environmental research and education and preservation of vital natural resources. A portion of our profits is donated to support these organizations so they can continue their important missions. These organizations include:

 For over 40 years, Ocean Conservancy has been advocating for a healthy ocean by supporting sustainable solutions based on science and cleanup efforts. Among many environmental achievements, Ocean Conservancy laid the groundwork for an international moratorium on commercial whaling, played an instrumental role in protecting fur seals from overhunting and banning the international trade of sea turtles. The organization created national marine sanctuaries and served as the lead non-governmental organization in the designation of 10 of the 13 marine sanctuaries.

 For 25 years, Rainforest Trust has been saving critical lands for conservation through land purchases and protected area designations. Rainforest Trust has played a central role in the creation of 73 new protected areas in 17 countries, including Falkland Islands, Costa Rica and Peru. Nearly 8 million acres have been saved thanks to Rainforest Trust's support of in-country partners across Latin America, with over 500,000 acres of critical lands purchased outright for reserves.

 Since 1980, Pacific Whale Foundation has been saving whales from extinction and protecting our oceans through science and advocacy. As an international organization, with ongoing research projects in Hawaii, Australia and Ecuador, PWF is an active participant in global efforts to address threats to whales and other marine life. A pioneer in non-invasive whale research, PWF was an early leader in educating the public, from a scientific perspective, about whales and the need for ocean conservation.

Thank you for choosing our products to achieve your educational goals.

With your purchase you support environmental causes around the world.

Biology
Practice Questions

UNIT 1. CELLULAR AND MOLECULAR BIOLOGY

Chapter 1.1: Eukaryotic Cell: Structure and Function

1. Facilitated transport can be differentiated from active transport, because:
   └ still passive

   A. active transport requires a symport
   B. facilitated transport displays saturation kinetics
   C. active transport displays sigmoidal kinetics
   D. active transport requires an energy source
   E. active transport only occurs in the mitochondrial inner membrane

2. The cell is the basic unit of function and reproduction, because:

   A. subcellular components cannot regenerate whole cells
   B. cells can move in space
   C. single cells can sometimes produce an entire organism
   D. cells can transform energy to do work
   E. a new cell can arise by the fusion of two cells

3. Which of the following is/are NOT able to readily diffuse through the plasma membrane of a cell without the aid of a transport protein?

   I. water
   II. small hydrophobic molecules
   III. small ions
   IV. neutral gas molecules

   A. I only   B. I and II only   C. I and IV only   D. III and IV only   E. I and III only

4. During a hydropathy analysis of a protein that has recently been sequenced, a researcher discovers that the protein has several regions that contain 20-25 hydrophobic amino acids. What conclusion would she draw from this finding?

   A. Protein would be specifically localized in the mitochondrial inner membrane
   B. Protein would be targeted to the mitochondrion
   C. Protein is likely to be an integral protein
   D. Protein is probably involved in glycolysis
   E. Protein is likely to be secreted from the cell

5. If a membrane bound vesicle that contains hydrolytic enzymes is isolated, most likely it is a:

   A. vacuole   B. microbody   C. chloroplast   D. phagosome   E. lysosome
   degrade stuff

6. A DNA damage checkpoint arrests cells in:

A. M/G2 transition

C. G1/S transition

B. G1/G2 transition

D. anaphase

E. S/G1 transition

7. If a segment of a double stranded DNA has a low ratio of guanine-cytosine (G-C) pairs relative to adenine-thymine (A-T) pairs, it is reasonable to assume that this nucleotide segment:

A. requires less energy to separate the two DNA strands than a comparable segment with a high C-G ratio

B. requires more energy to separate the two DNA strands than a comparable segment with a high C-G ratio

C. requires the same energy to separate the two DNA strands as a comparable segment with a high C-G ratio

D. contains more adenine than thymine

E. contains more cytosine than guanine

8. In general, phospholipids contain:

A. a glycerol molecule

C. unsaturated fatty acids

B. saturated fatty acids

D. a cholesterol molecule

E. a glucose molecule

9. The overall shape of a cell is determined by its:

A. cell membrane

C. nucleus

B. cytoskeleton

D. cytosol

E. endoplasmic reticulum

10. Which molecule generates the greatest osmotic pressure when placed into water?

A. 300 mM NaCl

C. 500 mM glucose

B. 250 mM CaCl$_2$

D. 600 mM urea

E. 100 mM KCl

11. Which cellular substituent is produced within the nucleus?

A. Golgi apparatus

B. lysosome

C. ribosome

D. rough endoplasmic reticulum

E. cell membrane

12. In the early stages of the cell cycle, progression from one phase to the next is controlled by:

A. the p53 transcription factor

B. anaphase promoting complexes

C. origin recognition complexes

D. the pre-replication complex

E. cyclin–CDK complexes

13. Which of the following is the correct sequence occurring during polypeptide synthesis?

A. DNA generates tRNA → tRNA anticodon binds to the mRNA codon in the cytoplasm → tRNA is carried by mRNA to the ribosomes, causing amino acids to join together in a specific order

B. DNA generates mRNA → mRNA moves to the ribosome → tRNA anticodon binds to the mRNA codon, causing amino acids to join together in their appropriate order

C. Specific RNA codons cause amino acids to line up in a specific order → tRNA anticodon attaches to mRNA codon → rRNA codon causes protein to cleave into specific amino acids

D. DNA regenerates mRNA in the nucleus → mRNA moves to the cytoplasm and attaches to the tRNA anticodon → operon regulates the sequence of amino acids in the appropriate order

E. DNA generates mRNA → mRNA anticodon binds to tRNA codon causing amino acids to join together in their appropriate order

14. Both prokaryotes and eukaryotes contain:

    I. a plasma membrane   II. ribosomes   III. peroxisomes

A. I only   C. I and II only

B. II only   D. II and III only   E. I, II and III

15. RNA is NOT expected to be found in which one of the following structures?

A. nucleus   C. prokaryotic cell

B. mitochondrion   D. ribosome   E. vacuole

16. Phosphotransferase is needed to form the mannose-6-phosphate tag that targets hydrolase enzymes to their lysosomal destination. Defective phosphotransferase causes I-cell disease, whereby the defective organelle which gives rise to this condition is the:

A. nucleus   C. Golgi apparatus

B. cell membrane   D. smooth ER   E. nucleolus

17. Mitochondria and chloroplasts are unusual organelles because they:

A. synthesize all of their ATP using substrate-level phosphorylation

B. contain cytochrome C oxidase

C. are totally devoid of heme-containing proteins

D. contain nuclear-encoded and organelle-encoded proteins

E. degrade macromolecules using hydrolytic enzymes

18. All of the following statements are true about cytoskeleton, EXCEPT that it:

A. is not required for mitosis   C. gives the cell mechanical support

B. maintains the cell's shape   D. is composed of microtubules and microfilaments

                                      E. is important for cell motility

19. Inside the cell, several key events in the cell cycle include:

    I. DNA damage repair and replication completion
    II. centrosome duplication
    III. assembly of the spindle and attachment of the kinetochores to the spindle

  A. I only
  B. I, II and III

  C. I and II only
  D. II and III only
  E. I and III only

20. A researcher labeled *Neurospora* mitochondria with a radioactive phosphatidylcholine membrane component and followed cell division by autoradiography in an unlabeled medium, allowing enough time for one cell division. What results did this scientist have to observe before concluding that pre-existing mitochondria give rise to new mitochondria?

  A. Daughter mitochondria are labeled equally
  B. Daughter mitochondria are all unlabeled
  C. One fourth of daughter mitochondria are labeled
  D. Some of the daughter mitochondria are unlabeled, while some are labeled
  E. Daughter mitochondria are all labeled

21. Which of the following involves the post-translational import of proteins?

  A. peroxisomes
  B. vacuoles

  C. Golgi complex
  D. lysosomes
  E. smooth ER

22. The width of a typical animal cell is closest to:

  A. 1 millimeter
  B. 20 micrometers

  C. 1 micrometer
  D. 10 nanometers
  E. 100 micrometers

23. Which organelle is identified by the sedimentation coefficient – S units (Svedberg units)?

  A. peroxisome
  B. nucleus

  C. mitochondrion
  D. nucleolus
  E. ribosome

24. Which of the following is NOT involved in osmosis?

  A. H_2O spontaneously moves from a hypertonic to a hypotonic environment
  B. H_2O spontaneously moves from an area of high solvent to low solvent concentration
  C. H_2O spontaneously moves from a hypotonic to a hypertonic environment
  D. Transport of H_2O
  E. Diffusion of H_2O

25. During cell division, cyclin B is marked for destruction by the:

   A. p53 transcription factor

   B. anaphase promoting complex

   C. CDK complex

   D. pre-replication complex

   E. maturation promoting factor

26. When a female mouse with a defect in mitochondrial protein required for fatty acid oxidation is crossed with a wild-type male, all progeny (both male and female) have the wild-type phenotype. Which statement is most likely correct?

   A. Mice do not exhibit maternal inheritance

   B. The defect is a result of an autosomal X-linked recessive trait

   C. The defect is a result of an X-linked recessive trait

   D. The defect is a result of a recessive mitochondrial gene

   E. The defect is a result of a nuclear gene mutation

27. The smooth ER is involved in:

   A. substrate-level phosphorylation

   B. exocytosis

   C. synthesis of phosphatidylcholine

   D. allosteric activation of enzymes

   E. synthesis of cytosolic proteins

28. What type of organelle is found in plants but not in animals?

   A. ribosomes   **B.** mitochondria   **C.** nucleus   **D.** plastids   **E.** nucleolus

29. Mitochondrial mutations are often limited to one tissue type. If an individual doesn't produce blood calcium decreasing hormone calcitonin, which tissue is most likely to carry the mitochondrial mutation?

   A. thyroid   **B.** kidney   **C.** parathyroid   **D.** liver   **E.** spleen

30. The rough ER is involved in:

   A. oxidative phosphorylation

   B. synthesis of plasma membrane proteins

   C. endocytosis

   D. post-translational modification of enzymes

   E. synthesis of lysosomal proteins

31. The best definition of active transport is the movement of:

   A. solutes across a semipermeable membrane down an electrochemical gradient

   B. solutes across a semipermeable membrane up a concentration gradient

   C. substances across a membrane in accordance with the Donnan equilibrium

   D. solutes via osmosis across a semipermeable membrane from high to low concentration

   E. H_2O via diffusion across a semipermeable membrane

32. Overexpression of cyclin D:

    A. increases contact inhibition
    B. decreases telomerase activity

    C. activates apoptosis
    D. promotes unscheduled entry into S phase
    E. promotes transition from G1 to S phase

33. A failure in which stage of spermatogenesis produces nondisjunction that results in a male having a XXY karyotype?

    A. prophase I
    B. metaphase

    C. prophase II
    D. telophase

    E. anaphase I

34. Protein targeting occurs during the synthesis of which type of proteins?

    A. nuclear proteins
    B. secreted proteins

    C. cytosolic proteins
    D. mitochondrial proteins
    E. chloroplast proteins

35. The difference between "free" and "attached" ribosomes is that:

    I. Free ribosomes are in the cytoplasm, while attached ribosomes are anchored to the endoplasmic reticulum

    II. Free ribosomes produce proteins in the cytosol, while attached ribosomes produce proteins that are inserted into the ER lumen

    III. Free ribosomes produce proteins that are exported from the cell, while attached ribosomes make proteins for mitochondria and chloroplasts

    A. I only
    B. II only

    C. I and II only
    D. I, II and III

    E. I and III only

36. The concentration of growth hormone receptors will significantly reduce after selective destruction of which structure?

    A. nucleolus
    B. nucleus

    C. cytosol
    D. plasma membrane

    E. peroxisomes

37. All of these processes are ATP-dependent, EXCEPT:

    A. export of Na⁺ from a neuron
    B. influx of Ca²⁺ into a muscle cell
    C. influx of K⁺ into a neuron
    D. exocytosis of neurotransmitter at a nerve terminus
    E. movement of urea across a cell membrane

38. The loss of function of p53 protein results in:

A. blockage in activation of the anaphase promoting complex
B. increase of contact inhibition
C. elimination of the DNA damage checkpoint
D. activation of apoptosis
E. suppression of spindle fiber assembly

39. Which of the following organelles is most closely associated with exocytosis of newly synthesized secretory protein?

A. peroxisome
B. ribosome
C. lysosome
D. Golgi apparatus
E. nucleus

40. Plant membranes are more fluid than animal membranes, because plant membranes:

A. contain large amounts of cholesterol
B. have higher amounts of unsaturated fatty acids compared to the membranes of animals
C. have lower amounts of unsaturated fatty acids compared to the membranes of animals
D. are only found in the inner membrane of the mitochondria
E. do not contain glycosylated proteins

41. Which of the following organelles are enclosed in a double membrane?

I. nucleus   II. chloroplast   III. mitochondrion

A. I and II only
B. II and III only
C. I and III only
D. I, II and III
E. I only

42. Proteins are marked and delivered to specific cell locations through:

A. specific protein transport channels
B. regulation signals released by the cell's cytoskeleton
C. post-translational modifications occurring in the Golgi
D. compartmentalization of the rough ER during protein synthesis
E. post-translational modifications occurring in the nucleus

43. The presence of which element differentiates a protein from a carbohydrate molecule?

A. carbon
B. hydrogen
C. nitrogen
D. oxygen
E. nitrogen and oxygen

44. What change occurs in the capillaries when arterial blood is infused with the plasma protein albumin?

    A. Decreased movement of H_2O from the capillaries into the interstitial fluid
    B. Increased movement of H_2O from the capillaries into the interstitial fluid
    C. Decreased movement of H_2O from the interstitial fluid into the capillaries
    D. Increased permeability to albumin
    E. Decreased permeability to albumin

45. The retinoblastoma protein controls:

    A. contact inhibition   pRb   **C.** activation of apoptosis
    B. transition from G1 to S phase   **D.** expression of cyclin D
                                            E. A and C

46. All of the following processes take place in the mitochondrion, EXCEPT:

    A. oxidation of pyruvate   **C.** electron transport chain
    B. Krebs cycle   **D.** reduction of FADH into $FADH_2$
                                           E. glycolysis

47. Which of the following is the most abundant lipid in the body?

    A. teichoic acid   **C.** glycogen
    B. peptidoglycan   **D.** triglycerides   **E.** glucose

L lipid RMS

48. What is the secretory sequence in the flow of newly-synthesized protein for export from the cell?

    A. Golgi → rough ER → smooth ER → plasma membrane
    B. Golgi → rough ER → plasma membrane
    C. smooth ER → rough ER → Golgi → plasma membrane
    D. rough ER → smooth ER → Golgi → plasma membrane
    E. rough ER → Golgi → plasma membrane

49. Digestive lysosomal hydrolysis would affect all of the following, EXCEPT:

    A. proteins   **B.** minerals   **C.** nucleotides   **D.** lipids   **E.** carbohydrates

50. Recycling of organelles within the cell is accomplished through autophagy by:

    A. mitochondria   **C.** nucleolus
    B. peroxisomes   **D.** lysosomes
                                           E. rough endoplasmic reticulum

51. All of the following are lipid derivatives, EXCEPT:

A. carotenoids   **B.** albumins   **C.** waxes   **D.** steroids   **E.** lecithin

52. Defective attachment of a chromosome to the spindle:

A. blocks activation of the anaphase promoting complex
B. activates exit from mitosis
C. prevents overexpression of cyclin D
D. activates sister chromatid separation
E. promotes transition from G1 to S phase

53. Which of the following stages is when human cells with a single unreplicated copy of the genome are formed?

A. mitosis   **B.** meiosis II   **C.** meiosis I   **D.** interphase   **E.** G1

54. Which organelle in the cell is the site of fatty acid, phospholipid and steroid synthesis?

A. endosome
B. peroxisome
C. rough endoplasmic reticulum
D. smooth endoplasmic reticulum
E. chloroplast

55. Which eukaryotic organelle is NOT membrane-bound?

A. nucleus   **B.** plastid   **C.** centriole   **D.** chloroplast   **E.** C and D

56. During which phase of mitotic division do spindle fibers split the centromere and separate the sister chromatids?

A. interphase   **B.** telophase   **C.** prophase   **D.** metaphase   **E.** anaphase

57. All of the following are correct about cyclic AMP (cAMP), EXCEPT:

A. the enzyme that catalyzes the formation of cAMP is located in the cytoplasm
B. membrane receptors are capable of activating the enzyme that forms cAMP
C. ATP is the precursor molecule in the formation of cAMP
D. adenylate cyclase is the enzyme that catalyzes the formation of cAMP
E. cAMP is a second messenger that triggers a cascade of intracellular reactions when a peptide hormone binds to the receptor on the cell membrane

58. Which molecule, and its associated protein kinase, ensures a proper progression of cell division?

A. oncogene
B. tumor suppressor
C. cyclin
D. histone
E. homeotic

---

59. Placed in a hypertonic solution, erythrocytes will undergo:

A. crenation → cell shriveling
B. expansion

C. plasmolysis and rupture
D. no change
E. shrinkage and then rapid expansion

60. Tiny organelles, abundant in the liver, that contain oxidases and detoxify substances (e.g. alcohol and hydrogen peroxide), are:

A. centrosomes
B. lysosomes

C. endosomes
D. rough endoplasmic reticulum
E. peroxisomes

difference: regular digestion vs. toxic substances

61. Microtubules can function independently or form protein complexes to produce structures like:

   I. flagella
   II. actin and myosin filaments in muscle cells → microfilaments
   III. mitotic spindle apparatus

A. III only
B. I and III only

C. II and III only
D. I and II only
E. I, II, and III

62. About a cell that secretes a lot of protein (e.g. a pancreatic exocrine cell), it can be assumed that this cell has:

   I. an abundance of rough endoplasmic reticulum
   II. a large Golgi apparatus
   III. a prominent nucleoli

A. I only
B. I and III only

C. II and III only
D. I, II and III
E. I and II only

63. A cell division where each of the two daughter cells receives a chromosome complement identical to that of a parent is:

A. mitosis
B. non-disjunction

C. meiosis
D. replication
E. crossing-over

64. Which of the following would be affected the LEAST by colchicines, which are known to interfere with microtubule formation?

A. organelle movement
B. meiosis

C. mitosis
D. cilia
E. pseudopodia for amoeboid motility

65. Which of the following processes is NOT an example of apoptosis?

A. Reabsorption of a tadpole's tail during metamorphosis into a frog
B. Formation of the endometrial lining of the uterus during the menstrual cycle
C. Formation of the synaptic cleft by triggering cell death in brain neuronal cells
D. Formation of fingers in the fetus by the removal of tissue between the digits
E. All are examples of apoptosis

66. Cells that utilize large quantities of ATP for active transport (e.g. epithelial cells of the intestine) have:

A. many mitochondria
B. high levels of DNA synthesis
C. high levels of adenylate cyclase
D. polyribosomes
E. many lysosomes

67. Clathrin is a protein that is collected on the cytoplasmic side of cell membranes and functions in the coordinated pinching off of membrane into receptor-mediated endocytosis. It can be predicted that a lipid-soluble toxin that inactivates clathrin results in:

A. increased ATP consumption
B. increased protein production on the rough endoplasmic reticulum
C. increased secretion of hormone into the extracellular fluid
D. increased ATP production
E. reduced delivery of polypeptide hormones to endosomes

68. Cells that respond to peptide hormones usually do so through a sequence of biochemical reactions involving membrane receptors and kinase activation. In order for cells to respond, the first and second messengers communicate, because:

A. peptide hormones pass through the cell membrane and elicit a response
B. the hormone-receptor complex moves into the cytoplasm as a unit
C. hormones alter cellular activities directly through gene expression
D. the G protein acts as a link between the first and second messengers
E. the presence of a cytosolic receptor binds the hormone

└ intermediate

Answer Key

1: D	11: C	21: C	31: B	41: D	51: B	61: B
2: A	12: E	22: B	32: D	42: C	52: A	62: D
3: E	13: B	23: E	33: E	43: C	53: B	63: A
4: C	14: C	24: A	34: B	44: A	54: D	64: E
5: E	15: E	25: B	35: C	45: B	55: C	65: B
6: C	16: C	26: E	36: D	46: E	56: E	66: A
7: A	17: D	27: C	37: E	47: D	57: A	67: E
8: A	18: A	28: D	38: C	48: E	58: C	68: D
9: B	19: B	29: A	39: D	49: B	59: A	
10: B	20: A	30: B	40: B	50: D	60: E	

Chapter 1.2: Molecular Biology of Eukaryotes

1. Which primers should be used to amplify the DNA shown in the figure below via PCR?

   A. 5'-CCCC-3' and 5'-AAAA-3'
   B. 5'-GGGG-3' and 5'-TTTT-3'
   C. 5'-AAAA-3' and 5'-GGGG-3'
   D. 5'-TTTT-3' and 5'-CCCC-3'
   E. none of the above

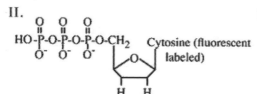

2. The major phenotypic expression of a genotype is in:

   A. rRNA   B. tRNA   C. mRNA   D. nucleic acids   E. proteins

3. Which molecule is used in DNA sequencing to cause termination when the template strand is G?

   I.

   HO-P-O-P-O-P-O-CH₂   Guanine (fluorescent labeled)

   III.

   HO-P-O-P-O-P-O-CH₂   Cytosine (fluorescent labeled)

   II.

   HO-P-O-P-O-P-O-CH₂   Cytosine (fluorescent labeled)

   IV.

   HO-P-O-CH₂   Cytosine (fluorescent labeled)

   A. molecule I & III   C. molecule II
   B. molecule III & IV   D. molecule IV   E. molecule III

4. Which of the following is NOT a part of post-translational modification of protein?

   A. addition of a 3' poly-A tail   C. methylation
   B. phosphorylation   D. glycosylation
   E. acetylation

5. A chromosome with its centromere in the middle is:

   A. acrocentric   C. metacentric
   B. telocentric   D. holocentric
   E. midcentromeric

6. Which of the following is NOT an example of an environmental factor that affects the way a gene is expressed?

    A. Heat shock proteins are synthesized in cells after a temperature increase

    B. *Drosophila* with specific genes develop bent wings when incubated at low temperatures and straight wings when incubated at high temperatures

    C. Himalayan hares change hair color after cooling of the naturally warm regions

    D. Shivering occurs after a decrease in body temperature

    E. All of the above

7. cDNA libraries contain:

    A. promoters

    B. intron portions of expressed genes

    C. exon portions of expressed genes

    D. non-expressed retrotransposons

    E. both introns and exons of expressed genes

8. Individual genes often encode for:

    I. enzymes with tertiary structure

    II. enzymes with quaternary structure

    III. complex polysaccharides

    A. I only

    B. II only

    C. I and II only

    D. I, II and III

    E. I and III only

9. What enzyme is often used to make a genomic library?

    A. RNA polymerase

    B. reverse transcriptase

    C. deoxyribonuclease

    D. DNA polymerase

    E. restriction endonuclease

10. Attachment of glycoprotein side chains and amino acid hydroxylation are part of post-translational modification. What is most likely the site for protein glycosylation?

    I. Lysosomes

    II. Golgi apparatus

    III. Rough endoplasmic reticulum

    A. II only   **B.** III only   **C.** I and II only   **D.** II and III only   **E.** I, II and III

11. The enzyme used for restoring the ends of the DNA in a chromosome is:

    A. telomerase   **B.** helicase   **C.** polymerase   **D.** gyrase   **E.** ligase

12. A cDNA library is made using:

    A. DNA from the region where the gene of interest is expressed

    B. mRNA from the region where the gene of interest is expressed

    C. mRNA from the region where the gene of interest is not expressed

    D. rRNA from the region where the gene of interest is expressed

    E. all of the above

13. Which of the following statements is called the *central dogma* of molecular biology?

  A. Information flow between DNA, RNA and protein is reversible

  B. Information flow in the cell is unidirectional, from protein to RNA to DNA

  C. The genetic code is ambiguous but not degenerate

  D. The DNA sequence of a gene can be predicted from the amino acid sequence of the protein

  E. Information flow in the cell is unidirectional, from DNA to RNA to protein

14. What is alternative splicing?

  A. Cleavage of peptide bonds to create different proteins

  B. Cleavage of DNA to make different genes

  C. Cleavage of hnRNA to make different mRNAs

  D. New method of splicing that does not involve snRNPs

  E. None of the above

15. This compound is synthesized by the nucleolus and is necessary for ribosomal function.

  A. ribozyme   **B.** riboflavin   **C.** liposome   **D.** rRNA   **E.** tRNA

16. *E. coli* RNA polymerase:

  I. synthesizes RNA in the 5' to 3' direction   III. copies a DNA template

  II. synthesizes RNA in the 3' to 5' direction   IV. copies an RNA template

  A. I and III only   **C.** I and IV only

  B. II and III only   **D.** II and IV only   **E.** I only

17. Within primary eukaryotic transcripts, introns are:

  A. often functioning as exons in other genes

  B. considerably different in size and number among different genes

  C. joined to form mature mRNA

  D. highly conserved in nucleotide sequence

  E. absent from the primary eukaryotic transcript because they are not transcribed

18. Which statement(s) is/are TRUE with respect to eukaryotic protein synthesis?

  I. exons of mRNA are spliced together before translation

  II. proteins must be spliced soon after translation

  III. prokaryotic ribosomes are smaller than eukaryotic ribosomes

  A. I only   **C.** I and II only

  B. II only   **D.** I, II and III only   **E.** I and III only

19. Which of the following would decrease the transcription of retrotransposons?

    I. Acetylation of histones associated with the retrotransposon

    II. Deacetylation of histones associated with the retrotransposon

    III. Methylation of retrotransposon DNA

    IV. Loss of methylation of retrotransposon DNA

A. I and II only   **C.** II and III only

B. I and III only   **D.** III and IV only   **E.** II and IV only

20. Which of the following is an exception to the principle of the *central dogma*?

A. yeast   **C.** bread mold

B. retroviruses   **D.** skin cells   **E.** onion cells

21. miRNA is generated from the cleavage of:

A. double stranded RNA   **C.** double stranded DNA

B. single stranded RNA   **D.** single stranded DNA   **E.** None of the above

22. DNA polymerase cannot fully replicate the 3' DNA end, which results in shorter DNA with every division cycle. The new strand synthesis mechanism that follows prevents the loss of the DNA coding region.

Which of the following structures is present at the location of the new strand synthesis?

A. kinetochore

B. centrosome

C. telomere

D. centromere

E. chromatid

23. The cell structure composed of a core particle of 8 histones is:

A. telomere   **C.** kinetochord

B. nucleosome   **D.** centrosome   **E.** spindle

24. Which statement is CORRECT about miRNA? (interference RNA):

    I. It is a backup system for tRNA in the regulation of translation

    II. It base pairs with mRNA and causes it to be cleaved

    III. It base pairs with mRNA and prevents its translation

A. II only   **C.** II and III only

B. I and II only   **D.** I, II and III   **E.** I and III only

25. The region of DNA in prokaryotes to which RNA polymerase binds most tightly is the:

A. promoter

B. poly C center

C. enhancer

D. operator site

E. minor groove

26. Genomic libraries contain:

I. promoters

II. intron portions of genes

III. exon portions of genes

IV. retrotransposons

A. I and II only

B. I and III only

C. I, II and III only

D. II and III only

E. I, II, III and IV

27. Which of these RNA molecules has a common secondary structure called cloverleaf and is relatively small?

A. hnRNA   **B.** mRNA   **C.** rRNA   **D.** tRNA   **E.** miRNA

28. A chromosome whereby the linear sequence of a group of genes is the reverse of the normal sequence has undergone:

A. translocation

B. inversion

C. duplication

D. deletion

E. position effect variegation

29. If a drug inhibits ribosomal RNA synthesis, which of these eukaryotic organelles would be most affected by this drug?

A. Golgi apparatus

B. lysosome

C. mitochondria

D. nucleus

E. nucleolus

30. What evidence shows that the AG gene is important for the formation of reproductive organs in *Arabidopsis* flowers?

A. The AG gene encodes a miRNA

B. RNA blot experiments show that the AG gene is strongly expressed in flowers, leaves and roots

C. The AG gene is in all flowering Arabidopsis plants

D. AG mutant flowers do not have reproductive organs

E. None of the above

31. Which of the following statements about the glycocalyx is FALSE?

A. May be composed of polysaccharide

B. May be composed of polypeptide

C. Protects from osmotic lysis

D. Is used to adhere to surfaces

E. May be responsible for virulence

32. Which of the following does NOT affect chromatin structure?

A. tandem repeats

B. DNA acetylation

C. histone acetylation

D. chromatin remodeling proteins

E. DNA methylation

33. Which enzyme maintains and regulates normal DNA coiling?

A. ligase

B. helicase

C. DNA polymerase I

D. DNA polymerase III

E. topoisomerase

34. RNA polymerase uses the two ribonucleotide triphosphates shown below to make 5'-CG-3'. Which of the indicated phosphorous atoms participates in phosphodiester bond formation?

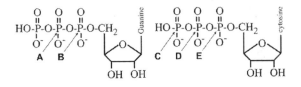

A. phosphorous atom A

B. phosphorous atom B

C. phosphorous atom C

D. phosphorous atom D

E. phosphorous atom E

35. When eukaryotic mRNA hybridizes with its corresponding DNA coding strand (i.e. heteroduplex analysis) and is visualized by electron microscopy, the looping strands of nucleic acid which are seen represent:

A. introns

B. exons

C. lariat structures

D. inverted repeats

E. overlapping genes

36. Which statement, if any, does NOT accurately describe an aspect of the nucleosome?

A. has an octet of proteins

B. has histone H1

C. is the first step in compacting the DNA in the nucleus

D. has DNA wrapped on the outside

E. has histone H3

37. Which of the following is NOT a chemical component of a bacterial cell wall?

A. N-acetylmuramic acid

B. peptidoglycan

C. teichoic acids

D. peptide chains

E. cellulose

38. Combinatorial control of gene transcription in eukaryotes is when:

    I. each transcription factor regulates only one gene

    II. a single transcription factor regulates a combination of genes

    III. presence or absence of a combination of transcription factors is required

A. I only

B. II only

C. II and III only

D. III only

E. I, II and III

39. Which of these post-transcriptional modifications has a mature eukaryotic mRNA undergone before being transported into the cytoplasm?

   A. addition of 3' G-cap and 5' poly-A-tail, removal of introns and splicing of exons

   B. RNA splicing together of exons and removal of introns

   C. RNA addition of 5' cap and 3' poly-A-tail

   D. addition of 5' G-cap and 3' poly-A-tail, removal of introns and splicing of exons

   E. RNA splicing together of introns and removal of exons

40. Which statement about gene expression is correct?

   A. The ribosome binding site lies at the 3' end of mRNA

   B. A second round of transcription can begin before the preceding transcript is completed

   C. Only one gene can be present within a given DNA sequence

   D. Mistakes in transcription are corrected by RNA polymerase

   E. Change in genotype always results in a changed phenotype

41. Considering that in vitro, the transcription factor SP1 binds nucleic acids with high affinity, where would the radio-labeled SP1 most likely NOT be found?

   A. Golgi apparatus   **C.** nucleolus

   B. mitochondria   **D.** ribosomes   **E.** nucleus

42. During splicing, snRNA base pairs with:

   A. mRNA sequences in the intron   **C.** DNA sequences in the intron

   B. hnRNA sequences in the exon   **D.** DNA sequences in the exon

                                     E. hnRNA sequences in the intron

43. Which of these macromolecules would be repaired rather than degraded?

   A. triglyceride   **C.** polypeptide

   B. polynucleotide   **D.** polysaccharide   **E.** proteins

44. The poly-A tail of RNA is:

   A. encoded in the DNA sequence of the gene

   B. added by the ribosome during translation

   C. base paired with tRNA during translation initiation

   D. enzymatically added soon after transcription is finished

   E. located in the cytoplasm of the cell

45. Which type of histone is not part of the nucleosome core particle?

   A. H3   **B.** H2B   **C.** H2A   **D.** H4   **E.** H1

46. During splicing, the phosphodiester bond at the upstream exon/intron boundary is hydrolyzed by:

   A. protein within the snRNP complex

   B. 2'-OH of a base within the intron

   C. 3'-OH of a base within the intron

   D. RNA polymerase III

   E. 3'-OH of a base within the exon

47. What is the expected charge, if any, on a histone that binds to DNA?

   A. neutral

   B. depends on the DNA conformation

   C. positive

   D. negative

   E. neutral or negative

48. RNA polymerase uses the two ribonucleotide triphosphates shown below to make 5'-CG-3'. Which of the indicated oxygen atoms participates in phosphodiester bond formation?

   A. oxygen atom A

   B. oxygen atom B

   C. oxygen atom C

   D. oxygen atom D

   E. either oxygen atom B or D

49. Chromosome regions with very few functional genes are:

   A. heterochromatin

   B. mid-repetitive sequences

   C. euchromatin

   D. chromatids

   E. nucleosomes

50. Which of the following is characteristic of prokaryotes only?

   A. primary transcripts of RNA have introns

   B. the processed RNA has a polyA

   C. the processed RNA has a 5'-cap

   D. transcription of the RNA occurs simultaneously with translation for the RNA

   E. the primary transcript is longer than the mRNA

Answer Key

1: A	11: A	21: A	31: C	41: A
2: E	12: B	22: C	32: A	42: E
3: C	13: E	23: B	33: E	43: B
4: A	14: C	24: C	34: B	44: D
5: C	15: D	25: A	35: A	45: E
6: D	16: A	26: E	36: B	46: B
7: C	17: B	27: D	37: E	47: C
8: A	18: E	28: B	38: D	48: C
9: E	19: C	29: E	39: D	49: A
10: D	20: B	30: D	40: B	50: D

Chapter 1.3: Cellular Metabolism and Enzymes

1. The atom responsible for generating a hydrogen bond that helps to stabilize the α-helical configuration of a polypeptide is:

    A. peptide bond atom
    B. atom found in the R-groups
    C. hydrogen of the carbonyl oxygen
    D. hydrogen of the amino nitrogen
    E. two of the above

2. The ATP molecule contains three phosphate groups, two of which are:

    A. bound as phosphoanhydrides
    B. bound to adenosine
    C. never hydrolyzed from the molecule
    D. cleaved off during most biochemical reactions
    E. equivalent in energy for the hydrolysis of each of the phosphates

3. Which attractive force is used by the side chains of nonpolar amino acids to interact with other nonpolar amino acids?

    A. ionic bonds   **C.** hydrophobic interaction
    B. hydrogen bonds   **D.** disulfide bonds   **E.** dipole-dipole

4. Fermentation yields less energy than aerobic respiration because:

    A. it requires a greater expenditure of cellular energy
    B. glucose molecules are not completely oxidized
    D. oxaloacetic acid serves as the final H^+ acceptor
    C. it requires more time for ATP production
    E. it occurs in H_2O

5. How would the reaction kinetics of an enzyme and its substrate change if an anti-substrate antibody is added?

    A. The antibody binds to the substrate, which increases the V_{max}
    B. The antibody binds to the substrate, which decreases K_m
    C. No change because K_m and V_{max} are independent of antibody concentration
    D. The antibody binds the substrate, which decreases V_{max}
    E. The antibody binds to the substrate, which increases K_m

6. Metabolism is:

    A. consumption of energy

    B. release of energy

    C. all conversions of matter and energy taking place in an organism

    D. production of heat by chemical reactions

    E. exchange of nutrients and waste products with the environment

7. During alcoholic fermentation, all of the following occurs, EXCEPT:

    A. release of CO_2

    B. oxidation of glyceraldehyde-3-phosphate

    C. oxygen is not consumed in the reaction

    D. ATP synthesis as a result of oxidative phosphorylation

    E. NADH is produced

8. When determining a protein's amino acid sequence, acid hydrolysis causes a partial destruction of tryptophan, conversion of asparagine into aspartic acid and conversion of glutamine into glutamic acid. Which of the following statements is NOT correct?

    A. Glutamine concentration is related to the level of aspartic acid

    B. Glutamic acid levels are an indirect indicator of glutamine concentration

    C. Tryptophan levels cannot be estimated accurately

    D. Asparagine levels cannot be estimated accurately

    E. Tryptophan and asparagine levels cannot be estimated accurately

9. What is the correct sequence of energy sources used by the body?

    A. fats → glucose → other carbohydrates → proteins

    B. glucose → other carbohydrates → fats → proteins

    C. glucose → other carbohydrates → proteins → fats

    D. glucose → fats → proteins → other carbohydrates

    E. fats → proteins → glucose → other carbohydrates

10. Enzymes act by:

    A. lowering the overall free energy change of the reaction

    B. decreasing the distance reactants must diffuse to find each other

    C. increasing the activation energy

    D. shifting equilibrium towards product formation

    E. decreasing the activation energy

11. For the following reaction, which statement is TRUE?

ATP + Glucose → Glucose-6-phosphate + ADP

A. reaction results in the formation of a phosphoester bond
B. reaction is endergonic
C. reaction is part of the Krebs cycle
D. free energy change for the reaction is approx. –4 kcal
E. reaction does not require an enzyme

12. Which of the following is a correct classification of cAMP, considering the fact that cAMP-dependent protein phosphorylation activates hormone-sensitive lipase?

A. DNA polymerase
B. lipoproteins
C. glycosphingolipids
D. second messenger
E. phospholipids

13. Which statement is NOT true about the Krebs cycle?

A. Krebs cycle occurs in the matrix of the mitochondria
B. Citrate is an intermediate in the Krebs cycle
C. Krebs cycle produces nucleotides such as NADH and $FADH_2$
D. Krebs cycle is linked to glycolysis by pyruvate
E. Krebs cycle is the single greatest direct source of ATP in the cell

14. The rate of V_{max} is directly related to:

I. Enzyme concentration
II. Substrate concentration
III. Concentration of a competitive inhibitor

A. I, II, and III
B. I and III only
C. I and II only
D. I only
E. II and III only

15. A reaction in which the substrate glucose binds to the enzyme hexokinase, and the conformation of both molecules changes, is an example of:

A. lock-and-key mechanism
B. induced fit mechanism
C. competitive inhibition
D. allosteric inhibition
E. covalent bond formation at active site

16. You are studying an enzyme that catalyzes a reaction that has a free energy change of +5 kcal. If you double the amount of your enzyme in a reaction mixture, what would be the free energy change for the reaction?

A. –10 kcal   B. –5 kcal   C. 0 kcal   D. +5 kcal   E. +10 kcal

17. Glucokinase and hexokinase catalyze the first glycolysis reaction; glucokinase has a higher K_m. Which of the following is a correct statement, if K_m is equal to [Substrate] = $1/2V_{max}$?

A. hexokinase is always functional and is not regulated by negative feedback

B. hexokinase and glucokinase are not isozymes

C. glucokinase is not a zymogen

D. glucokinase becomes active from high levels of fructose

E. none of the above

18. α-helices and β-pleated sheets are characteristic of which level of protein folding?

A. primary

B. secondary

C. tertiary

D. quaternary

E. secondary & tertiary

19. Coenzymes are:

A. minerals such as Ca^{2+} and Mg^{2+}

B. small inorganic molecules that work with an enzyme to enhance reaction rate

C. linking together of two or more enzymes

D. small molecules that do not regulate enzymes

E. small organic molecules that work with an enzyme to enhance reaction rate

20. Hemoglobin is an example of a protein that:

A. is initially inactive in the cell

B. has a quaternary structure

C. carries out a catalytic reaction

D. has only tertiary structure

E. has a signal sequence

21. The site of the TCA cycle in eukaryotic cells, as opposed to prokaryotes, is:

A. mitochondria

B. endoplasmic reticulum

C. cytosol

D. nucleolus

E. intermembrane space of the mitochondria

22. All of the following are metabolic waste products, EXCEPT:

A. lactate   **B.** pyruvate   **C.** CO_2 **D.** H_2O **E.** ammonia

23. When measuring the reaction velocity as a function of substrate concentration, what is likely to occur if the enzyme concentration changes?

   A. V_{max} changes, while K_m remains constant

   B. V_{max} remains constant, while V changes

   C. V_{max} remains constant, while K_m changes

   D. V_{max} remains constant, while V and K_m change

   E. Not possible to predict without experimental data

24. A holoenzyme is:

   A. inactive enzyme without its cofactor   **C.** active enzyme with its organic moiety

   B. inactive enzyme without its coenzyme   **D.** active enzyme with its coenzyme

                                                 E. active enzyme with its cofactor

25. *Clostridium butyricum* is a heterotrophic anaerobe that grows on glucose and converts it to butyric acid as a product. If the free energy for this reaction is –50 kcal, the maximum number of ATP that this organism can synthesize from one molecule of glucose is approximately:

   A. 5 ATP   **B.** 36 ATP   **C.** 7 ATP   **D.** 38 ATP   **E.** 0 ATP

26. Which amino acid is directly affected by dithiothreitol (DTT) known to reduce and break disulfide bonds?

   A. methionine   **C.** glutamine

   B. leucine   **D.** cysteine   **E.** proline

27. Which of the following choices represents a correct pairing of aspects for cellular respiration?

   A. Krebs cycle – cytoplasm

   B. fatty acid degradation – lysosomes

   C. electron transport chain – inner mitochondrial membrane

   D. glycolysis – inner mitochondrial membrane

   E. ATP synthesis – outer mitochondrial membrane

28. An apoenzyme is an:

   A. active enzyme with its organic moiety

   B. inactive enzyme without its inorganic cofactor

   C. active enzyme with its cofactor

   D. active enzyme with its coenzyme

   E. inactive enzyme without its cofactor

29. Several different forces are involved in the stabilization of the tertiary structure of a protein. Which of the following is most likely involved in this stabilization?

A. glycosidic bonds   **C.** peptide bonds

B. disulfide bonds   **D.** anhydride bonds   **E.** phosphodiester bonds

30. In the non-oxidative branch of the pentose phosphate pathway, transketolase is a reaction catalyst enzyme and its activity depends on a prosthetic group. Which bond is used by a prosthetic group to attach to its target?

A. van der Waals interactions   **C.** ionic bond

B. covalent bond   **D.** hydrogen bond   **E.** dipole-dipole interactions

31. The process of $C_6H_{12}O_6 + O_2 \rightarrow CO_2 + H_2O$ is completed in the:

A. plasma membrane   **C.** ribosome

B. cytoplasm   **D.** nucleus   **E.** mitochondria

32. In a hyperthyroidism patient, the oxidative metabolism rate measured through the basal metabolic rate (BMR) will be:

A. indeterminable   **C.** above normal

B. below normal   **D.** normal   **E.** between below normal to normal

33. Cofactors are:

    I. small inorganic molecules that work with an enzyme to enhance reaction rate

    II. small organic molecules that work with an enzyme to enhance reaction rate

    III. small molecules that regulate enzyme activity

A. I only   **C.** II and III only

B. I and II only   **D.** I, II and III   **E.** I and III only

34. All proteins:

A. are post-translationally modified   **C.** have catalytic activity

B. have a primary structure   **D.** contain prosthetic groups

                                      E. contain disulfide bonds

35. After being gently denatured with the denaturant removed, proteins can recover significant activity because recovery of structure depends on?

A. 4° structure of the polypeptide   **C.** 2° structure of the polypeptide

B. 3° structure of the polypeptide   **D.** 1° structure of the polypeptide

                                        E. interactions between polypeptide and its prosthetic groups

36. All of the following statements about glycolysis are true, EXCEPT:

    A. end-product can be lactate, ethanol, CO_2, and pyruvate

    B. $FADH_2$ is produced during glycolysis

    C. a molecule of glucose is converted into two molecules of pyruvate

    D. net total of two ATPs is produced

    E. NADH is produced

37. Vitamins are:

    A. necessary components in the human diet

    B. present in plants but not in animals

    C. absent in bacteria within the gastrointestinal tract

    D. all water soluble

    E. inorganic components of the diet

38. Hemoglobin is a protein that contains a:

    A. site where proteolysis occurs   **C.** bound zinc atom

    B. phosphate group at its active site   **D.** prosthetic group

    E. serine phosphate at its active site

39. Which of these amino acids is nonoptically active because it does not contain four different groups bonded to the α carbon?

    A. valine   **B.** aspartic acid   **C.** glutamate   **D.** cysteine   **E.** glycine

40. Which of the following statements is TRUE for glycolytic pathway?

    A. glucose produces a net of two molecules of ATP and two molecules of NADH

    B. glucose produces one molecule of pyruvate

    C. O_2 is a reactant for glycolysis

    D. glucose is partially reduced

    E. pyruvate is the final product of the Krebs cycle and is the immediate for the next series of reactions in cellular respiration

41. Which of these metabolic processes take(s) place in the mitochondria?

    I. Krebs cycle   II. glycolysis   III. electron transport chain

    A. II only   **C.** II and III only

    B. I and III only   **D.** I, II and III   **E.** I and II only

42. Which statement below best describes the usual relationship of the inhibitor molecule to the allosteric enzyme in feedback inhibition of enzyme activity?

    A. The inhibitor is the substrate of the enzyme
    B. The inhibitor is the product of the enzyme-catalyzed reaction
    C. The inhibitor is the final product of the metabolic pathway
    D. The inhibitor is a metabolically unrelated signal molecule
    E. The inhibitor binds to a tertiary protein

43. *Clostridium butyricum* is an obligate anaerobe that grows on glucose and converts it to butyric acid. If the ΔG for this reaction is –50 kcal, the synthesis of ATP occurs through:

    A. substrate-level phosphorylation
    B. oxidative phosphorylation
    C. neither substrate-level nor oxidative phosphorylation
    D. both substrate-level and oxidative phosphorylation
    E. electron transport cascade

44. While covalent bonds are the strongest bonds that form protein structure, which of the following are connected by a peptide bond?

    A. ammonium group and ester group
    B. two amino groups
    C. the α carbons
    D. two carboxylate groups
    E. amino group and carboxylate group

45. All of the following statements apply to oxidative phosphorylation, EXCEPT:

    A. it can occur under anaerobic conditions
    B. it produces two ATPs for each $FADH_2$
    C. it involves O_2 as the final electron acceptor
    D. it takes place on the inner membrane of the mitochondrion
    E. it involves a cytochrome electron transport chain

46. Like other catalysts, enzymes:

    I. increase the rate of reactions without affecting ΔG
    II. shift the chemical equilibrium from more reactants to more products
    III. do not alter the chemical equilibrium between reactants and products

    A. I only
    B. I and II only
    C. I and III only
    D. III only
    E. II and III only

47. Enzyme activity can be regulated by:

    I. zymogen proteolysis
    II. changes in substrate concentration
    III. post-translational modification

A. I only
B. II and III only
C. I, II and III
D. I and II only
E. I and III only

48. Which interactions stabilize parallel and non-parallel beta-pleated sheets?

A. hydrophobic interactions
B. hydrogen bonds
C. van der Waals interactions
D. covalent bonds
E. dipole-dipole interactions

49. Glycogen is:

A. degraded by glycogenesis
B. synthesized by glycogenolysis
C. unbranched molecule
D. found in both plants and animals
E. the storage polymer of glucose

50. If $[S] = 2\ K_m$, what portion of active sites of the enzyme is filled by substrate?

A. 3/4
B. 2/3
C. 1/2
D. 1/3
E. 1/4

51. In allosteric regulation, how is enzyme activity affected by the binding of a small regulatory molecule to an enzyme?

A. It is inhibited
B. It is stimulated
C. Can be either stimulated or inhibited
D. Is neither stimulated nor inhibited
E. The rate is increased to twice K_m and then plateaus

52. All biological reactions:

A. are exergonic
B. have an activation energy
C. are endergonic
D. occur without a catalyst
E. are irreversible

53. In eukaryotes, the energy is trapped in a high-energy phosphate group during oxidative phosphorylation that occurs in the:

A. nucleus
B. mitochondrial matrix
C. inner mitochondrial membrane
D. outer mitochondrial membrane
E. cytoplasmic face of the plasma membrane

54. All of the following statements about enzymes are true, EXCEPT:

    A. They function optimally at a particular temperature

    B. They function optimally at a particular pH

    C. They may interact with non-protein molecules to achieve biological activity

    D. Their activity is not affected by a genetic mutation

    E. They are almost always proteins

55. The Gibbs free-energy change (ΔG) of a reaction is determined by:

    I. intrinsic properties of the reactants and products

    II. concentrations of the reactants and products

    III. temperature of the reactants and products

    A. I only   **C.** II and III only

    B. I and III only   **D.** I, II and III   **E.** I and II only

56. Allosteric enzymes:

    I. are regulated by metabolites that bind at sites other than the active site

    II. have quaternary structure

    III. show cooperative binding of substrate

    A. I only   **C.** II and III only

    B. I and II only   **D.** I, II and III   **E.** III only

57. Which symptom is characteristic of a patient exposed to monoamine oxidase inhibitors, given that they prevent the breakdown of catecholamines (i.e. epinephrine)?

    A. decreased blood flow to skeletal muscles   **C.** dilated pupils

    B. excessive digestive activity   **D.** decreased heart rate

    E. increased peristalsis along the GI tract

58. The ΔG for hydrolysis of ATP to ADP and Pi is:

    A. Greater than +7.3 kcal/mole   **C.** –7.3 kcal/mole

    B. +7.3 kcal/mole   **D.** –0.5 kcal/mole   **E.** none of the above

59. The active site of an enzyme is where:

    I. prosthetic group is bound

    II. proteolysis occurs for zymogens

    III. non-competitive inhibitors bind

    A. I only   **C.** I and II only

    B. II only   **D.** II and III only   **E.** I, II and III

60. The hydrolysis of ATP → ADP + phosphate allows glucose-6-phosphate to be synthesized from glucose and phosphate because:

 A. heat produced from ATP hydrolysis drives glucose-6-phosphate synthesis
 B. enzymatic coupling of these two reactions allows the energy of ATP hydrolysis to drive the synthesis of glucose-6-phosphate
 C. energy of glucose phosphorylation drives ATP splitting
 D. all of the above
 E. None of the above

61. Which of the following statements is TRUE?

 A. Protein function can be altered by modification after synthesis
 B. Covalent bonds stabilize the secondary structure of proteins
 C. A protein with a single polypeptide subunit has quaternary structure
 D. Integral proteins contain high amounts of acidic amino acids
 E. Protein denaturation is always reversible

62. What does the alcoholic fermentation pathway have in common with the oxidation of pyruvate under aerobic conditions?

 A. no commonality
 B. triose sugar is a product of each reaction
 C. ethyl alcohol is a product of each reaction
 D. CO_2 is a product of each reaction
 E. NADH is a product of each reaction

63. Which one of the following statements is TRUE?

 A. All disaccharides must contain fructose
 B. Most polysaccharides contain ribose
 C. All polysaccharides are energy-generating molecules
 D. Polysaccharides are only found in animal cells
 E. Glucose and fructose have different chemical properties even with the same molecular formula

Answer Key

1: D	11: A	21: A	31: E	41: B	51: C	61: A
2: A	12: D	22: B	32: C	42: C	52: B	62: D
3: C	13: E	23: A	33: D	43: A	53: C	63: E
4: B	14: D	24: E	34: B	44: E	54: D	
5: E	15: B	25: C	35: D	45: A	55: D	
6: C	16: D	26: D	36: B	46: C	56: D	
7: D	17: C	27: C	37: A	47: C	57: E	
8: A	18: B	28: E	38: D	48: B	58: C	
9: B	19: E	29: B	39: E	49: E	59: C	
10: E	20: B	30: B	40: A	50: B	60: B	

Chapter 1.4: Specialized Cells and Tissues

1. What is the role of Ca^{2+} in muscle contractions?

   A. Breaking the cross-bridges as a cofactor in the hydrolysis of ATP
   B. Re-establishing the polarity of the plasma membrane following an action potential
   C. Transmitting the action potential across the neuromuscular junction
   D. Spreading the action potential through the T tubules
   E. Binding to the troponin complex, which leads to exposure of the myosin-binding sites

2. As sodium moves from the extracellular to the intracellular space, which anion must follow it for electrical neutrality?

   A. chloride   B. potassium   C. magnesium   D. lithium   E. calcium

3. Which of the following statements about caffeine is FALSE?

   A. It inhibits the signaling pathway normally stimulated by epinephrine
   B. It is a signal molecule
   C. It acts in different ways in different tissues
   D. It is a common ingredient in headache remedies
   E. It indirectly leads to an increased rate of conversion of glycogen into glucose

---

Use the graph to answer questions **4–8**

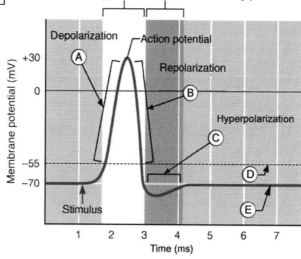

4. From the figure above, the arrow A points to which ion entering the axon?

   A. Ca^{2+} C. Cl^-
   B. K^+ D. Na^+
              E. Mg^{2+}

5. From the figure above, the arrow B points to which ion leaving the axon?

   A. Ca^{2+} C. Cl^-
   B. K^+ D. Na^+
              E. Mg^{2+}

---

6. From the figure above, the arrow C points to the slow close of:

A. ATPase

B. Mg^{2+} voltage gated channels

C. Cl^- voltage gated channels

D. Na^+ voltage gated channels

E. K^+ voltage gated channels

7. From the figure above, the arrow D points to:

A. hyperpolarization

B. threshold

C. overshoot

D. resting potential

E. summation

8. From the figure above, the arrow E points to:

A. hyperpolarization

B. hypopolarization

C. overshoot

D. resting potential

E. summation

9. What is the effect of chemical X that denatures all enzymes in the synaptic cleft?

A. inhibition of depolarization of the presynaptic membrane

B. prolonged depolarization of the presynaptic membrane

C. prolonged depolarization of the postsynaptic membrane

D. inhibition of depolarization of the postsynaptic membrane

E. inhibition of communication between the presynaptic and postsynaptic neurons

10. A "resting" motor neuron is expected to:

A. release high levels of acetylcholine

B. have high permeability to Na^+

C. have equal permeability to Na^+ and K^+

D. exhibit a resting potential that is more negative than the threshold potential

E. have a higher concentration of Na^+ on the inside of the cell than on the outside

11. Which of the following results from administration of digitalis known to block Na^+/K^+ ATPase?

A. decrease of intracellular $[K^+]$

B. increase of extracellular $[Na^+]$

C. decrease of intracellular $[Ca^{2+}]$

D. decrease of intracellular $[Na^+]$

E. increase of intracellular $[K^+]$

12. If a neuron's membrane potential goes from –70 mV to –90 mV, this is an example of:

A. depolarization

B. repolarization

C. hyperpolarization

D. Na^+ channel inactivation

E. hypopolerization

13. Which of the following tissues is/are example(s) of connective tissue?

      I. bone   II. cartilage   III. nervous

A. II only
C. II and III only
B. I only
D. I, II and III
E. I and II only

14. Which of these factors does NOT affect the resting membrane potential?

A. The Na^+/K^+ pump
B. Active transport across the plasma membrane of the axon
C. Equal distribution of ions across the plasma membrane of the axon
D. Selective permeability for ions across the plasma membrane of the axon
E. The relative concentration of ions across the plasma membrane

15. The first event to occur when a resting axon reaches its threshold potential is:

A. closing of K^+ gates
C. hyperpolarization of the membrane potential
B. activation of the Na^+/K^+ pump
D. closing of Na^+ gates
E. opening of Na^+ gates

16. Multiple sclerosis is a demyelinating disease that most severely affects which of the following muscle fibers?

A. C-pain – temperature and mechanoreception (velocity 0.75-3.5 m/s)
B. gamma – intrafusal muscle spindle (velocity 4-24 m/s)
C. beta – touch (velocity 25-75 m/s)
D. alpha – extrafusal muscle spindle (velocity 75-130 m/s)
E. beta – pressure (velocity 25-75 m/s)

17. The measure of the post synaptic response after stimulation by the presynaptic cell is called synaptic:

A. area   **B.** strength   **C.** speed   **D.** summation   **E.** refractory

18. What molecule must a cell have to be able to respond to a signal?

A. paracrine   **B.** receptor   **C.** autocrine   **D.** responder   **E.** A and C

19. Muscle contraction takes place due to:

A. myosin filaments and actin filaments sliding past each other
B. shortening of only actin filaments
C. shortening of only myosin filaments
D. simultaneous shortening of both myosin and actin filaments
E. myosin filaments elongating, while actin filaments shorten

20. When an organism dies, its muscles remain in the contracted state of *rigor mortis* for a brief period. Which of the following most directly contributes to this phenomenon?

A. no ATP is available to move cross-bridges
B. no ATP is available to break bonds between the thick and thin filaments
C. no calcium to bind to troponin
D. no oxygen supplied to muscles
E. glycogen remaining in the muscles

21. Which of the following would be expected to occur in a cell exposed to ouabain, known to block the activity of the Na^+/K^+ ATPase?

A. increase in ATP consumption
B. spontaneous depolarization
C. increase in extracellular $[Na^+]$
D. increase in intracellular $[K^+]$
E. increase in both extracellular $[Na^+]$ and intracellular $[K^+]$

22. The region on neurons that brings graded potential to the neuronal cell body is:

A. axon
B. dendrite
C. T-tubule
D. node of Ranvier
E. axon terminus

23. Which statement correctly describes the role of the myelin sheath in action potential transmission?

A. Saltatory conduction dissipates current through specialized leakage channels
B. Oligodendrocytes cover the nodes of Ranvier to prevent backflow of current
C. Protein fibers cover the axon and prevent leakage of current across the membrane
D. Nodes of Ranvier insulate the axon
E. Lipids insulate the axons, while membrane depolarization occurs within the nodes

24. Which structure in humans is analogous to an electrical device that allows current to flow only in one direction?

A. dendrite
B. axon process
C. myelin sheath
D. synaptic cleft
E. spinal nerve

25. In the communication link between a motor neuron and a skeletal muscle:

A. motor neuron is the presynaptic, and the skeletal muscle is the postsynaptic cell
B. motor neuron is the postsynaptic, and the skeletal muscle is the presynaptic cell
C. action potentials are possible on the motor neuron, but not on the skeletal muscle
D. action potentials are possible on the skeletal muscle, but not on the motor neuron
E. motor neuron fires action potentials, but the skeletal muscle is not electrochemically excitable

26. Which of these organelles undergoes self-replication?

A. DNA   **C.** nucleolus

B. mitochondria   **D.** ribosomes   **E.** nucleus

27. The resting membrane potential of a neuron would be closest to which of the following?

A. 70 mV   **B.** 120 mV   **C.** −70 mV   **D.** 0 mV   **E.** 30mV

28. A molecule that binds to the three-dimensional structure of another molecule's receptor is:

A. responder   **C.** ligand

B. receptor   **D.** ion channel   **E.** filament

29. *Myasthenia gravis* is a severe autoimmune disease of neuromuscular junctions where a patient's body produces antibodies against the acetylcholine receptors of the muscle membrane (i.e. the sarcolemma), which causes their removal by phagocytosis. Which of these processes would be directly affected by this condition?

A. Acetylcholine synthesis

B. Calcium release by the endoplasmic reticulum

C. Sarcomere shortening during muscle contraction

D. Density of the receptors on the presynaptic membrane

E. Action potential conduction across the sarcolemma

30. Which of the following are shared by skeletal, cardiac, and smooth muscle?

A. A bands and I bands   **C.** gap junctions

B. transverse tubules   **D.** thick and thin filaments   **E.** motor units

31. Procaine is a local anesthetic used during many dental procedures that inhibits the propagation of an action potential along a neuron by:

A. increasing the myelin deposit of Schwann cells

B. removing Schwann cells covering the axon

C. increasing Cl^- movement out of the neuron in response to an action potential

D. stimulating Ca^{2+} voltage gated channels at the synapse

E. blocking Na^+ voltage gated channels

32. The type of glial cell that myelinates an axon in the CNS is the:

A. astrocyte   **C.** Schwann Cell

B. oligodendrocyte   **D.** choroid plexus   **E.** chondrocyte

33. The myelin sheath around axons of the peripheral nervous system is produced by:

A. axon hillock

B. nerve cell body

C. nodes of Ranvier

D. Schwann cell

E. oligodendrocytes

34. All of these statements are true for muscles, EXCEPT:

A. Tetanus is a condition of sustained contraction due to an overlap of twitch impulses

B. Tonus is the state of partial contraction which occurs in a resting muscle

C. Isometric contraction means that the length of the muscle is constant

D. Isotonic contraction means that the length of the muscle shortens

E. Resting muscle is completely relaxed

35. Tissues are composed of cells, and a group of tissues functioning together make up:

A. organs

B. membranes

C. organ systems

D. organelles

E. organisms

36. If a neuronal membrane, which is normally slightly passively permeable to K^+, becomes impermeable, but the Na^+/K^+-ATPase remains active, the neuron's resting potential would become:

A. less positive, because $[K^+]$ increases inside the neuron

B. more negative, because $[K^+]$ increases outside the neuron

C. more positive, because $[K^+]$ increases inside the neuron

D. more positive, because $[K^+]$ increases outside the neuron

E. more negative, because $[K^+]$ increases inside the neuron

37. The asymmetric concentration gradient of Na^+ and K^+ across the membrane is maintained by the:

A. voltage gated channels

B. Na^+/K^+ ATPase

C. mitochondria

D. constitutive ion channels

E. Schwann cells

38. Which of the following statements about the insulin receptor is FALSE?

A. Once activated, it undergoes autophosphorylation

B. It requires binding by two insulin molecules to be activated

C. It catalyzes the phosphorylation of the insulin response substrate

D. It is located entirely within the cytoplasm

E. None of the above

39. Which of these processes results in hyperpolarization?

   A. excessive outflow of K^+

   B. excessive outflow of Na^+

   C. excessive influx of K^+

   D. excessive influx of Na^+

   E. excessive outflow of Na^+ and excessive influx of K^+

40. The deficiency of which vitamin is associated with neural tube defects?

   A. B_6 **B.** calcium   **C.** folic acid   **D.** B_{12} **E.** B_7

41. Which membrane-bound protein channel must be inhibited to cause a blockage of nerve conduction by a hydrophobic local anesthetic such as lidocaine?

   A. Cl^- channel   **C.** K^+ channel

   B. Ca^{2+} channel   **D.** Na^+ channel   **E.** Mg^{2+} channel

42. Which of the following is NOT a class of molecules used as a neurotransmitter?

   A. peptides   **C.** amino acids

   B. catecholamines   **D.** neuropeptides   **E.** enzymes

43. Which of the following is FALSE?

   A. Cells are bombarded with numerous signals but they respond to only a few

   B. A cell's receptors determine whether or not the cell will respond to a signal

   C. Receptor proteins are very specific

   D. There are only a few kinds of signal receptor proteins

   E. None of the above

44. Which neurons are involved in piloerection of hair standing on its end?

   A. sympathetic motor neurons

   B. sympathetic sensory neurons

   C. parasympathetic motor neurons

   D. parasympathetic sensory neurons

   E. sympathetic and parasympathetic sensory neurons

45. Which statement best summarizes the relationship between the cytoplasm and cytosol?

   A. Cytoplasm includes the cytosol, a watery fluid inside the cell

   B. Cytoplasm is within the nucleolus, while cytosol is within the cell outside the nucleus

   C. Cytoplasm is within the cell outside the nucleus, while cytosol is within the nucleolus

   D. Cytoplasm is within the nucleus, while cytosol is within the nucleolus

   E. Cytoplasm is within the nucleolus, while cytosol is within the mitochondria

46. During depolarization of a muscle cell, Ca^{2+} is released from the sarcoplasmic reticulum and binds to which of the following structures?

A. muscle ATPase

B. troponin C

C. myosin heads

D. actin thin filaments

E. myosin thick filaments

47. Which of the following is TRUE about saltatory conduction?

A. Current passes through the myelin sheath

B. Voltage-gated Ca^{2+} channels are concentrated at the nodes of Ranvier

C. Myelinated axons exhibit greater conduction velocity than non-myelinated axons

D. It is much slower than conduction along non-myelinated axons

E. The direction of depolarization is reversed

48. The specialized region of the neuron that connects to the axon and sums the graded inputs prior to propagation of the all or none depolarization is:

A. axon hillock

B. nerve cell body

C. node of Ranvier

D. glial cell

E. axon terminus

49. Myelin covers axons and is responsible for:

A. initiating the action potential

B. allowing pumping of Na^+ out of the cell

C. maintaining the resting potential

D. determining the threshold of the neuron

E. allowing faster conduction of impulses

50. Which tissue forms coverings, linings and glands?

A. connective tissue

B. cellular matrix

C. endothelium

D. epithelial

E. peritoneum

51. What is the function of the nodes of Ranvier within the neuron?

A. To provide a binding site for acetylcholine

B. To provide a space for Schwann cells to deposit myelin

C. To regenerate the anterograde conduction of the action potential

D. To permit the axon hillock to generate a stronger action potential

E. To permit the axon process to generate an action potential with greater amplitude

52. An organ is defined as a structure that has a recognizable shape, has specific functions and is composed of two or more different types of:

A. tissues B. cells C. germ layers D. mesoderm E. endoderm

53. What can be deduced about the conduction velocity of Purkinje fibers?

A. Fast, ion independent channel
B. Slow, ion independent channel
C. Fast, Na^+ dependent channel
D. Slow, Na^+ dependent channel
E. none of the above

54. Which of the following is NOT classified as one of the four primary (basic) types of tissue?

A. connective B. blood C. muscle D. nervous E. epithelial

55. The release of a neurotransmitter into the synaptic cleft results from the influx of:

A. Na^+ ions C. Mg^{2+} ions
B. K^+ ions D. Cl^- ions E. Ca^{2+} ions

56. Which connective tissue stores triglycerides and provides cushioning and support for organs?

A. glycogen C. connective
B. muscle D. adipose E. endothelial

57. Which of the following non-excitable cells utilize voltage gated Na^+ channels?

A. cardiac cells C. smooth muscle
B. arterioles D. skeletal muscle E. nerve cells

58. The cells lining the air sacs in the lungs make up:

A. simple columnar epithelium
B. highly elastic connective tissue
C. stratified squamous epithelium
D. pseudostratified ciliated columnar epithelium
E. simple squamous epithelium

Answer Key

1: E	11: A	21: B	31: E	41: D	51: C
2: A	12: C	22: B	32: B	42: E	52: A
3: A	13: E	23: E	33: D	43: D	53: C
4: D	14: C	24: D	34: E	44: A	54: B
5: B	15: E	25: A	35: A	45: A	55: E
6: E	16: D	26: B	36: C	46: B	56: D
7: B	17: D	27: C	37: B	47: C	57: B
8: D	18: B	28: C	38: D	48: A	58: E
9: C	19: A	29: E	39: A	49: E	
10: D	20: B	30: D	40: C	50: D	

Chapter 1.5: Microbiology

1. What is the *major* distinction between prokaryotic and eukaryotic cells?

    A. prokaryotic cells don't have DNA, and eukaryotic cells do
    B. prokaryotic cells cannot obtain energy from their environment
    C. eukaryotic cells are smaller than prokaryotic cells
    D. prokaryotic cells have not prospered, while eukaryotic cells are evolutionary "successes"
    E. prokaryotic cells don't have a nucleolus, but eukaryotic cells do

2. Which of these statements describes the actions of penicillin?

    A. it is a reversible competitive inhibitor
    B. it is an irreversible competitive inhibitor
    C. it activates transpeptidase that digests the bacterial cell wall
    D. it is an effective antiviral agent
    E. it acts as a noncompetitive inhibitor

3. Which of the following statements is applicable to all viruses?

    A. They have RNA genome
    B. They have DNA genome
    C. They have chromosomes
    D. They cannot replicate outside of host cell
    E. They have reverse transcriptase

4. The replica plating technique of Joshua and Esther Lederberg demonstrated that:

    A. mutations are usually beneficial
    B. mutations are usually deleterious
    C. streptomycin caused the formation of streptomycin-resistant bacteria
    D. streptomycin revealed the presence of streptomycin-resistant bacteria
    E. the frequency of mutations is proportional to the concentration of streptomycin

5. Operons:

    A. are a common feature of the eukaryote genome
    B. often coordinate the production of enzymes that function in a single pathway
    C. have multiple translation start and stop sites that are used by the ribosome
    D. usually undergo alternative splicing
    E. both B and C

6. Which of the following is TRUE for the life cycle of sexually-reproducing *Neurospora* fungus?

A. Only mitosis occurs

B. Fertilization and meiosis are separated

C. Meiosis quickly follows fertilization

D. Fertilization immediately follows meiosis

E. Mitosis quickly follows fertilization

7. Which of the following is found in prokaryotic cells?

A. mitochondria   C. nuclei

B. chloroplasts   D. enzymes   E. extensive endomembrane system

8. Which organelle is the site of protein modification and carbohydrate synthesis?

A. Golgi apparatus   C. peroxisomes

B. lysosomes   D. smooth ER   E. nucleolus

9. If a suspension of Hfr cells is mixed with an excess of F^- cells, what is most likely to occur?

A. Most of the F^- cells are transformed into F^+ cells

B. The F^- cells produce sex pili that attach to the Hfr cells

C. Hfr chromosomal DNA is transferred to F^- cells by conjugation

D. Hfr cells replicate the F factor independently of their chromosomes

E. Most of the F^+ cells become F^- cells

10. What carcinogen and mutagen test looks for an increased reversion frequency in a His⁻ bacteria strain?

A. *Salmonella* reversion test   C. mutagen test

B. auxotrophic reversion test   D. amber test   E. Ames test

11. The type of bacteria NOT able to grow on minimal media due to mutations affecting metabolism:

A. auxotrophs   C. heterotrophs

B. chemotrophs   D. prototrophs   E. all are able to grow

12. Which of the following statements is TRUE?

A. Endospores are for reproduction

B. Endospores allow a cell to survive environmental changes

C. Endospores are easily stained with Gram stain

D. Cell produces one endospore and keeps growing

E. Cell produces many endospores and keeps growing

13. Most fungi spend the biggest portion of their life cycle as:

   A. neither haploid nor diploid   **C.** diploid

   B. both haploid and diploid   **D.** polyploidy   **E.** haploid

14. An aerobic bacteria culture that has been exposed to cyanide gas is infected by a bacteriophage strain. However, the replication of viruses does not occur. What is cyanide's action mechanism?

   A. Binding to viral nucleic acid

   B. Denaturing bacteriophage enzymes

   C. Inhibiting aerobic ATP production

   D. Destroying bacteriophage binding sites on the bacterial cell wall

   E. Denaturing viral enzymes needed for replication

15. Recipient cells acquire genes from free DNA molecules in the surrounding medium by:

   A. generalized transduction   **C.** transduction

   B. conjugation   **D.** recombination   **E.** transformation

16. Viruses can have a genome that is:

   A. single-stranded DNA   **C.** double-stranded RNA

   B. single-stranded RNA   **D.** double-stranded DNA   **E.** all of the above

17. Which assumption must be true to map the order of bacterial genes on the chromosome in a Hfr strain?

   A. Bacterial genes are polycistronic

   B. A given Hfr strain always transfers its genes in the same order

   C. The rate of chromosome transfer varies between bacteria of the same strain

   D. The inserted F factor and bacterial genes are replicated by different mechanisms

   E. All of the above statements are true

18. Which of the following statements about prokaryotic cells is generally FALSE?

   A. They have a semirigid cell wall

   B. They are motile by means of flagella

   C. They possess 80S ribosomes

   D. They reproduce by binary fission

   E. They lack membrane-bound nuclei

19. All of the following events play a role in the life cycle of a typical retrovirus EXCEPT:

A. injection of viral DNA into the host cell

B. integration of viral DNA into the host genome

C. reverse transcriptase gene is transcribed and mRNA is translated inside the host cell

D. viral DNA incorporated into the host genome may be replicated along with the host DNA

E. none of the above

20. All of the following are true for viruses EXCEPT:

A. Genetic material may be either single-stranded or double-stranded RNA

B. Virus may replicate in a bacterial or eukaryotic host

C. Virus may replicate without a host

D. The protein coat of the virus does not enter a host bacterial cell

E. Genetic material may be either single-stranded or double-stranded DNA

21. DNA transfer from a bacterial donor cell to a recipient cell by cell-to-cell contact is:

A. conjugation

B. transformation

C. transduction

D. recombination

E. transposons

22. On the overnight agar plates with *E. coli*, replication of a virus is marked by:

A. no visible change

B. bacterial colonies on the agar surface

C. growth of a smooth layer of bacteria across the plate

D. growth of bacteria across the entire plate except for small clear patches

E. absence of any growth on the plate

23. Which of the following about gram-negative cell walls is FALSE?

A. They protect the cell in a hypotonic environment

B. They have an extra outer layer of lipoproteins, lipopolysaccharides and phospholipids

C. They are toxic to humans

D. They have a thinner outer membrane

E. They are sensitive to penicillin

24. All of the following may be present in a mature virus found outside the host cell EXCEPT:

A. core proteins

B. both RNA and DNA

C. protein capsid

D. phospholipid bilayer envelope

E. none of the above are necessary

25. All of the following are correct about *lac* operon EXCEPT:

    A. Repressor protein binds to the operator, halting gene expression
    B. Promoter is the binding site of RNA polymerase
    C. There is not a gene that encodes for a repressor protein
    D. There are 3 structural genes that code for functional proteins
    E. *Lac* operon is found in eukaryotes

26. Phage DNA integrated into the chromosome is:

    A. lytic phage
    B. specialized transducing phage
    C. lysogenic phage
    D. prophage
    E. insertion sequence

27. Many RNA copies of the retrovirus RNA genome are made by:

    A. host cell DNA polymerases
    B. reverse transcriptase
    C. host cell RNA polymerases
    D. host cell ribosomes
    E. none of the above

28. In the laboratory, *E. coli* are grown at the temperature of 37°C because:

    A. *E. coli* strain is a 37°C temperature-sensitive mutant
    B. *E. coli* reproduce most rapidly at this temperature
    C. lower temperatures inhibit conjugation
    D. *E. coli* are obligate aerobes
    E. *E. coli* obtain energy from the temperature of the growth medium

29. Some bacteria are able to propel themselves through liquid by means of:

    A. flagellum   **C.** centrosome
    B. centriole   **D.** peptidoglycan   **E.** cell wall

30. When most viruses infect eukaryotic cells:

    A. their capsid enters the host cell
    B. they replicate independently of the host cell during a lysogenic infection
    C. they enter the cell via endocytosis
    D. they do not need tail fibers to recognize the host cell
    E. they replicate as the host cell replicates during a lytic infection

31. Which organelle(s) is/are NOT present in bacteria?

    I. peroxisomes   III. ribosomes
    II. nucleolus   IV. flagellum

A. I only

B. II only

C. I and II only

D. I, II, and III

E. III and IV only

32. A bacterial cell carrying a prophage is:

A. virulent

B. temperate

C. exconjugant

D. transformant

E. lysogen

33. All of the following are true of prokaryotic translation EXCEPT:

A. mRNA is not spliced before initiation

B. N-terminal amino acid of nascent polypeptides is formylated

C. mRNA chain being translated may not be fully transcribed before translation begins

D. Hydrogen bonds between amino acids and mRNA codons are necessary for translation

E. Translation and transcription both happen in the same location within the cell

34. Prokaryotes are about how many times smaller in diameter than a typical eukaryote?

A. two   **B.** ten   **C.** 100   **D.** 10,000   **E.** zero

35. Which of these structures are found in prokaryotes?

    I. A cell wall containing peptidoglycan
    II. A plasma membrane with cholesterol
    III. Ribosomes

A. I only

B. II only

C. I and II only

D. I and III only

E. II and III only

36. Which of the following is characteristic of viruses?

A. membrane-bound organelles

B. genetic material not made of nucleic acids

C. peptidoglycan cell wall

D. phospholipid bilayer membrane

E. protein coat

37. An F-plasmid that can integrate into the bacterial chromosome by homologous recombination is:

A. episome

B. viron

C. endoconjugate

D. lytic

E. virulent

38. The RNA genome of retrovirus is converted to double-stranded DNA by:

    A. host cell DNA polymerases

    B. reverse transcriptase

    C. host cell RNA polymerases

    D. host cell ribosomes

    E. none of the above

39. Which of the following statements is true about T4 infection of *E. coli*?

    A. T4 mRNA is translated by bacterial ribosomes while still being transcribed from DNA

    B. One of the first genes expressed during viral infection is a lysozyme that facilitates cell lysis

    C. The final stage for lytic cycles of infection is the viral assembly after the virus leaves the cell

    D. T4 buds via endocytosis through the plasma membrane to leave the cell

    E. T4 exits the cell through protein pores in the plasma membrane to leave the cell

40. Which of the following statements about a gram-positive cell wall is FALSE?

    A. It maintains the shape of the cell

    B. It is sensitive to lysozyme

    C. It protects the cell in a hypertonic environment

    D. It contains teichoic acids

    E. It is sensitive to penicillin

41. Which statement is CORRECT about the lipopolysaccharide layer outside the peptidoglycan cell wall of a Gram negative bacterium?

    A. It allows the bacterium to attach to solid objects

    B. It does not contain a phospholipid membrane

    C. It protects the bacterium against antibiotics

    D. It absorbs and holds Gram stain

    E. It appears deep purple from Gram stain

42. DNase added to a bacterial cell causes hydrolysis of the cell's DNA, preventing protein synthesis and cell death. Regardless, some viruses pre-treated with DNase continue to produce new proteins following infection because:

    A. viral genome contains multiple copies of their genes

    B. viruses are homozygous for necessary genes

    C. viral genome contains multiple reading frames

    D. icosahedral protein coat of the virus denatures DNase

    E. viral genome is comprised of RNA

43. In a mating between Hfr and F⁻ cells, the F⁻ recipient:

    A. becomes Hfr   **C.** remains F⁻

    B. becomes F′   **D.** becomes F⁺   **E.** cannot establish lysogeny

44. The unique feature about the methionine residue used for prokaryotic initiation of translation is that it is:

    A. formylated   **C.** methylated

    B. hydrophilic   **D.** acetylated   **E.** hydrophobic

45. Which of the following statements best describes what takes place when a bacterial cell is placed in a 35% solution of a large polysaccharide (e.g. dextran)?

    A. NaCl moves into the cell from a higher to a lower concentration

    B. The cell undergoes plasmolysis

    C. H_2O moves out of the cell

    D. H_2O moves into the cell

    E. No change result because solution is isotonic

46. If the Gram stain method is used to stain a Gram positive bacterium, it appears?

    A. deep purple, because of a thicker peptidoglycan cell wall

    B. deep purple, because of a thinner peptidoglycan cell wall

    C. red or pink, because of a thicker peptidoglycan cell wall

    D. red or pink, because of a thinner peptidoglycan cell wall

    E. red or pink, because of the absence of a peptidoglycan cell wall

47. Which enzyme replicates the F factor in F⁺ bacteria prior to conjugation?

    A. DNA polymerase   **C.** DNA ligase

    B. reverse transcriptase   **D.** integrase   **E.** RNA polymerase

48. In a mating between Hfr and F⁻ cells, the Hfr donor:

    A. becomes F′

    B. remains Hfr

    C. becomes F⁺

    D. becomes F⁻

    E. loses part of the chromosome

49. Which statement(s) is/are TRUE regarding retrotransposons?

    I. They are never found between genes in the human genome
    II. They comprise close to half of the human genome
    III. They can cause mutations by inserting themselves into genes

A. I and II only   **C.** I and III only
B. II only   **D.** I, II and III   **E.** II and III only

50. Which of the following is TRUE about plasmids?

    I. They are small organelles in the bacterial cytoplasm
    II. They are transcribed and translated simultaneously
    III. They are replicated by bacterial enzymes

A. I only   **C.** I and III only
B. II and III only   **D.** I, II, and III   **E.** I and II only

51. Which of the following have a cell wall?

A. protoplasts   **B.** fungi
C. L forms   **D.** viruses   **E.** mycoplasmas

52. Which of the following statements about a prokaryotic cell is correct?

A. It contains a range of different organelles
B. It has a nucleolus within the cytoplasm
C. It contains cell walls composed of chitin
D. It has a nuclear membrane that encloses a nucleus
E. It uses glycolysis to produce ATP

53. Which of these statements applies to both a bacteriophage and retrovirus?

A. They are capable of infecting human cells
B. They act as immunosuppressive agents
C. They integrate their genetic material into the genome of the host cell
D. They have genes that encode for reverse transcriptase
E. They have a genome of RNA

54. Integration of the phage DNA into the bacterial chromosome is facilitated by:

A. DNA polymerase encoded in the host genome
B. topoisomerase encoded in the phage
C. several proteins, some of them encoded in the phage and some in the host genome
D. site-specific recombinase encoded in the host genome
E. site-specific recombinase encoded in the phage

55. After being digested by restriction enzymes, how is DNA ligated into a plasmid in two different directions?

   A. Both ends of a DNA fragment produced by a restriction enzyme are identical if rotated 180°

   B. The existing DNA strands serve as primers for DNA polymerase

   C. DNA ligase enzymes are able to link any two pieces of DNA together

   D. Plasmid DNA is single-stranded so the ligated strands form double-stranded segments

   E. DNA polymerase links any two pieces of DNA together

56. Virulent phage:

   I. is capable only of lytic growth

   II. can only undergo a process called lysogeny

   III. needs several different proteins to be incorporated into bacterial chromosome

   A. I and II only   **C.** II only

   B. III only   **D.** I only   **E.** I, II, and III

57. Which statement best describes the promoter in an operon?

   A. It activates the repressor-inducer complex to permit transcription

   B. It is a molecule that inactivates the repressor and turns on the operon

   C. It is the binding site for RNA polymerase

   D. It is the binding site for the repressor

   E. It is the binding site for DNA polymerase

Answer Key

1: E	11: A	21: A	31: C	41: C	51: B
2: B	12: B	22: D	32: E	42: E	52: E
3: D	13: E	23: E	33: D	43: C	53: C
4: D	14: C	24: B	34: B	44: A	54: E
5: B	15: E	25: E	35: D	45: C	55: A
6: C	16: E	26: D	36: E	46: A	56: D
7: D	17: B	27: C	37: A	47: A	57: C
8: A	18: C	28: B	38: B	48: B	
9: C	19: A	29: A	39: A	49: E	
10: E	20: C	30: C	40: C	50: B	

Chapter 1.6: Photosynthesis

1. What mechanism is used by C_4 plants to conserve water?

    A. developing deep roots
    B. performing the Calvin cycle during the day
    C. developing water storages in their leaves and stems
    D. performing the Calvin cycle early in the morning
    E. closing the stomata during heat and dryness

2. What mechanism is used by CAM plants to conserve water?

    A. performing the Calvin cycle during the day
    B. developing water storages in their leaves and stems
    C. including carbon dioxide into RuBP
    D. opening their stomata during the night
    E. closing their stomata during the night

3. Which of the choices is an autotroph?

    A. beech tree   **B.** octopus   **C.** zooplankton   **D.** coral   **E.** dog

4. Which of the following is NOT a true statement about ATP?

    A. ATP provides energy for the mechanical functions of cells
    B. Used ATP is discarded by the cell as waste
    C. ATP consists of ribose, adenine, and three phosphate groups
    D. ADP is produced when ATP releases energy
    E. ATP conversion to ADP is an exothermic reaction

5. Photosynthesis plays a role in the metabolism of a plant through:

    A. breaking down sugars into H_2O and O_2
    B. converting H_2O into CO_2
    C. consuming CO_2 and synthesizing sugars
    D. converting O_2 into cellulose
    E. converting O_2 into sugars

6. From the figure below, all of the following are parts of an ADP molecule EXCEPT:

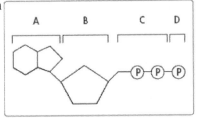

   A. structure A
   B. structure B
   C. structure C
   D. structure D
   E. structures A and B

7. Within the inner membrane of a chloroplast, stacks of thylakoids are surrounded by:

   A. stroma fluid   **B.** grana   **C.** chloroplast   **D.** thylakoids   **E.** cristae

8. What happens during photosynthesis?

   A. Autotrophs consume carbohydrates
   B. Autotrophs produce carbohydrates
   C. Heterotrophs consume ATP
   D. Heterotrophs produce ATP
   E. Heterotrophs produce sugars

9. The role of stomata is to facilitate:

   A. gas exchange   **C.** Calvin cycle
   B. H_2O release   **D.** H_2O uptake   **E.** Chloroplast formation

10. Plants gather the sun's energy using:

   A. glucose   **C.** thylakoids
   B. chloroplasts   **D.** mitochondria   **E.** pigments

11. Which structures are used by plants to obtain most of their water?

   A. Stomata   **C.** Roots
   B. Flowers   **D.** Chloroplasts   **E.** Shoots

12. Which structure in the image illustrates a single thylakoid?

   A. Structure A
   B. Structure B
   C. Structure C
   D. Structures A and B
   E. Structures B and C

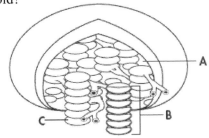

13. Membranous structures found in chloroplasts are:

A. organelles   **C.** stomata

B. grana   **D.** plasma   **E.** thylakoids

14. Chlorophyll is located within the chloroplast in the:

A. thylakoid space

B. thylakoid membrane

C. stroma

D. ATP

E. space between the inner and outer membrane of the thylakoid

15. Grana is/are:

A. stacks of membranous sacs

B. chloroplast pigments

C. fluid found in the chloroplasts

D. the space between the inner and outer membrane of chloroplasts

E. the membrane-bound compartment of chloroplasts

16. Which chemical shown in the figure is an electron carrier molecule?

A. NADP$^+$

B. Oxygen

C. H_2O

D. Carbon dioxide

E. Sugars

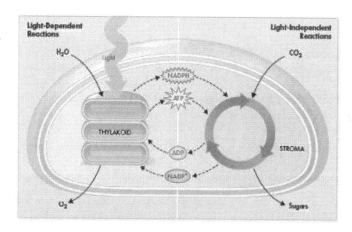

17. Which of the following is most likely to result if a shade-tolerant plant would receive a minimal amount of water while receiving its necessary amount of light?

A. Accelerated plant growth   **C.** Increased output of oxygen

B. Low ATP production   **D.** Increased consumption of carbon dioxide

                                     E. Plant death

18. Why are electron carriers needed for transporting electrons from one part of the chloroplast to another?

    A. High-energy electrons get their energy from electron carriers

    B. High-energy electrons are not soluble in cytoplasm

    C. High-energy electrons are easily oxidized

    D. High-energy electrons would be destroyed

    E. High-energy electrons are highly reactive

19. Which equation best describes the reaction of photosynthesis?

    A. $6O_2 + 6CO_2 \rightarrow C_6H_{12}O_6 + 6H_2O$

    B. $C_6H_{12}O_6 + 6H_2O \rightarrow 6CO_2 + 6O_2$

    C. $C_6H_{12}O_6 \rightarrow 6CO_2 + 6H_2O + 6O_2$

    D. $6H_2O + 6CO_2 \rightarrow C_6H_{12}O_6 + 6O_2$

    E. $6 CO_2 + 6 O_2 + 6 H_2O \rightarrow C_6H_{12}O_6$

20. What is the result when there is an increase in the intensity of light that a plant receives?

    A. The rate of photosynthesis increases and then levels off

    B. The rate of photosynthesis does not change

    C. The rate of photosynthesis increases indefinitely with light intensity

    D. The rate of photosynthesis decreases indefinitely with light intensity

    E. The rate of photosynthesis increases indefinitely to the square root of the light intensity

21. In redox reactions of photosynthesis, electron transfer goes as follows:

    A. $H_2O \rightarrow CO_2$

    B. $C_6H_{12}O_6 \rightarrow O_2$

    C. $O_2 \rightarrow C_6H_{12}O_6$

    D. $CO_2 \rightarrow H_2O$

    E. $C_6H_{12}O_6 \rightarrow H_2O$

22. Which of the following is NOT a step in the light-dependent reactions?

    A. ATP synthase allows H^+ ions to pass through the thylakoid membrane

    B. ATP and NADPH are used to produce high-energy sugars

    C. High-energy electrons move through the electron transport chain

    D. Pigments in photosystem II absorb light

    E. Electrons move from photosystem II to photosystem I

23. Which of the following is the product of the light reaction of photosynthesis?

    A. sugar

    B. carbon monoxide

    C. water

    D. carbon dioxide

    E. oxygen

24. Photosystems I and II are found in:

    A. cell membrane

    B. the Calvin cycle

    C. thylakoid membrane

    D. stroma

    E. matrix of the mitochondrion

25. Which two components from the light reactions are required by the C_3 cycle?

    A. NADH and RuBP   **C.** ATP and NADH

    B. NADPH and ATP   **D.** glucose and $NADP^+$ **E.** water and $NADP^+$

26. Which pathway is the correct flow of electrons during photosynthesis?

    A. Photosystem I $\rightarrow$ Calvin cycle $\rightarrow$ $NADP^+$

    B. Light $\rightarrow$ Photosystem I $\rightarrow$ Photosystem II

    C. H_2O $\rightarrow$ Photosystem I $\rightarrow$ Photosystem II

    D. O_2 $\rightarrow$ ADP $\rightarrow$ Calvin cycle

    E. H_2O $\rightarrow$ $NADP^+$ $\rightarrow$ Calvin cycle

27. Light-dependent reactions of photosynthesis produce:

    A. H_2O and RuBP   **C.** NADH and ATP

    B. ATP and CO_2 **D.** NADPH and ATP   **E.** H_2O and O_2

28. The Calvin cycle takes place in the:

    A. chlorophyll molecules   **C.** photosystems

    B. thylakoid membranes   **D.** stroma   **E.** matrix

29. A wavelength of which light is reflected by a blue-colored plant?

    A. violet   **B.** yellow   **C.** blue   **D.** lime   **E.** purple

30. What are the three parts of an ATP molecule?

    A. Adenine, ribose and three phosphate groups

    B. NADH, NADPH and $FADH_2$

    C. Adenine, thylakoid and a phosphate group

    D. Stroma, grana and chlorophyll

    E. Adenine, ribose, and a phosphate group

31. When observing a plant, its visible color is:

 A. wavelength emitted by the plant
 B. wavelength from the return of excited electrons to their ground state
 C. wavelength being reflected by that plant
 D. wavelength absorbed by the pigment of the plant
 E. wavelength from the excited electrons of the plant

32. Energy is released from ATP when:

 A. a phosphate group is removed
 B. ATP is exposed to sunlight
 C. adenine binds to ribose
 D. a phosphate group is added
 E. a base is removed

33. The wavelength energy least utilized by photosynthesis is:

 A. orange **B.** blue **C.** violet **D.** cyan **E.** green

34. Organisms, such as plants, that make their own food are called:

 A. symbiotic **C.** autotrophs
 B. parasitic **D.** heterotrophs **E.** omnivores

35. During the fall foliage, which pigments are responsible for the yellow, red and orange colors of leaves?

 A. carotenoids **C.** melanin
 B. chlorophyll *b* **D.** chlorophyll *a* **E.** anthocyanin

36. Which of the following organisms is a heterotroph?

 A. sunflower **C.** alga
 B. flowering plant **D.** mushroom **E.** deciduous trees

37. In addition to chlorophyll *a*, plants also have chlorophyll *b* and carotenoids accessory pigments because:

 A. there is not enough of chlorophyll *a* produced for plant's energy needs
 B. plants must have leaves of different colors
 C. these pigments reflect more energy
 D. these pigments protect plants from UV radiation
 E. these pigments absorb energy wavelengths that chlorophyll *a* does not

38. Plants get the energy they need for photosynthesis by absorbing:

    A. chlorophyll *b*

    B. energy from the sun

    C. high-energy sugars

    D. chlorophyll *a*

    E. both chlorophyll *a* and *b*

39. A discrete packet of light is called:

    A. photon   **B.** neutron   **C.** proton   **D.** wavelength   **E.** electron

40. Most plants appear green because chlorophyll:

    A. does not absorb violet light   **C.** absorbs violet light

    B. does not absorb green light   **D.** absorbs green light

                                 E. generate wavelengths of green light

41. In visible light, shorter wavelengths carry:

    A. more energy   **C.** more red color

    B. more photons   **D.** less energy   **E.** less photons

42. Interconnected sacs of membrane suspended in a thick fluid are:

    A. stroma   **C.** chlorophyll

    B. thylakoids   **D.** grana   **E.** stomata

43. When a photon is absorbed by a molecule, one of its electrons is raised to a(n):

    A. higher state   **C.** lower state

    B. excited state   **D.** ground state   **E.** valence

44. What is the function of $NADP^+$ in photosynthesis?

    A. Photosystem   **C.** Electron carrier

    B. Pigment   **D.** High-energy sugar   **E.** Absorb photons

45. A molecule releases energy gained from absorption of a photon through:

    I. fluorescence   II. heat   III. light

    A. I only   **C.** III only

    B. II only   **D.** II and III only   **E.** I, II and III

46. Photosynthesis uses sunlight to convert water and carbon dioxide into:

A. oxygen and high-energy sugars
B. ATP and oxygen

C. high-energy sugars and proteins
D. oxygen and carbon
E. ATP and high-energy sugars

47. Photosystems are found in:

A. grana
B. stroma

C. stomata
D. thylakoid membranes

E. matrix of mitochondria

48. The light-dependent reactions take place:

A. within the thylakoid membranes
B. in the outer membrane of the chloroplast
C. in the stroma of the chloroplast
D. within the mitochondria membranes
E. in the inner membrane of the chloroplast

49. Which compound is located in the photosystem's reaction center?

A. FADPH
B. rhodopsin

C. ATP
D. ADP

E. chlorophyll *a*

50. What are the products of the light-dependent reactions?

A. CO_2 gas, O_2 gas and NADPH
B. ATP, CO_2 gas and NADPH
C. ATP, NADPH and O_2 gas
D. O_2 gas and glucose
E. ATP, NADPH and glucose

51. Light reactions of photosynthesis occur in:

A. thylakoid membranes
B. matrix

C. grana
D. stomata

E. cytosol

52. Which of the following activities happens within the stroma?

A. Electrons move through the electron transport chain
B. The Calvin cycle produces sugars
C. ATP synthase produces ATP
D. Photosystem I absorbs light
E. Photosystem II produces sugars

53. Oxygen released by a photosystem comes from:

   A. ATP   **C.** Chlorophyll *a*

   B. carbon dioxide   **D.** water   **E.** light

54. The Calvin cycle is another name for the:

   A. photosynthesis reaction   **C.** light-independent reactions

   B. electron transport chain   **D.** light-dependent reactions

                               E. Photosystem I

55. Electrons for the light reactions originate from:

   A. NADPH   **B.** water   **C.** light   **D.** carbon dioxide   **E.** ATP

56. Which mechanism takes place in both light reactions of photosynthesis and cellular respiration?

   A. C_3 cycle   **C.** Glycolysis

   B. Beta oxidation   **D.** Krebs cycle   **E.** Electron transport chain

57. During photosynthesis, an H^+ ion gradient is formed across:

   A. mitochondrial inner membrane   **C.** thylakoid membrane

   B. inner chloroplast membrane   **D.** mitochondrial outer membrane

                               E. stroma membrane

58. Which of the following is the correct matching of molecules and products in the Calvin cycle?

   A. ATP + NADPH + carbon dioxide ⇒ sugar

   B. light + water + carbon dioxide ⇒ sugar and oxygen

   C. ATP + NADPH + carbon dioxide ⇒ sugar and oxygen

   D. light + water + carbon dioxide ⇒ sugar

   E. carbon dioxide + light ⇒ oxygen and water

Answer Key

1: E	11: C	21: A	31: C	41: A	51: A
2: D	12: C	22: B	32: A	42: B	52: B
3: A	13: E	23: E	33: E	43: B	53: D
4: B	14: B	24: C	34: C	44: C	54: C
5: C	15: A	25: B	35: A	45: E	55: B
6: D	16: A	26: E	36: D	46: A	56: E
7: A	17: B	27: D	37: E	47: D	57: C
8: B	18: E	28: D	38: B	48: A	58: A
9: A	19: D	29: C	39: A	49: E	
10: E	20: A	30: A	40: B	50: C	

UNIT 2. ECOLOGY

Chapter 2.1: Energy Flow, Nutrient Cycles, Ecosystems, Biomes

1. The lowest level of environmental complexity that includes living and nonliving factors is:

A. ecosystem   **B.** biosphere   **C.** biome   **D.** community   **E.** population

2. How does an area's weather differ from the area's climate?

A. Weather does not change very much and an area's climate may change many times
B. Weather is the area's daily conditions while climate is the area's average conditions
C. Weather involves temperature and precipitation while climate involves only temperature
D. Weather depends on where it is located on Earth while the area's climate does not
E. Weather involves temperature and precipitation while climate involves only precipitation

3. One type of symbiosis is:

A. parasitism   **C.** competition
B. predation   **D.** succession   **E.** none of the above

4. Climate zones are the result of differences in:

A. thickness of the ozone layer   **C.** angle of the sun's rays
B. greenhouse gases   **D.** heat transport   **E.** altitude of the observer

5. The greenhouse effect is:

A. an unnatural phenomenon that causes heat energy to be radiated back into the atmosphere
B. the result of the differences in the angle of the sun's rays
C. primarily related to the levels of ozone in the atmosphere
D. a phenomenon that has only occurred for the last 50 years
E. a natural phenomenon that maintains Earth's temperature range

6. The tendency for warm air to rise and cool air to sink results in:

A. regional precipitation   **C.** ocean upwelling
B. the seasons   **D.** global wind patterns   **E.** regional temperature

7. An ecosystem in which water either covers the soil or is present at or near the surface of the soil for at least part of the year is called a:

A. estuary   **C.** mangrove swamp   **B.** salt marsh   **D.** pond   **E.** wetland

8. Which of the following is a biological aspect of an organism's niche?

    A. composition of soil

    B. amount of sunlight

    C. predators

    D. the water in the area

    E. availability of minerals

9. An organism's niche is:

    A. the range of temperatures that the organism needs to survive

    B. a full description of the place an organism lives

    C. the range of physical and biological conditions in which an organism lives and the way it obtains what it needs to survive and reproduce

    D. all the physical factors in the organism's environment

    E. all the biological factors in the organism's environment

10. No two species can occupy the same niche in the same habitat at the same time:

    A. unless the species require different biotic factors

    B. because of the competitive exclusion principle

    C. unless the species require different abiotic factors

    D. because of the interactions that shape the ecosystem

    E. unless the species require different biotic and the same abiotic factors

11. Plants are:

    A. omnivores   **C.** primary consumers

    B. herbivores   **D.** primary producers   **E.** detritivores

12. How do most primary producers make their own food?

    A. By breaking down remains into carbon dioxide

    B. By converting water into carbon dioxide

    C. By using chemical energy to make carbohydrates

    D. By using heat energy to make nutrients

    E. By using light energy to make carbohydrates

13. Compared to land, the open oceans:

    A. are nutrient-poor environments   **C.** have less zooplankton

    B. are rich in silica and iron   **D.** contain abundant oxygen

                                     E. are nutrient-rich environments

14. Several species of warblers can live in the same spruce tree ONLY because they:

    A. can find different temperatures within the tree
    B. occupy different niches within the tree
    C. have different habitats within the tree
    D. don't eat food from the tree
    E. can find different amounts of direct sunlight within the tree

15. All the interconnected feeding relationships in an ecosystem make up a food:

    A. web   **B.** network   **C.** chain   **D.** framework   **E.** scheme

16. A symbiotic relationship in which both species benefit is:

    A. predation   **C.** commensalism
    B. parasitism   **D.** omnivorism   **E.** mutualism

17. A wolf pack hunts, kills, and feeds on a moose. In this interaction, the wolves are:

    A. predators   **C.** prey
    B. mutualists   **D.** hosts   **E.** symbionts

18. The total amount of living tissue within a given trophic level is:

    A. energy mass   **C.** organic mass
    B. biomass   **D.** trophic mass   **E.** abiotic

19. An interaction in which an animal feeds on plants is:

    A. symbiosis   **C.** herbivory
    B. predation   **D.** carnivory   **E.** parasitism

20. A symbiotic relationship in which one organism is harmed and another benefits is:

    A. synnecrosis   **C.** mutualism
    B. predation   **D.** parasitism   **E.** commensalism

21. What animals eat both producers and consumers?

    A. autotrophs   **C.** omnivores
    B. chemotrophs   **D.** herbivores   **E.** heterotrophs

22. Ecosystem services include:

    A. food production   **C.** solar energy
    B. production of oxygen   **D.** all of the above   **E.** none of the above

23. Organisms that can capture energy and produce food are known as:

A. omnivores   **C.** herbivores

B. heterotrophs   **D.** consumers   **E.** autotrophs

24. What is one difference between primary and secondary succession?

A. Secondary succession begins with lichens and primary succession begins with trees

B. Primary succession modifies the environment while secondary succession does not

C. Secondary succession begins on soil while primary succession begins on newly exposed surfaces

D. Primary succession is rapid and secondary succession is slow

E. Both primary succession and secondary succession are rapid

25. A term that means the same thing as *consumer* is:

A. carbohydrate   **C.** autotroph

B. heterotroph   **D.** producer   **E.** detritivore

26. Primary succession would most likely occur after:

A. severe storm   **C.** earthquake

B. farm land is abandoned   **D.** forest fire   **E.** lava flow

27. Which of the following organisms is a detritivore?

A. fungus   **B.** snail   **C.** crow   **D.** caterpillar   **E.** mouse

28. Matter can recycle through the biosphere because:

A. biological systems do not deplete matter but transform it

B. biological systems use only carbon, oxygen, hydrogen, and nitrogen

C. matter does not change into new compounds

D. matter is assembled into chemical compounds

E. biological systems do not change matter into new compounds

29. A tropical rain forest may not return to its original climax community after which type of disturbances?

A. volcanic eruption   **C.** burning of a forest fire

B. flooding after a hurricane   **D.** clearing and farming   **E.** earthquake

30. A collection of all the organisms that live in a particular place, together with their nonliving environment is a(n):

  A. ecosystem   **C.** population

  B. biome   **D.** community   **E.** biomass

31. Which biome is characterized by very low temperatures, little precipitation, and permafrost?

  A. tropical dry forest   **C.** temperate forest

  B. tundra   **D.** desert   **E.** savannah

32. A bird stalks, kills, and then eats an insect. Based on its behavior, which pair of ecological terms describes the bird?

  A. herbivore ↔ decomposer

  B. autotroph ↔ herbivore

  C. carnivore ↔ consumer

  D. producer ↔ heterotroph

  E. herbivore ↔ consumer

33. Which two biomes have the least precipitation?

  A. boreal forest and temperate woodland

  B. tundra and desert

  C. tropical savanna and tropical dry forest

  D. tundra and temperate shrubland

  E. tropical rain forest and temperate grassland

34. What goes in box 5 of the food web in the figure provided?

  A. decomposers

  B. carnivores

  C. scavengers

  D. herbivores

  E. omnivores

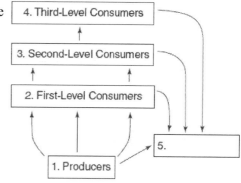

35. The average, year-after-year conditions of temperature and precipitation in a particular region are referred to as:

  A. zonation   **C.** weather   **E.** weather and climate

  B. microclimate   **D.** climate

36. Nitrogen fixation is carried out primarily by:

A. producers
B. consumers

C. humans
D. plants

E. bacteria

37. The rate at which organic matter is created by producers is called:

A. primary succession
B. nitrogen fixation

C. primary productivity
D. nutrient limit

E. secondary succession

38. Which landforms are NOT classified into a major biome?

A. coastlines
B. islands

C. prairies
D. mountain ranges

E. deserts

39. The North Pole and the South Poles are:

A. not classified into major biomes
B. part of aquatic ecosystems

C. classified as tundra biomes
D. not home to any animals
E. classified as temperate biomes

40. The type of interaction in which one organism captures and feeds on another is:

A. mutualism
B. symbiosis

C. competition
D. parasitism

E. predation

41. Are you likely to find zooplankton in the aphotic, benthic zone of an ocean?

A. No, zooplankton cannot chemosynthesize in the dark without oxygen in the water
B. No, zooplankton feed on phytoplankton that cannot photosynthesize without light
C. No, zooplankton cannot photosynthesize in the dark without oxygen in the water
D. Yes, zooplankton is chemosynthetic autotrophs
E. Yes, zooplankton can photosynthesize in the dark

42. Carbon cycles through the biosphere in all of the following processes EXCEPT:

A. decomposition of plants
B. burning of fossil fuels

C. transpiration
D. photosynthesis

E. decomposition of animals

43. The nutrient availability of aquatic ecosystems is the:

A. number of different animal species living in the water
B. amount of rainfall the water receives
C. number of other organisms present in the water
D. amount of nitrogen, oxygen, and other elements dissolved in the water
E. amount of salinity of the water

44. What type of organism forms the base of many aquatic food webs?

A. phytoplankton   **C.** secondary consumers

B. mangrove trees   **D.** plants   **E.** zooplankton

45. The movements of energy and nutrients through living systems are different because:

A. nutrients flow in two directions while energy recycles

B. energy forms chemical compounds while nutrients are lost as heat

C. energy flows in one direction while nutrients recycle

D. energy is limited in the biosphere while nutrients are always available

E. nutrients are lost as heat while energy is limited in the biosphere

46. Boreal forest biomes are:

A. found near the equator

B. also known as taiga

C. home to more species than all other biomes combined

D. hot and wet year-round

E. made up of mostly hardwoods

47. Freshwater ecosystems that often originate from underground sources in mountains or hills are:

A. lakes   **C.** estuaries

B. wetlands   **D.** ponds   **E.** rivers and streams

48. Which is one way a wetland differs from a lake or pond?

A. Water does not always cover a wetland as it does a lake or pond

B. Wetlands are salty while lakes and ponds are fresh water

C. Water flows in a lake or pond but never flows in a wetland

D. Wetlands are nesting areas for birds while lakes and ponds are not nesting areas

E. None of the above are differences

49. A wetland that contains a mixture of fresh water and salt water is:

A. pond   **B.** river   **C.** stream   **D.** estuary   **E.** wetland

50. The permanently dark zone of the ocean is called the:

A. aphotic zone   **C.** photic zone

B. intertidal zone   **D.** coastal zone   **E.** intercostals zone

51. Each of the following is an abiotic factor in the environment EXCEPT:

A. temperature   **B.** rainfall   **C.** plant life   **D.** soil type   **E.** pH

52. Estuaries are commercially important because:

  A. fossil fuels are found in estuaries   **C.** hotels are often built in estuaries

  B. lumber trees grow in estuaries   **D.** abundant fish species live in estuaries

                                            E. the shoreline of estuaries attracts tourists

53. Animals that get energy by eating the carcasses of other animals that have been killed by predators or have died of natural causes are:

  A. detritivores   **C.** omnivores

  B. heterotrophs   **D.** autotrophs   **E.** scavengers

54. Which of the following statements is NOT true about the open ocean?

  A. The open ocean begins at the low-tide mark and extends to the end of the continental shelf

  B. Most of the photosynthetic activity on Earth occurs in the open ocean within the photic zone

  C. The open ocean has low levels of nutrients

  D. Organisms in the deep ocean are exposed to frigid temperatures

  E. Organisms in the deep ocean are exposed to total darkness

55. The branch of biology dealing with interactions among organisms and between organisms and their environment is:

  A. paleontology   **C.** microbiology

  B. ecology   **D.** entomology   **E.** zoology

56. Which of the following descriptions about the organization of an ecosystem is correct?

  A. Species make up communities which make up populations

  B. Species make up populations which make up communities

  C. Communities make up species which make up populations

  D. Populations make up species which make up communities

  E. Communities make up populations which make up species

57. The photic zone:

  A. is deep, cold, and permanently dark

  B. extends to where the light intensity is reduced to 50% compared to the surface

  C. extends to the bottom of the open ocean

  D. extends to a depth of about 600 feet

  E. extends to where chemosynthetic bacteria are the producers

Answer Key

1: A	11: D	21: C	31: B	41: B	51: C
2: B	12: E	22: D	32: C	42: C	52: D
3: A	13: A	23: E	33: B	43: D	53: E
4: C	14: B	24: C	34: A	44: A	54: A
5: E	15: A	25: B	35: D	45: C	55: B
6: D	16: E	26: E	36: E	46: B	56: B
7: E	17: A	27: B	37: C	47: E	57: D
8: C	18: B	28: A	38: B	48: A	
9: C	19: C	29: D	39: A	49: D	
10: B	20: D	30: A	40: E	50: A	

Chapter 2.2: Populations, Communities, Conservation Biology

1. A developer wants to build a new housing development in or around a large city. Which of the following plans would be LEAST harmful to the environment?

    A. Building a neighborhood in a meadow at the edge of the city
    B. Filling a wetland area and building oceanfront condominiums
    C. Clearing a forested area outside of the city to build houses
    D. Building apartments at the site of an abandoned factory in the city
    E. Building apartments on an abandoned farm

2. Assemblages of different populations that live together in a defined area are called:

    A. communities
    B. species
    C. ecosystems
    D. habitats
    E. biodiversity

3. There are 160 Saguaro cactus plants per square kilometer in a certain area of Arizona desert. To which population characteristic does this information refer?

    A. age structure
    B. population density
    C. growth rate
    D. geographic range
    E. death rate

4. Using resources in a way that does not cause long-term environmental harm is:

    A. subsistence hunting
    B. biological magnification
    C. monoculture
    D. sustainable development
    E. none of the above

5. What does the range of a population teach the observer that density does not?

    A. The deaths per unit area
    B. The births per unit area
    C. The areas inhabited by a population
    D. The number that live in an area
    E. The migrations per unit area

6. Which ecological inquiry method is an ecologist using when she enters an area periodically to count the population numbers of a certain species?

    A. modeling
    B. experimenting
    C. hypothesizing
    D. questioning
    E. observing

7. Which of the following is NOT one of the factors that play a role in population growth rate?

    A. demography
    B. emigration
    C. death rate
    D. Immigration
    E. birth rate

8. Which of the following is an example of population density?

   A. number of deaths per year
   B. number of bacteria per square millimeter
   C. number of births per year
   D. number of frogs in a pond
   E. immigration rate per year

9. An example of a biotic factor is:

   A. sunlight   **C.** competing species
   B. soil type   **D.** average temperature   **E.** average monthly rainfall

10. An example of a nonrenewable resource is:

   A. wood   **B.** fish   **C.** sunlight   **D.** wind   **E.** coal

11. A mathematical formula designed to predict population fluctuations in a community is:

   A. ecological model
   B. ecological observation
   C. biological experiment
   D. biological system
   E. population experiment

12. The 1930s, Dust Bowl in the Great Plains was caused by:

   A. using renewable resources
   B. poor farming practices
   C. deforestation
   D. contour plowing
   E. using nonrenewable resources

13. The movement of organisms into a range is:

   A. population shift   **C.** immigration
   B. carrying capacity   **D.** emigration   **E.** bottleneck effect

14. Which of the following is NOT a basic method used by ecologists to study the living world?

   A. modeling   **C.** experimenting
   B. observing   **D.** animal training   **E.** hypothesizing

15. If immigration and emigration numbers remain equal, which is the most important contributing factor to a slowed growth rate?

    A. decreased death rate   **C.** increased birthrate

    B. constant birthrate   **D.** constant death rate   **E.** decreased birthrate

16. When farming, overgrazing, climate change, and/or seasonal drought change farmland into land that cannot support plant life, it is:

    A. deforestation   **C.** extinction

    B. monoculture   **D.** depletion   **E.** desertification

17. Which factor might NOT contribute to an exponential growth rate in a given population?

    A. reduced resources   **C.** higher birthrates

    B. less competition   **D.** lower death rates   **E.** increased longevity

18. Which of the following is NOT considered a sustainable-development strategy for management of Earth's resources?

    A. selective harvesting of trees   **C.** desertification

    B. crop rotation   **D.** contour plowing   **E.** none of the above

19. Which are two ways a population can decrease in size?

    A. Emigration and increased birthrate   **C.** Increased death rate and immigration

    B. Decreased birthrate and emigration   **D.** Immigration and emigration

                                         E. Increased birthrate and death rate

20. Farmers can reduce soil erosion by:

    A. plowing up roots

    B. grazing cattle on the land

    C. contour plowing

    D. increasing irrigation

    E. planting monocultures

21. Which age structure is most likely for a population that has not completed the demographic transition?

    A. 10 percent of people aged 50–54   **C.** 15 percent of people below age 15

    B. 5 percent of people aged 10–14   **D.** 50 percent of people below age 15

                                         E. 20 percent of people above age 60

22. Which two factors increase population size?

　A. births and immigration

　B. deaths and emigration

　C. births and emigration

　D. deaths and immigration

　E. none of the above

23. In a logistic growth curve, exponential growth is the phase in which the population:

　A. growth begins to slow down 　　**C.** reaches carrying capacity

　B. growth stops 　　**D.** death rate exceeds growth rate

　　　　　　　　　　　　　　　　E. grows quickly

24. An example of sustainable resource use is the use of predators and parasites to:

　A. control pest insects 　　**C.** harm natural resources

　B. eat unwanted plants 　　**D.** pollinate plants 　　**E.** feed live stock

25. The graph in the figure below shows the growth of a bacterial population. Which of the following correctly describes the growth curve?

　A. demographic 　　**C.** logistic

　B. exponential 　　**D.** limiting

　　　　　　　　　　E. linear

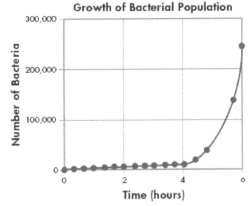

26. DDT was used to:

　A. form ozone 　　**C.** kill insects

　B. feed animals 　　**D.** fertilize soil 　　**E.** form greenhouse gases

27. All renewable resources:

　A. are living

　B. can be recycled or reused

　C. are unlimited in supply

　D. can regenerate or be replenished

　E. once were living

28. Density-dependent limiting factors include:

A. blizzards   **C.** disease

B. damming of rivers   **D.** earthquakes   **E.** average temperature

29. If a population grows larger than the carrying capacity of the environment:

A. birthrate must fall   **C.** birthrate may rise

B. death rate must fall   **D.** birthrate and death rate may rise

                              E. death rate may rise

30. One property that makes DDT hazardous over the long run is that DDT is:

A. deadly to herbivores

B. subject to biological magnification

C. volatile pesticide

D. insecticide

E. organic pesticide

31. As the population gets larger, it grows more quickly because the size of each generation of offspring is larger than the generation before it resulting in:

A. multiple growth   **C.** growth density

B. exponential growth   **D.** logistic growth   **E.** linear growth

32. The gray-brown haze often found over large cities is:

A. vapor   **C.** greenhouse gases

B. particulates   **D.** ozone layer   **E.** smog

33. Compounds that contribute to the formation of acid rain contain:

A. nitrogen and sulfur

B. ammonia and nitrates

C. carbon dioxide and oxygen

D. calcium and phosphorus

E. carbon dioxide and ammonia

34. The various growth phases through which most populations go are represented on:

A. normal curve   **C.** logistic growth curve

B. population curve   **D.** exponential growth curve

                              E. linear growth curve

35. Water lilies do not grow in desert sand because water availability to these plants in a desert is:

A. competition factor

B. logistic growth curve

C. limiting factor

D. carrying capacity

E. none of the above

36. The sulfur and nitrogen compounds in smog combine with water to form:

A. acid rain

B. chlorofluorocarbons

C. ozone

D. ammonia

E. greenhouse gases

37. Air and water pollution have been reduced by:

A. raising more cattle for food

B. increasing biological magnification

C. using fossil fuels in factories

D. using only unleaded gasoline

E. using more effective pesticides

38. Which would least likely to be affected by a density-dependent limiting factor?

A. population with a high immigration rate

B. large, dense population

C. population with a high birthrate

D. small, scattered population

E. population with a low death rate

39. Raising cattle and farming rice contribute to air pollution by:

A. releasing ozone into the atmosphere

B. producing smog which reacts to form dangerous ozone gas

C. producing particulates into the air

D. releasing sulfur compounds that form acid rain

E. releasing the greenhouse gas methane into the atmosphere

40. For most populations that are growing, as resources become less available, the population:

A. enters a phase of exponential growth

B. reaches carrying capacity

C. increases more rapidly

D. declines rapidly

E. enters a phase of linear growth

41. Which of the terms best describes the number of different species in the biosphere or in a particular area?

A. species diversity

B. genetic diversity

C. ecosystem diversity

D. biodiversity

E. ecological diversity

42. Which density-dependent factors other than the predator/prey relationship affect the populations of moose and wolves on an island?

A. A hurricane for both moose and wolves
B. Food availability for the moose and disease for the wolf
C. Extreme temperatures for the moose and flooding for the wolves
D. Parasitic wasps for the wolves and clear-cut forest for the moose
E. A drought for both moose and wolves

43. How are species diversity and genetic diversity different?

A. Species diversity measures the number of individuals of a species while genetic diversity measures the total variety of species
B. Conservation biology is concerned with species diversity but not with genetic diversity
C. Species diversity is evaluated only in ecosystems while genetic diversity is evaluated in the entire biosphere
D. Species diversity measures the number of species in the biosphere while genetic diversity measures the variety of genes in the biosphere, including genetic variation within species
E. Species diversity results from natural selection while genetic diversity results from genetic engineering

44. Which of the following is NOT likely to be a limiting factor on the sea otter population living in the ocean?

A. prey availability   **C.** disease
B. predation   **D.** competition   **E.** drought

45. Introduced species can threaten biodiversity because they can:

A. crowd out native species   **C.** cause desertification of land
B. reduce fertility of native species   **D.** cause biological magnification
                                             E. mate with native species

46. Which of the following is a density-independent limiting factor?

A. parasitism and disease   **C.** predator/prey relationships
B. eruption of a volcano   **D.** struggle for food
                                     E. available water or sunlight

47. What would reduce competition within a species' population?

A. higher population density   **C.** higher birthrate
B. fewer resources   **D.** fewer individuals
                                     E. lower death rate

48. A major factor that negatively affects biodiversity is:

A. nonrenewable resources

B. contour plowing

C. habitat fragmentation

D. biological magnification

E. preservation of ecosystems

49. All of the following are threats to biodiversity EXCEPT:

A. habitat fragmentation

B. desertification

C. habitat preservation

D. biological magnification of toxic compounds

E. all of the above are threats

50. The graph in the figure below shows the changes in a mosquito population. What caused the changes seen in the graph?

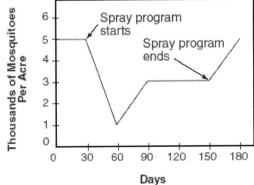

A. increase in resources

B. density-dependent limiting factor

C. reduction in resources

D. increase in predation

E. density-independent limiting factor

51. Each of the following is a density-dependent limiting factor EXCEPT:

A. crowding

B. disease

C. competition

D. temperature

E. predation

52. The "hot spot" strategy seeks to protect species in danger of extinction due to:

A. human activity

B. biological magnification

C. captive breeding programs

D. expanding national parks

E. preservation of ecosystems

53. About 500 years ago, the world's population started to:

A. level off

B. grow more rapidly

C. reach carrying capacity

D. decrease

E. grow sporadically

54. Protecting an entire ecosystem ensures that:

  A. interactions among many species are preserved
  B. governments will set aside land
  C. existing parks and reserves will expand
  D. captive breeding programs will succeed
  E. wetlands will not be disturbed

55. The goals of biodiversity conservation include all of the following EXCEPT:

  A. Ensuring that local people benefit from conservation efforts
  B. Preserving habitats and ecosystems
  C. Introducing exotic species into new environments
  D. Protecting individual species
  E. Protecting populations

56. Countries in the first stage of demographic transition have:

  A. slowly growing population
  B. more old people than young people
  C. low death rate and a low birthrate
  D. high death rate and a low birthrate
  E. high death rate and a high birthrate

57. Demography is the scientific study of:

  A. disease
  B. human populations
  C. parasitism
  D. modernized countries
  E. urban development

58. The amount of land and water necessary to provide the resources for a person's living and to neutralize that person's waste is that person's:

  A. habitat
  B. ecological sustainability
  C. biodiversity
  D. ecological footprint
  E. environmental sustainability

59. Population density refers to the:

  A. number of people in each age group
  B. carrying capacity
  C. number of individuals in community
  D. area inhabited by a population
  E. number of individuals per unit area

60. The anticipated human population by the year 2050 is about:

A. 78 million   **C.** 9 billion

B. 11trillion   **D.** 3.8 billion   **E.** 900 million

61. Demographic transition is change from high birthrates and high death rates to:

A. indefinite growth

B. low birthrates and high death rates

C. low birthrates and low death rates

D. exponential growth

E. high birthrates and low death rates

62. Imported plants and animals in Hawaii have:

A. increased crop yields

B. improved soil fertility

C. increased the native bird species

D. caused native species to die out

E. caused habitat fragmentation

63. A benefit of monoculture farming practices is:

A. pest resistance of the crops

B. the ability to grow a lot of food

C. the ability to spend less money on fertilizer

D. the use of less water for irrigation

E. disease resistance of the crops

64. An example of a density-independent limiting factor is:

A. disease   **C.** mutualism

B. predation   **D.** competition   **E.** unusual weather

65. In countries like India, the human population is growing:

A. logistically   **C.** exponentially

B. linearly   **D.** transitionally   **E.** demographically

66. Most of the worldwide human population is growing exponentially because:

A. human populations do not conform to the logistic model

B. the food supply is limitless

C. human populations have not reached their exponential curve

D. most countries have not yet completed the demographic transition

E. competition for resources is absent

67. Success at solving an environmental problem is more likely when researchers follow the basic principles of ecology because:

    A. ecology uses scientific research to identify the cause of the problem and the best practices to solve the problem

    B. ecologists are very good at influencing government officials into changing laws to improve the environment

    C. ecological solutions to problems are usually very easy to implement and can be done quickly

    D. most people in the world are more interested in saving the environment than in their own comfort and convenience

    E. ecology is concerned with the preservation of natural resources

68. Which environmental problem can be identified by ecologists through the data in the graph?

    A. habitat fragmentation

    B. desertification

    C. the hole in the ozone layer

    D. ecological diversity

    E. global warming

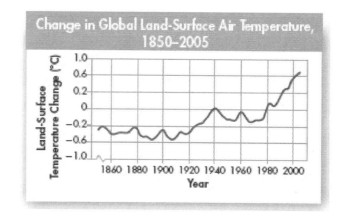

69. The sum total of the genetically based variety of all organisms in the biosphere is called:

    A. ecosystem diversity   **C.** genetic diversity

    B. biodiversity   **D.** species diversity   **E.** none of the above

70. An ecological hot spot is an area where:

    A. many habitats and species are at high risk of extinction

    B. species diversity is too high

    C. habitats show a high amount of biodiversity

    D. hunting is encouraged

    E. species show a high amount of biodiversity

Answer Key

1: D	11: A	21: D	31: B	41: A	51: D	61: C
2: A	12: B	22: A	32: E	42: B	52: A	62: D
3: B	13: C	23: E	33: A	43: D	53: B	63: B
4: D	14: D	24: A	34: C	44: E	54: A	64: E
5: C	15: E	25: B	35: C	45: A	55: C	65: C
6: E	16: E	26: C	36: A	46: B	56: E	66: D
7: A	17: A	27: D	37: D	47: D	57: B	67: A
8: B	18: C	28: C	38: D	48: C	58: D	68: E
9: C	19: B	29: E	39: E	49: C	59: E	69: B
10: E	20: D	30: B	40: B	50: E	60: C	70: A

UNIT 3. GENETICS

Chapter 3.1: DNA and Protein Synthesis

1. DNA and RNA differ because:

A. only DNA contains phosphodiester bonds
B. only RNA contains pyrimidines
C. DNA is in the nucleus and RNA is in the cytosol
D. RNA is associated with ribosomes and DNA is associated with histones
E. RNA contains a phosphate group in its ribose ring

2. In the 1920s, circumstantial evidence indicated that DNA was the genetic material. Which of the following experiments led to the acceptance of this hypothesis?

A. Griffith's experiments with *Streptococcus pneumoniae*
B. Avery, MacLeod and McCarty's work with isolating the transforming principle
C. Hershey and Chase's experiments with viruses and radioisotopes
D. A, B and C were used to support this hypothesis
E. Darwin's theory of natural selection

3. Within a guanine molecule, the N-glycosidic bond is relatively unstable and can be hydrolyzed through depurination. Which of these molecules is most likely to undergo depurination?

A. sterols   **C.** phospholipids
B. lipids   **D.** proteins   **E.** DNA

4. What process duplicates a single gene?

A. Unequal recombination at repeated sequences that flank the gene
B. Equal recombination at repeated sequences that flank the gene
C. Unequal recombination within the single gene
D. Equal recombination within the single gene
E. All of the above

5. Which element is NOT found within nucleic acids?

A. nitrogen   **C.** phosphorus
B. oxygen   **D.** sulfur   **E.** carbon

6. Which RNA molecule is translated?

   A. miRNA   **B.** tRNA   **C.** rRNA   **D.** mRNA   **E.** C and D

7. The aging of normal cells is associated with:

   A. loss of telomerase activity
   B. a decrease in contact inhibition
   C. an increase in mutation rate
   D. activation of the maturation promoting factor
   E. extranuclear inheritance

8. Protein synthesis in eukaryotic cells initiates in which of following structures?

   A. nucleus
   B. Golgi
   C. cytoplasm
   D. rough endoplasmic reticulum
   E. smooth endoplasmic reticulum

9. When a gene is duplicated on one chromatid, the gene on the other chromatid is:

   A. also duplicated   **C.** transposed to another site
   B. inverted   **D.** maintained as a single gene   **E.** deleted

10. Experiments designed by Avery, McLeod and McCarty to identify the transforming principle were based on:

      I. purifying each of the macromolecule types from a cell-free extract
      II. removing each of the macromolecules from a cell, then testing its type
      III. selectively destroying the different macromolecules in a cell-free extract

   A. I only   **B.** II only   **C.** III only   **D.** I, II and III   **E.** I and II only

11. What is the term for a blotting method where proteins are transferred from a gel to membranes and probed by antibodies to specific proteins?

   A. Eastern blotting
   B. Western blotting
   C. Northern blotting
   D. Southern blotting
   E. Both Northern and Western blotting

12. The figure shows a nucleotide. At what position will the incoming nucleotide be attached in the figure?

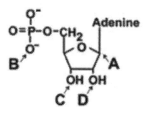

A. position A
B. position B
C. position C
D. position D
E. none of the above

13. All of the following is correct about DNA, EXCEPT:

A. the strands are anti-parallel
B. the basic unit is a nucleotide
C. the sugar molecule is deoxyribose
D. guanine binds to cytosine via three hydrogen bonds
E. adenine and guanine are pyrimidines

14. Tumor-suppressor genes normally control:

A. cell differentiation
B. necrosis
C. cell proliferation or activation of apoptosis
D. sister chromatid separation
E. protein degradation

15. Select the correct statement for aminoacyl tRNA synthetase.

A. It binds several different amino acids
B. It is an enzyme that uses energy from ATP to attach a specific amino acid to a tRNA
C. It is a tRNA that covalently binds amino acids
D. It synthesizes tRNA
E. It synthesizes rRNA

16. Griffith's experiment with pneumococcus demonstrated that:

A. smooth bacteria can survive heating
B. DNA, not protein, is the genetic molecule
C. materials from dead organisms can affect and change living organisms
D. nonliving viruses can change living cells
E. the virus injects its DNA into the host cell

17. In the genetic code, deciphered by Noble laureate Marshall W. Nirenberg, each amino acid is coded for by three nucleotides (codons). How many possible codons exist in nature that code for 20 amino acids found in polypeptides?

A. 4   B. 20   C. 27   D. 64   E. 16

---

18. What mechanism is used to target proteins to organelles (e.g. chloroplast, mitochondrion)?

   A. Addition of phosphate groups to the protein
   B. Synthesizing the proteins as zymogens
   C. Adding prosthetic groups to the protein
   D. Cysteine bond formation
   E. Signal sequence at the N-terminus of the polypeptide

19. When DNA is treated with 2-aminopurine, adenine is replaced by guanine on one strand. During replication, the complementary strand will have a substitution of:

   A. guanine for adenine
   B. adenine for guanine
   C. cytosine for thymine
   D. thymine for cytosine
   E. adenine for cytosine

20. Before the genetic code was determined experimentally by Nobel laureate Marshall W. Nirenberg, why was it hypothesized that each codon would contain at least three bases?

   A. Three bases are needed to produce a stable codon structure
   B. There were three known nucleotide bases
   C. There were more proteins than nucleotide bases
   D. Three bases can form $4^3 = 64$ pairs, which is enough to encode 20 amino acids
   E. There were twenty known amino acids

21. DNA of bacteria grown in heavy medium (^{15}N) were isolated and added to an *in vitro* synthesis system. Then the bacteria are grown in light medium (^{14}N). After several hours, a sample of DNA was taken and analyzed for differing densities. How many DNA densities were found in the sample after 2 generations?

   A. 1   **B.** 2   **C.** 4   **D.** 8   **E.** 12

22. Which experimental procedure(s) simultaneously measure(s) the level of all mRNA in a tissue?

   I. Northern blot   II. *In situ* hybridization   III. Microarray experiment

   A. I only   **C.** III only
   B. II only   **D.** I, II and III   **E.** I and III only

23. In order to show that DNA is the "transforming principle," Avery, MacLeod and McCarty showed that DNA could transform nonvirulent strains of pneumococcus. Their hypothesis was strengthened by their demonstration that:

   A. enzymes that destroyed proteins also destroyed transforming activity

   B. enzymes that destroyed nucleic acids also destroyed transforming activity

   C. enzymes that destroyed complex carbohydrates also destroyed transforming activity

   D. the transforming activity was destroyed by boiling

   E. other strains of bacteria were also transformed successfully

24. Which of these are components of the codon-anticodon hybridization on the ribosome that determines the fidelity of protein synthesis?

   A. mRNA & tRNA   **C.** tRNA & rRNA

   B. mRNA & rRNA   **D.** DNA & RNA polymerase   **E.** RNA polymerase

25. Which procedure measures the level of mRNA from only a single gene?

   I. Northern blot

   II. *In situ* hybridization

   III. Microarray experiment

   A. II only   **C.** II and III only

   B. I and II only   **D.** I, II and III   **E.** I and III only

26. Which stage of cell division is the stage when chromosomes replicate?

   A. prophase   **B.** telophase   **C.** anaphase   **D.** metaphase   **E.** interphase

27. If the transcript's sequence is 5'-CUAAGGGCUAC-3', what is the sequence of the DNA template?

   A. 3'-GUAGCCCUUAG-5'

   B. 3'-GTACGCCTTAG-5'

   C. 5'-GTAACCCTTAG-3'

   D. 5'-GUTACCUGUAG-3'

   E. 5'-GTAGCCCTTAG-3'

28. Duplicated genes:

   A. are more common in prokaryote genomes than in eukaryote genomes

   B. are closely related but diverged in sequence and function over evolutionary time

   C. never encode for important proteins, such as transcription factors

   D. encode for proteins that catalyze different steps of a biochemical pathway

   E. all of the above

29. The Hershey–Chase experiment:

    A. proved that DNA replication is semiconservative

    B. used ^{32}P to label protein

    C. used ^{35}S to label DNA

    D. supported the hypothesis that DNA is the transforming molecule

    E. both A and C

30. Which of these statements is NOT correct about DNA replication?

    A. DNA polymerase synthesizes and proofreads the DNA

    B. RNA primers are necessary for the hybridization of the polymerase

    C. Ligase relaxes positive supercoils that accumulate as the replication fork opens

    D. DNA polymerase adds Okazaki fragments in a 5' → 3' direction

    E. DNA polymerase adds deoxynucleotides in a 5' → 3' direction

31. Which of the following structures represent a peptide bond between adjacent amino acids?

    A. structure A   **C.** structure C

    B. structure B   **D.** structure D

                            E. structures A and D

32. All of these statements apply to proteins, EXCEPT:

    A. they regulate cell membrane trafficking   **C.** they can be hormones

    B. they catalyze chemical reactions   **D.** they undergo self-replication

                                         E. they bind antigens

33. Eukaryote RNA polymerase usually:

    A. binds to the TATAA promoter sequence and initiates transcription

    B. needs general transcription factors to bind to the promoter and initiate basal-level transcription

    C. needs specific regulatory transcription factors to bind to the promoter and initiate basal-level transcription

    D. transcribes tRNA genes

    E. transcribes mRNA genes

34. If a particular RNA sequence has a cytosine content of 25%, what is its adenine content?

    A. 50%

    B. 37.5%

    C. 12.5%

    D. 25%

    E. cannot be determined

35. If a portion of prokaryotic mRNA has the base sequence 5′-ACUACUA<u>U</u>GCGUCGA-3′, what could result from a mutation where the underlined base is changed to A?

    I. truncation of the polypeptide

    II. inhibition of initiation of translation

    III. no effect on protein synthesis

    A. I and II only   **C.** II and III only

    B. I and III only   **D.** III only   **E.** II only

36. Which statement is INCORRECT about the genetic code?

    A. Many amino acids are specified by more than one codon

    B. Most codons specify more than one amino acid

    C. There are multiple stop codons

    D. Codons are 3 bases in length

    E. The start codon inserts methionine at the amino end of the polypeptide

37. In bacteria, the enzyme that removes the RNA primers is called:

    A. DNA ligase   **C.** reverse transcriptase

    B. primase   **D.** DNA polymerase I   **E.** helicase

38. Okazaki fragments are:

    A. synthesized in a 5'→ 3' direction by DNA polymerase I

    B. covalently linked by DNA polymerase I

    C. components of the leading strand

    D. components of DNA synthesized to fill in gaps after excision of the RNA primer

    E. synthesized in a 5'→ 3' direction by DNA polymerase III

39. Peptide bond synthesis is catalyzed by:

    A. tRNA in the cytoplasm   **C.** ribosomal RNA

    B. ribosomal proteins   **D.** mRNA in the ribosome

                                         E. none of the above

40. Which of the following statements does NOT apply to protein synthesis?

   A. The process does not require energy

   B. rRNA is required for proper binding of the mRNA message

   C. tRNA molecules shuttle amino acids that are assembled into the polypeptide

   D. The amino acid is bound to the 3' end of the tRNA

   E. The mRNA is synthesized from $5' \rightarrow 3'$

41. All of the following statements about PCR are correct, EXCEPT:

   A. PCR can be used to obtain large quantities of a particular DNA sequence

   B. PCR does not require knowledge of the terminal DNA sequences of the region to be amplified

   C. PCR uses a DNA polymerase to synthesize DNA

   D. PCR uses short synthetic oligonucleotide primers

   E. PCR involves heating the DNA sample to denature complementary base pairing

42. The shape of a tRNA is determined primarily by:

   A. its number of bases

   B. proteins that bind it

   C. tRNA and aminoacyl tRNA synthetase interactions

   D. intramolecular base pairing

   E. hydrophobic interactions

43. In prokaryotic cells, methylated guanine contributes to:

   A. increased rate of DNA replication

   B. decreased rate of DNA replication

   C. correcting the separation of DNA strands

   D. proofreading the replicated strands

   E. correcting mismatched pairs of bases

44. In the polymerization reaction by DNA polymerase, what is the function of magnesium?

Primer (free 3' OH) + 5' PPP $\longrightarrow$ Primer 3'O-P-5' + PPi

Mg^{+2}, 4 dNTPs, DNA polymerase

   A. cofactor

   B. monovalent metal ion

   C. substrate

   D. enzyme

   E. coenzyme

45. The structure of the ribosome is created by:

   I. internal base pairing of rRNA

   II. ribosomal proteins

   III. internal base pairing of mRNA

   IV. internal base pairing of tRNA

   A. I only

   B. II only

   C. I and II only

   D. I, II and IV only

   E. I, II and III only

46. Which statement is true for tRNA?

A. It has some short double-stranded segments
B. It has a poly-A tail
C. It is produced in the nucleolus
D. It is a long molecule of RNA
E. It is the template for protein synthesis

47. Which chemical group is at the 5' end of a single polynucleotide strand?

A. diester group
B. purine base
C. hydroxyl group
D. phosphate group
E. nitrogen group

48. The drug aminoacyl-tRNA is an analog of puromycin. Both have an amino group capable of forming a peptide bond, but puromycin lacks a carboxyl group to form another peptide bond. What is a possible effect of adding puromycin to bacteria undergoing protein synthesis?

A. Inhibition of initiation of protein synthesis
B. Inhibition of entry of aminoacyl-tRNA into the P site during elongation
C. Inability to form a complete ribosome
D. Substitution of puromycin for another amino acid in the protein, yielding a normal length protein
E. Termination of protein synthesis via covalent attachment of puromycin

49. In *E. coli* cells, DNA polymerase I:

    I.   synthesizes most of the Okazaki fragments
    II. simultaneously copies both strands of DNA
    III. degrades the RNA primer portion of Okazaki fragments

A. I only   **B.** II only   **C.** III only   **D.** I and III only   **E.** I and II only

50. The overall fidelity of DNA replication is very high. During DNA synthesis, the error rate is on the order of one mismatched nucleotide per:

A. 100   **B.** 1,000   **C.** 10,000   **D.** 1,000,000   **E.** 10,000.000

51. All of the following are contained within a molecule of DNA, EXCEPT:

A. nitrogenous bases
B. phosphodiester bonds
C. polypeptide bonds
D. deoxyribose sugars
E. phosphate groups

52. In *E. coli* cells, DNA polymerase III:

    A. synthesizes most of the Okazaki fragments

    B. removes RNA primer

    C. is the only DNA polymerase used by *E. coli* during replication

    D. degrades the RNA portion of an Okazaki fragment

    E. synthesizes DNA in the 3' to 5' direction

53. Which molecule belongs to a different chemical category than all others?

    A. uracil   **B.** guanine   **C.** adenine   **D.** thymine   **E.** cysteine

54. Enzyme that cleaves DNA at sequence-specific site is:

    A. restriction endonuclease   **C.** DNA polymerase

    B. exonuclease   **D.** ligase   **E.** integrase

55. *E. coli* RNA polymerase initiated transcription and synthesized one phosphodiester bond. Which molecule shown below is RNA polymerase made from?

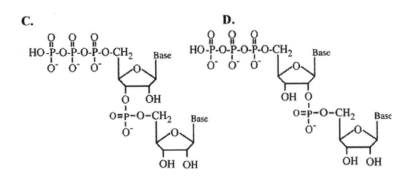

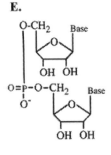

56. What is the rate that PCR increases the amount of DNA during each additional cycle?

    A. additively   **C.** linearly

    B. exponentially   **D.** systematically   **E.** gradually

57. Which of the following is present in RNA, but absent in DNA of a cell?

    A. additional hydroxyl group   **C.** thymine

    B. hydrogen bonds   **D.** double helix   **E.** phosphodiester bonds

58. After the new DNA strands are synthesized, which enzyme is needed to complete the process of DNA replication?

    A. primase   **C.** helicase

    B. ligase   **D.** reverse transcriptase   **E.** both B and D

59. When a base is paired with its complementary strand, which of these strands would have the highest melting point?

    A. TTAGTCTC   **C.** AGCTTCGT

    B. TTTTAAAA   **D.** CGCGTATA   **E.** GCCAGTCG

60. A technique that looks for possible gene functions by mutating wildtype genes is:

    A. contig building   **C.** reverse genetics

    B. transgenetics   **D.** gene therapy   **E.** gene mapping

61. How many high-energy phosphate bonds are needed for the translation of a 50-amino acid polypeptide (starting with mRNA, tRNA, amino acids, and all the necessary enzymes)?

    A. 49   **B.** 50   **C.** 101   **D.** 199   **E.** 150

62. The mRNA in *E. coli* cells is composed primarily of:

    A. four bases – A, T, C, G

    B. phosphodiester linkages connecting deoxyribonucleotide molecules

    C. two strands that base pair in anti-parallel orientation

    D. processed RNA molecule containing introns

    E. phosphodiester linkages connecting ribonucleotide molecules

63. Which of the following statements about mismatch repair of DNA is correct?

    A. DNA is scanned for any base-pairing mismatches after methyl groups are added to guanines

    B. Errors in replication made by DNA polymerase are corrected on the unmethylated strand

    C. All abnormal bases are removed by the proofreading mechanism

    D. Repairs from high-energy radiation damage are made

    E. Mismatch repair occurs on each strand of DNA during replication

64. What is the first amino acid of each protein of eukaryotic cells?

A. methionine
B. glutamate

C. valine
D. proline

E. isoleucine

65. DNA in *E. coli* is composed of:

    I. four bases – A, T, C, G
    II. phosphodiester linkages that connect deoxyribonucleotide molecules
    III. two strands that base pair in anti-parallel orientation
    IV. phosphodiester linkages that utilize the 3'-OH

A. I and II only
B. I and III only

C. I, II and III only
D. I, II, III and IV

E. II, III and IV only

66. If a peptide has the sequence val-ser-met-pro and the tRNA molecules used in its synthesis have the corresponding sequence of anticodons 3'-CAG-5', 3'-UCG-5', 3'-UAC-5', 3'-UUU-5', what sequence of the DNA codes for this peptide?

A. 5'–CAGTCGTACTTT–3'
B. 5'–TTTCATGCTGAC–3'

C. 5'–GACGCTCATTTT–3'
D. 5'–UUUCAUGCUGAC–3'
E. 5'–CAGUCGUACUUU–3'

67. The site of the DNA template that RNA polymerase binds to during transcription is:

A. promoter
B. leader sequence

C. enhancer
D. domain

E. transcription factor

68. Which dipeptide is synthesized by a ribosome?

A. isoleucine-glycine
B. cytosine-guanine

C. proline-thymine
D. uracil-glutamic acid

E. both A and B

69. Which of the following is the correct order of events in the delivery of a protein to its cellular destination?

A. Signal sequence binds to a docking protein → transmembrane-gated channel opens →
    → protein enters the organelle
B. Membrane channel is formed → signal sequence binds to docking protein →
    → chaperonins unfold protein → protein enters the organelle → protein refolds
C. Chaperonins unfold protein → signal sequence binds to docking protein →
    → membrane channel is formed → protein enters the organelle → protein refolds
D. Membrane channel is formed → chaperonins unfold protein → signal sequence binds to
    a docking protein → protein enters the organelle → protein refolds
E. Signal sequence binds to a docking protein → membrane channel is formed →
    → chaperonins unfold protein → protein enters the organelle → protein refolds

70. Select the correct mRNA sequences depending on the direction RNA polymerase transcribes.

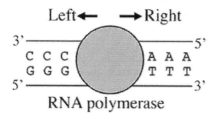

RNA sequence if RNA polymerase goes left:

RNA sequence if RNA polymerase goes right:

A. 5'-GGG-3'   5'-AAA-3'

B. 5'-GGG-3'   5'-UUU-3'

C. 5'-CCC-3'   5'-AAA-3'

D. 5'-CCC-3'   5'-UUU-3'

E. None of the above

71. Which of the following statements is TRUE for the base composition of DNA?

A. In double stranded DNA, the number of G bases equals the number of T bases

B. In double stranded DNA, the number of A bases equals the number of T bases

C. In double stranded DNA, the number of C bases equals the number of T bases

D. In each single strand, the number of A bases equals the number T bases

E. In double stranded DNA, the number of G bases equals the number of A bases

72. The figure shows a replication fork in *E. coli*. Which of the indicated sites is the 3' end of the lagging strand?

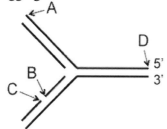

A. site A   **C.** site C

B. site B   **D.** site D   **E.** sites C and D

73. Ribosomal subunits are isolated from bacteria grown in a "heavy" ^{13}C and ^{15}N medium and added to an *in vitro* system that actively synthesizes protein. Following translation, a sample is removed and centrifuged. Which of these would be the best illustration of centrifugation results?

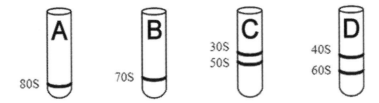

A. Test tube A
B. Test tube B
C. Test tube C
D. Test tube D
E. Test tubes A and C

74. Which statement is TRUE?

A. polypeptides are synthesized by addition of amino acids to the amino terminus
B. prokaryotic RNA usually undergoes nuclear processing
C. RNA polymerase has a proof-reading activity
D. prokaryotic RNA contains introns
E. 3' end of mRNA corresponds to the carboxyl terminus of the protein

75. A codon for histidine is 5'-CAU-3'. The anticodon in the tRNA that brings histidine to the ribosome is:

A. 5'-CAU-3'
B. 5'-GUA-3'
C. 5'-UAC-3'
D. 5'-AUG-3'
E. none of the above

76. During translation elongation, the existing polypeptide chain is transferred to which site as the ribosome moves in the 3' direction?

A. tRNA occupying the A site
B. tRNA occupying the P site
C. ribosomal rRNA
D. signal recognition particle
E. none of the above

77. Which of the following is likely the primer to amplify the DNA fragments *via* PCR?

5'-ATCGGTATGTAACGCTCACCTGT-3'

A. 5'-ACAG-3'
B. 5'-AGAC-3'
C. 5'-TAGC-3'
D. 5'-GACT-3'
E. 5'-CTGT-3'

78. Which of the following statements about the genetic code is FALSE?

    A. It is mostly the same for *E. coli* and humans

    B. It is redundant

    C. It is ambiguous

    D. It has one codon for starting translation

    E. All of the above are true statements

79. What portion of the polypeptide chain is responsible for establishing and maintaining the force that is used to stabilize the secondary structure?

    A. C-terminus   **C.** carbonyl oxygen

    B. N-terminus   **D.** R-groups   **E.** both A and B

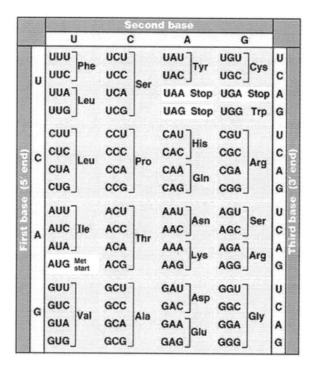

80. A ribosome has made a tri-peptide, MET-ARG-SER, which is attached to the tRNA in the P site. Using the genetic code table, what codon is in the E site of the ribosome?

    A. AUG   **C.** UCA

    B. CGU   **D.** UGA   **E.** It cannot be determined

81. A ribosome has made a tri-peptide, MET-ARG-SER, which is attached to the tRNA in the P site. Using the genetic code table, what codon is in the A site of the ribosome?

    A. AUG   **C.** UCA

    B. CGU   **D.** UGA   **E.** It cannot be determined

Answer Key

1: D	11: B	21: B	31: A	41: B	51: C	61: D	71: B
2: D	12: C	22: C	32: D	42: D	52: A	62: E	72: B
3: E	13: E	23: B	33: B	43: E	53: E	63: B	73: C
4: A	14: C	24: A	34: E	44: A	54: A	64: A	74: E
5: D	15: B	25: B	35: A	45: C	55: C	65: D	75: D
6: D	16: C	26: E	36: B	46: A	56: B	66: B	76: B
7: A	17: D	27: E	37: D	47: D	57: A	67: A	77: A
8: C	18: E	28: B	38: E	48: E	58: B	68: A	78: C
9: E	19: C	29: D	39: C	49: C	59: E	69: E	79: C
10: C	20: D	30: C	40: A	50: D	60: C	70: D	80: B

Chapter 3.2: Mendelian Genetics, Inheritance Patterns

1. Which of the following is a characteristic that makes an organism unsuitable for genetic studies?

A. Large number of chromosomes

B. Short generation time

C. Ease of cultivation

D. Ability to control crosses

E. Availability of a variation for traits

2. People with the sex-linked genetic disease, hemophilia, suffer from excessive bleeding because their blood will not clot. Tom, Mary and their 4 daughters do not exhibit symptoms of hemophilia. However, their son exhibits symptoms of hemophilia because:

A. Tom is heterozygous

B. Tom is homozygous

C. Mary is heterozygous

D. Mary is homozygous

E. All of the above are equally probable

3. Which factor would NOT favor an *r-selection* over a *K-selection* reproductive strategy?

A. Commercial predation by humans

B. Limited space

C. Shorter growing season

D. Frequent and intense seasonal flooding

E. None of the above

4. Several eye colors are characteristic for *Drosophila melanogaster*. Red eyes are dominant over sepia or white eyes. What percent of offspring of a sepia eyed fly will have sepia eyes if that fly mated with a red eyed fly that was a cross of red eyed and sepia eyed parents?

A. 0%   **B.** 25%   **C.** 50%   **D.** 75%   **E.** 100%

5. Color blindness mutations in humans result from:

A. fragile X syndrome

B. chromosome nondisjunction

C. reciprocal translocation

D. dosage compensation

E. unequal crossing-over

6. A small subpopulation of flies with a slightly advantageous modification in structure was found extinct after a locally-isolated decimating fire. A geneticist would most likely attribute the loss of this advantageous gene to:

A. differential reproduction

B. natural selection

C. Hardy-Weinberg principle

D. genetic drift

E. high genetic mutation rate

7. Which of the following methods was NOT used by Mendel in his study of the genetics of the garden pea?

    A. Maintenance of true-breeding lines

    B. Cross-pollination

    C. Microscopy

    D. Production of hybrid plants

    E. Quantitative analysis of results

8. If you cross AAbbCc x AaBbCc where A, B and C are unlinked genes, what is the probability of obtaining an offspring with AaBbCc genotype?

    A. 1/4   **B.** 1/16   **C.** 1/64   **D.** 1/32   **E.** 1/8

9. What is the risk of having a child affected by a disease with an autosomal recessive inheritance if both the mother and father are carriers for that disease?

    A. 0%   **B.** 25%   **C.** 50%   **D.** 75%   **E.** 66%

10. All of the following are necessary conditions for the Hardy-Weinberg equilibrium, EXCEPT:

    A. forward mutation rate equals backward mutation rate

    B. random emigration and immigration

    C. large gene pool

    D. random mating

    E. no inbreeding

11. For the multi-step progression of cancer, the major mutational target(s) is/are:

    A. telomerase   **C.** tumor suppressor gene

    B. X-linked traits   **D.** trinucleotide repeats

    E. transcription factors

12. Since the gene responsible for color blindness is found on the X chromosome, what is the chance that a son of a colorblind man and a woman-carrier will be colorblind?

    A. 75%   **B.** 100%   **C.** 25%   **D.** 50%   **E.** 66%

13. If two strains of true-breeding plants that have different alleles for a certain character are crossed, their progeny are:

    A. P generation   **C.** F_2 generation

    B. F_1 generation   **D.** F_1 crosses

    E. F_2 progeny

14. The arabidopsis plant has 5 pairs of homologous chromosomes. Suppose an Arabidopsis is heterozygous for 5 mutations, and each mutation is on a different chromosome. How many genetically distinct gametes will this plant make after meiosis?

   A. 5   **B.** 10   **C.** 32   **D.** 64   **E.** 25

15. An unknown inheritance pattern has the following characteristics:

- 25% probability of having a homozygous unaffected child
- 25% probability of having a homozygous affected child
- 50% probability of having a heterozygous child

Which of these Mendel's inheritance patterns best matches the above observations?

   A. autosomal recessive   **C.** X-linked recessive

   B. autosomal dominant   **D.** X-linked dominant

                               E. cannot be determined without more information

16. What is the frequency of heterozygotes within a population in Hardy-Weinberg equilibrium if the frequency of the dominant allele D is three times that of the recessive allele d?

   A. 7.25%   **B.** 12.75%   **C.** 33%   **D.** 37.5%   **E.** 50%

17. A recessive allele may appear in a phenotype due to:

   A. gain-of-function mutation   **C.** senescence

   B. acquired dominance   **D.** processivity

                               E. the loss of heterozygosity

18. Which of the following observations would support the theory of maternal inheritance for the spunky phenotype?

   A. Spunky female x wild-type male → progeny all spunky

   B. Wild-type female x spunky male → progeny all spunky

   C. Wild-type female x spunky male → progeny 1/2 spunky, 1/2 wild-type

   D. Spunky female x wild-type male → progeny 1/2 spunky, 1/2 wild-type

   E. Spunky female x wild-type male → progeny all wild-type

19. Mendel concluded that each pea has two units for each characteristic, and each gamete contains one unit. Mendel's "unit" is now referred to as:

   A. genome   **C.** codon

   B. hnRNA   **D.** transcription factor

                    E. gene

20. Which of the following is most likely to lead to a complete loss of gene function?

    A. Missense mutation that causes the nonpolar methionine to be replaced with glycine

    B. GC base pair being converted to an AT base pair in the promoter

    C. A mutation in the third codon of the open reading frame

    D. A base pair change that does not affect the amino acid sequence

    E. All of the above

21. All of the following effects are possible to occur after a mutation, EXCEPT:

    A. abnormal lipid production

    B. abnormal protein production

    C. gain of enzyme function

    D. loss of enzyme function

    E. no change in protein production

22. Which cross must produce all green, smooth peas if green (G) is dominant over yellow (g) and smooth (S) is dominant over wrinkled (s)?

    A. GgSs x GGSS

    B. GgSS x ggSS

    C. Ggss x GGSs

    D. GgSs x GgSs   **E.** None of the above

23. Retinoblastoma is inherited as:

    A. a multifactorial trait

    B. X-linked recessive

    C. Mendelian dominant

    D. Mendelian recessive

    E. an extranuclear trait

24. Tay-Sachs disease is a rare autosomal recessive genetic disorder. If a male heterozygous carrier and a female heterozygous carrier have a first child who is homozygous for the wild type, what is the chance that the second child develops Tay-Sachs?

    A. 1/3   **B.** 1/2   **C.** 1/16   **D.** 1/8   **E.** 1/4

25. Mendel's crossing of spherical-seeded pea plants with wrinkled-seeded pea plants resulted in progeny that all had spherical seeds. This indicates that the wrinkled-seed trait is:

    A. codominant

    B. dominant

    C. recessive

    D. penetrance

    E. both A and C

26. The end result of mitosis is the production of:

    A. two (1N) cells identical to the parent cell

    B. two (2N) cells identical to the parent cell

    C. four (1N) cells identical to the parent cell

    D. four (2N) cells identical to the parent cell

    E. four (1N) unique cells that are genetically different than the parent cell

27. What is the probability (p) of randomly rolling a pair of 4s using two six-sided dice?

    A. 1/12   **B.** 1/16   **C.** 1/6   **D.** 1/72   **E.** 1/36

28. The degree of genetic linkage is often measured by the:

    A. frequency of nonsense mutations
    B. histone distribution
    C. frequency of missense mutations
    D. probability of crossing over
    E. AT/GC ratio

29. Cancers associated with defects in mismatch repair are passed down via:

    A. dominant inheritance   **C.** X-linked inheritance
    B. maternal inheritance   **D.** epigenetic inheritance
                             E. recessive inheritance

30. Given that color blindness is a recessive trait inherited through a sex-linked gene on the X chromosome, what is the probability that a daughter born to a colorblind father and a mother who carries the trait will be a carrier?

    A. 0%   **B.** 25%   **C.** 50%   **D.** 100%   **E.** 12.5%

31. What is the probability that a cross between a true-breeding pea plant with a dominant trait and a true-breeding pea plant with a recessive trait will result in all F_1 progeny having the dominant trait?

    A. 1/2   **B.** 1/4   **C.** 0   **D.** 1   **E.** 1/9

32. The tall allele is dominant to the short allele. True breeding tall plants were crossed to true breeding short plants. The F_1 plants were self crossed to produce F_2 progeny. What are the phenotypes of the F_1 and F_2 progeny?

    A. All F_1 and 1/4 of the F_2 are short
    B. All F_1 are short and 1/4 of the F_2 are tall
    C. All F_1 and 3/4 of the F_2 are tall
    D. All F_1 are tall and 3/4 of the F_2 plants are short
    E. All of the above are equally probable

33. All of these DNA lesions result in a frame shift mutation, EXCEPT:

    A. 1 inserted base pairs   **C.** 4 inserted base pair
    B. 2 substituted base pairs   **D.** 2 deleted base pairs
                                 E. 5 deleted base pairs

34. If two species with the AaBbCc genotype reproduce, what is the probability that their progeny have the AABBCC genotype?

A. 1/2   **B.** 1/4   **C.** 1/16   **D.** 1/64   **E.** 1/8

35. At the hypoxanthine guanine phosphoribosyl transferase (HPRT) locus, a normal amount of mRNA is present, but no protein is observed. This phenotype is caused by:

A. frameshift mutation

B. mutation in the gene altering the restriction pattern, but not affecting the protein; for instance, the mutated nucleotide is in the third codon position of the open reading frame

C. point mutation leading to an amino acid substitution important for enzyme function

D. gene deletion or mutation affecting the promoter

E. nonsense mutation affecting message translation

36. Hemophilia is a recessive X-linked trait. Knowing that females with Turner's syndrome have a high incidence of hemophilia, it can be concluded that these females have:

A. lost an X and gained a Y   **C.** gained an X

B. lost an X   **D.** gained a Y   **E.** none of the above

37. What is the pattern of inheritance for a rare recessive allele?

A. Every affected person has an affected parent

B. Unaffected parents can produce children who are affected

C. Unaffected mothers have affected sons and daughters who are carriers

D. Every affected person produces an affected offspring

E. None of the above

38. True-breeding plants with large purple flowers were crossed with true-breeding plants with small white flowers. The F_1 progeny all had large purple flowers. The F_1 progeny were crossed to true breeding plants with small white flowers. Among 1000 progeny:

Number of progeny	Flower size	Flower color
250	small	white
250	small	purple
250	large	white
250	large	purple

Most likely, the genes for flower size and color:

A. are unlinked   **C.** are linked and separated by no more than 25 centimorgans

B. are sex-linked   **D.** require determination of the cross of the F_2 progeny

E. cannot be determined

39. Mitosis does NOT serve the purpose of:

    A. replenishment of erythrocytes   **C.** organ repair

    B. formation of scar tissue   **D.** tissue growth

                                          E. transduction

40. If tall height and brown eye color are dominant, what is the probability for a heterozygous tall, heterozygous brown-eyed mother and a homozygous tall, homozygous blue-eyed father to have a tall child with blue eyes? Note: the genes for eye color and height are unlinked.

    A. 3/4   **B.** 1/8   **C.** 1/4   **D.** 1/2   **E.** None of the above

41. Recombination frequencies:

    A. arise from completely random genetic exchanges

    B. are the same for cis and trans heterozygotes

    C. decrease with distance

    D. are the same for all genes

    E. are the same for all chromosomes

42. How many different gametes can be produced from the genotype *AaBbCc*, assuming independent assortment?

    A. 4   **B.** 6   **C.** 8   **D.** 16   **E.** 3

43. What is the pattern of inheritance for a rare dominant allele?

    A. Every affected person has an affected parent

    B. Unaffected parents can produce children who are affected

    C. Unaffected mothers have affected sons and daughters who are carriers

    D. Every affected person produces an affected offspring

    E. All of the above

44. People that are homozygous for a recessive autosomal mutation accumulate harmful amounts of lipids. Jane and her parents are not afflicted. However, Jane's sister accumulates lipids. What is the probability that Jane is heterozygous for the mutation?

    A. 1/4   **B.** 1/3   **C.** 2/3   **D.** 1/2   **E.** ¾

45. A genetic disease that shows an earlier onset and more severe symptoms with every generation is an example of:

    A. codominance   **C.** heterozygous advantage

    B. penetrance   **D.** gain-of-function mutations   **E.** anticipation

46. For a trait with two alleles, if the frequency of the recessive allele is 0.6 in a certain population, what is the frequency of individuals expressing the dominant phenotype?

    A. 0.48   **B.** 0.64   **C.** 0.16   **D.** 0.36   **E.** 0.12

47. The maximum recombination frequency between two genes is:

    A. 100%   **B.** 80%   **C.** 50%   **D.** 10%   **E.** 1%

48. In mice, short hair is dominant over long hair. If a short-haired individual is crossed with a long-haired individual and both long and short-haired offspring result, what can be concluded?

    A. Short-haired individual is homozygous
    B. Short-haired individual is heterozygous
    C. Long-haired individual is homozygous
    D. Long-haired individual is heterozygous
    E. More information is required

49. The end result of meiosis in males is the production of:

    A. two (1N) cells genetically identical to the parent cell
    B. two (2N) cells genetically identical to the parent cell
    C. four (1N) cells genetically identical to the parent cell
    D. four (1N) unique cells genetically different from the parent cell
    E. four (2N) cells genetically identical to the parent cell

50. Which of the following is a correct example of transversion, a mutation where purine gets converted to pyrimidine?

    A. uracil → thymine   **C.** thymine → adenine
    B. cytosine → thymine   **D.** guanine → adenine
                              E. guanine → cytosine

51. Which of the following is a type of genetic mutation?

    I. insertion   II. frameshift   III. nonsense   IV. missense

    A. I and II only   **C.** II and IV only
    B. I, II and III only   **D.** I, II, III and IV   **E.** II, III and IV only

52. Two reciprocal crossing over events appear in the progeny at an approximate ratio of:

    A. 4:1   **B.** 3:1   **C.** 2:1   **D.** 2:3   **E.** 1:1

53. In dogs, phenotype A (erect ears and barking, while following a scent) is caused by dominant alleles; phenotype B (droopy ears and silent, while following a scent) is caused by recessive alleles. A dog that is homozygous dominant for both traits is mated with a dog that is homozygous recessive for both traits. If the two genes are unlinked, which of the following is the expected F_1 phenotypic ratio?

    A. 9:3:3:1   **B.** 1:1   **C.** 16:0   **D.** 1:2:1   **E.** None of the above

54. Mutations:

    A. always cause severe mutant phenotypes
    B. never cause severe mutant phenotypes
    C. are not inherited by progeny
    D. may cause premature termination of translation
    E. are none of the above

55. What is the risk of having a child affected by a disease with autosomal dominant inheritance if both the mother and father have one mutant gene for that disease?

    A. 0%   **B.** 25%   **C.** 50%   **D.** 12.5%   **E.** 75%

56. Given the recombinant frequencies below, what is the sequence of linked genes D, E, F and G?

    GE: 23%   ED: 15%   EF: 8%
    GD: 8%   GF: 15%   DF: 7%

    A. FGDE   **B.** EFGD   **C.** GFDE   **D.** GDFE   **E.** DEFG

Questions **57** through **63** are based on the following:

The pedigree illustrated by the schematic shows the inheritance of albinism, a homozygous recessive condition manifested in a total lack of pigment. Specify the following genotypes using *A* and *a* to indicate dominant and recessive alleles, respectively.

Note: solid figures are albino individuals.

57. Individual A-1 in the pedigree shown is:

    A. *AA*   **C.** *Aa*
    B. *aa*   **D.** any of the above
                E. none of the above

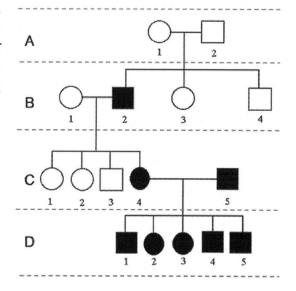

58. Individual A-2 in the pedigree shown is:

  A. *AA*   **C.** *Aa*
  B. *aa*   **D.** any of the above   **E.** none of the above

59. Individual B-1 in the pedigree above is:

  A. *AA*   **C.** *Aa*
  B. *aa*   **D.** any of the above   **E.** none of the above

60. Individual B-2 in the pedigree above is:

  A. *AA*   **C.** *Aa*
  B. *aa*   **D.** any of the above   **E.** none of the above

61. Individual C-3 in the pedigree above is:

  A. *AA*   **C.** *Aa*
  B. *aa*   **D.** any of the above   **E.** none of the above

62. Individual C-4 in the pedigree above is:

  A. *AA*   **C.** *Aa*
  B. *aa*   **D.** any of the above   **E.** none of the above

63. Individual D-4 in the pedigree above is:

  A. *AA*   **C.** *Aa*
  B. *aa*   **D.** any of the above   **E.** none of the above

64. In cocker spaniels, black color (B) is dominant over red (b), and solid color (S) is dominant over spotted (s). If the genes are unlinked and the offspring of BBss and bbss individuals are mated with each other, what fraction of their offspring will be black and spotted?

  A. 1/16   **B.** 9/16   **C.** 1/9   **D.** 3/16   **E.** 3/4

65. Why do genes that cause disease often appear to skip generations in an X-linked recessive inheritance?

  A. Disease is primarily transmitted through unaffected carrier females
  B. Males with an affected gene are carriers, but don't show the disease
  C. X-linked diseases are only expressed in males
  D. All X-linked diseases display incomplete penetrance
  E. none of the above

66. The "calico" coat pattern of a female cat is a result of:

A. endoreduplication

B. unequal crossing-over

C. random X chromosome inactivation

D. Turner syndrome

E. trisomy of the X chromosome

67. Which of these statements is/are true for an autosomal dominant inheritance?

  I. A single allele of the mutant gene is needed to exhibit the phenotype

  II. Transmission to the son by the father is not observed

  III. Autosomal dominant traits do not skip generations

A. II only

B. I, II and III

C. I only

D. I and III only

E. II and III only

68. What chromosomal abnormality results in some XY individuals being phenotypically females?

A. fragile X syndrome

B. Barr body formation

C. dosage compensation

D. mosaicism

E. deletion of the portion of the Y chromosome containing the testis-determining factor

69. Which event would likely NOT disrupt the Hardy-Weinberg equilibrium?

A. An entire population is exposed to intense cosmic radiation

B. Massive volcano eruption kills 1/6 of a large homogeneous population

C. Old and infirm within the population are selectively targeted by a predator

D. Population experiences emigration

E. Mutations occurring within organisms in the population

70. Which event has a risk of the joining together of two recessive alleles, resulting in genetic defect?

A. genetic mutation

B. transformation

C. inbreeding

D. crossing over

E. trisomy

71. To engineer polyploid plants in plant breeding, genetic engineers use drugs that:

A. insert new DNA into plants' genome

B. damage the DNA and cause mutations

C. rearrange the sequences of codons on the DNA strand

D. alter the number of chromosomes

E. shorten the length of DNA strands

72. To create mules, people have bred horses with donkeys. This is called:

A. crossing over   **C.** hybridization

B. genetic engineering   **D.** inbreeding   **E.** genetic mutation

73. Bacteria transformed by a plasmid can be distinguished from untransformed bacteria by the:

A. genetic marker   **C.** presence of DNA strands

B. trisomy   **D.** absence of cytosine   **E.** hybridization

74. When treated by penicillin, a bacterial culture transformed with recombinant plasmids that contain a gene for resistance to this antibiotic will:

A. undergo lysis   **C.** die

B. rapidly replicate DNA   **D.** survive   **E.** alternate generations

75. Which of the following takes place during transformation?

A. bacterial DNA undergoes mutation

B. bacterial DNA is inserted into a plasmid

C. two bacteria exchange genetic material via pilli

D. a prokaryotic cell becomes eukaryotic

E. a bacterial cell takes in foreign DNA material

76. Recombinant DNA experiments utilize plasmids because:

A. they contain foreign DNA

B. their genetic material can't be cut with restriction enzymes

C. they are unable to replicate inside the bacteria

D. they are used to transform bacteria

E. they are a natural part of bacterial genome

77. The figure shown illustrates:

A. PCR making a copy of the DNA

B. enzyme cutting the DNA

C. use of hybridization in genetic engineering

D. DNA sequencing via gel electrophoresis

E. insertion of genetic marker

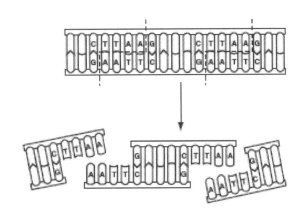

78. Between which two nucleotides is the DNA being cut in the figure shown?

A. thymine and adenine   **C.** adenine and cytosine

B. adenine and guanine   **D.** thymine and guanine

                                      E. thymine and cytosine

79. What is the product of combining DNA from different sources?

A. recombinant DNA

B. clone

C. hybrid DNA

D. mutant

E. plasmid

80. Plasmids are easily inserted into bacteria but not into yeast because yeast are:

A. mutants

B. protists

C. eukaryotes

D. prokaryotes

E. RNA carriers

81. Which pairing represents two transgenic organisms?

A. Apple tree hybrid and a polyploid cherry tree

B. Human growth hormone producing bacteria and genetically modified soy beans

C. Apple tree hybrid and human growth hormone producing bacteria

D. Genetically modified soy beans and a polyploid cherry tree

E. Polyploid cherry tree and human growth hormone producing bacteria

82. Which task requires recombinant DNA technology?

A. Creating bacteria that produce human growth hormone

B. Creating a polyploid cherry tree

C. Crossing two types of orange trees to create new orange fruits

D. Crossing a donkey with a horse to breed a mule

E. Paternity testing

83. A sheep named Dolly was different from animals produced by sexual reproduction because:

A. her DNA is identical to the DNA of her offspring

B. she was carried by a surrogate mother, while her DNA came from two other individuals

C. all of her cells' DNA is identical

D. her DNA was taken from a somatic cell of an adult individual

E. her DNA was taken from a gamete of neonatal individual

84. Transformation of a plant cell is considered successful when:

A. plasmid cannot enter the cell

B. cell produces daughter cells that subsequently produce other daughter cells

C. cell goes through programmed cell death (apoptosis)

D. cell's enzymes destroy the plasmid after it enters the cell

E. foreign DNA is integrated into one of the plant cell's chromosomes

85. Genetically modified (GM) crops are designed to produce a higher yield per plant by producing:

   A. less food but with higher nutritional content
   B. the same amount of food but with higher nutritional content
   C. more food per acre
   D. less food per acre
   E. none of the above

86. Gene therapy is considered successful when:

   A. the person's cells express the newly introduced gene
   B. the replacement gene is integrated into viral DNA
   C. the virus with the replacement gene enters the person's cells
   D. the person's cells replicate the newly introduced gene
   E. the replacement gene is integrated into the person's genome

87. Which choice correctly describes the process of establishing parental relationships through DNA fingerprinting of certain genes?

   A. Mitochondrial DNA links a daughter to the mother, while plasmid DNA links to the father
   B. Mitochondrial DNA links a son to the mother, while Y chromosome links to the father
   C. X chromosome links a daughter to the mother, while Y chromosome links to the father
   D. X chromosome links a son to the mother, while mitochondrial DNA links to the father
   E. Y chromosome links a son to the mother, while X chromosome links a girl to the father

Answer Key

1: A	11: C	21: A	31: D	41: A	51: D	61: C	71: D
2: C	12: D	22: A	32: C	42: C	52: E	62: B	72: C
3: B	13: B	23: C	33: B	43: A	53: C	63: B	73: A
4: C	14: C	24: E	34: D	44: C	54: D	64: E	74: D
5: E	15: A	25: C	35: E	45: E	55: E	65: A	75: E
6: D	16: D	26: B	36: B	46: B	56: D	66: C	76: D
7: C	17: E	27: E	37: B	47: C	57: C	67: D	77: B
8: E	18: A	28: D	38: A	48: B	58: C	68: E	78: B
9: B	19: E	29: A	39: E	49: D	59: C	69: B	79: A
10: B	20: B	30: C	40: D	50: E	60: B	70: C	80: C
							81: B
							82: A
							83: D
							84: E
							85: C
							86: A
							87: B

UNIT 4. ORGANISMAL BIOLOGY

Chapter 4.1: Plants: Structure, Function, Reproduction

1. In angiosperms, pollen grains are produced in the:

    A. ovule   **B.** sepal   **C.** stigma   **D.** carpel   **E.** anther

2. In most plants, which organs are adapted to capture sunlight for photosynthesis?

    A. leaves   **B.** flowers   **C.** roots   **D.** stems   **E.** stomata

3. Which of the following is true?

    A. Plants take carbon dioxide from the atmosphere and do not use oxygen
    B. Most plants get the water they need from the soil
    C. Plants have adaptations that minimize light absorption
    D. Plants require less water on sunny days
    E. The leaves provide the majority of water used by plants

4. In angiosperms, xylem consists of tracheids and:

    A. vessel elements   **C.** sieve tube elements
    B. parenchyma   **D.** albuminous cells   **E.** companion cells

5. In angiosperms, flowers are adaptations for reproduction. Flowers that are pollinated by animals do NOT:

    A. produce nectar to attract pollinators
    B. pollinate more efficiently than wind-pollinated plants
    C. grow larger in size
    D. exist as brightly colored structures
    E. exist as small structures

6. Without gas exchange, a plant would be unable to:

    A. make minerals   **C.** make food
    B. reproduce   **D.** absorb sunlight   **E.** flower

7. Which of the following is NOT part of the female structure of a flower?

    A. ovary   **B.** stigma   **C.** style   **D.** filament   **E.** ovules

8. Living on land required that plants:

A. have cell walls   **C.** conserve water

B. exchange gases   **D.** have photosynthetic pigments   **E.** utilize oxygen

9. If some of the xylem of a young maple tree were destroyed, it would most likely interfere with the tree's ability to:

A. absorb nutrients from the soil   **C.** conduct water to the roots

B. conduct water to the leaves   **D.** conduct sugars to the roots

                                     E. undergo photosynthesis

10. Vascular tissue in plants consists of:

A. epidermis   **C.** meristems

B. parenchyma and collenchyma   **D.** xylem and phloem   **E.** rhizoid

11. The ancestors of land plants likely evolved from:

A. prokaryotes that carried on photosynthesis

B. a protist that lived on land

C. an organism similar to green algae

D. mosses that lived in water

E. mosses that lived on land

12. When a plant reproduces vegetatively:

A. only root tissue can be used to produce new offspring

B. offspring are produced by mitosis alone

C. meiosis produces a new gametophyte

D. offspring will differ from the parent

E. mitosis produces a new gametophyte

13. Seeds that are spread by animals usually are contained in:

A. lightweight structures   **C.** unripened ovaries

B. thin coatings that are easily digested   **D.** environmentally resistant structures

                                     E. fleshy and nutritious fruits

14. To observe mitosis, which of the following should be examined under a compound microscope?

A. xylem from a tree trunk   **C.** epidermis of a leaf

B. phloem from the stem of a plant   **D.** tip of a shoot   **E.** rhizoid

15. What would have most likely occurred if plants had not begun to live on land?

    A. Animals would not undergo cellular respiration
    B. There would be no CO_2 in the atmosphere
    C. Animals also would not live on land
    D. There would be no green algae in the oceans
    E. There would be no oxygen in the atmosphere

16. The vascular cylinder of a root consists of:

    A. phloem and xylem   **C.** xylem only
    B. phloem and ground tissue   **D.** phloem only   **E.** xylem and ground tissue

17. Vegetative reproduction can be a disadvantage when:

    A. the parent plant must reproduce quickly
    B. the parent plant is not competing with other organisms for resources
    C. conditions are favorable for growth
    D. conditions in the physical environment suddenly change
    E. the parent plant reproduces slowly

18. The layer of cells that encloses the vascular tissue in the central region of a root is:

    A. apical meristem   **B.** epidermis   **C.** cortex   **D.** meristems   **E.** endodermis

19. What is a basic difference between a sporophyte and a gametophyte?

    A. A sporophyte is part of the diploid phase, while a gametophyte is part of the haploid phase of the plant life cycle
    B. A sporophyte is much smaller than the gametophyte
    C. A sporophyte is a reproductive structure, while a gametophyte is not
    D. A sporophyte undergoes sexual reproduction, while a gametophyte undergoes asexual reproduction
    E. A gametophyte is part of the diploid phase, while a sporophyte is part of the haploid phase of the plant life cycle

20. The outer covering of a plant consists of:

    A. meristematic tissue   **C.** vascular tissue
    B. dermal tissue   **D.** ground tissue   **E.** albuminous cells

21. How have the sporophyte and gametophyte changed as plants have evolved?

    A. The gametophyte has become smaller, while the sporophyte has become larger
    B. Many more recently evolved plants no longer have a sporophyte
    C. The sporophyte has become larger, while the gametophyte has not changed
    D. More recently evolved plants are mostly haploid gametophytes
    E. The gametophyte has become larger, while the sporophyte has become smaller

22. A period during which the embryo of a seed is alive but not growing is:

A. germination   **B.** dormancy   **C.** dispersal   **D.** fertilization   **E.** necrosis

23. Which of the following correctly describes seeds and fruits in angiosperms?

A. Fruits provide nutrients for the seed
B. Fruits do not protect the seed, so they are not favored by natural selection
C. Fruits surround no more than one seed apiece
D. Fruits develop from the endosperm of the seed
E. Fruits are matured ovaries that contain seeds

24. All of the following are reasons why green algae are classified as plants, EXCEPT:

A. They have genes similar to genes in land plants
B. They have chlorophyll *a*
C. They have cellulose in their cell walls
D. They are single-celled organisms
E. They have chlorophyll *b*

25. In bryophytes, which of the following structures is like a root?

A. Stalk   **C.** Rhizoid
B. Capsule   **D.** Sporophyte   **E.** stomata

26. A seed that is dispersed to an area far away from the parent plant always faces less:

A. unfavorable conditions for germination   **C.** chance of self-pollination
B. favorable conditions for germination   **D.** competition for space
     E. competition with the parent plants

27. Root pressure:

A. is produced in the vascular cylinder by active transport
B. is produced within the medulla of the root
C. is produced within the cortex of the stem
D. causes a plant's roots to increase in size
E. forces water in xylem downward

28. Which of the following statements about green algae is true?

A. Like other plants, they have specialized structures
B. They are multicellular plants
C. Evidence suggests they were the first plants
D. They are found in dry areas on land
E. None of the above

29. The plant embryo in a seed begins to grow again during:

A. pollination   **C.** dormancy

B. germination   **D.** fertilization   **E.** B and C

30. A period of dormancy can allow seeds to germinate:

A. in extreme temperatures   **C.** under cold conditions

B. without water   **D.** under ideal conditions

     E. under warm conditions

31. One of the three main functions of stems is to:

A. store sugars   **C.** facilitate photosynthesis

B. store carbon dioxide   **D.** carry nutrients and H_2O between roots and leaves

     E. store water

32. Which hormone stimulates increase in size, especially in stems and fruits?

A. germination   **C.** auxin

B. ethylene   **D.** cytokinin   **E.** gibberellin

33. Of the following locations, you would most likely find mosses growing in:

A. exposed ground with direct sun   **C.** sandy soil near a beach

B. shaded ground near a small pond   **D.** hot and dry soil

     E. mineral-poor soil

34. When environmental conditions are unfavorable, why is it an advantage for green algae to reproduce sexually?

A. Sexual reproduction produces offspring with greater genetic variety and durable zygotes that help green algae survive unfavorable conditions

B. Sexual reproduction produces multicellular algae instead of unicellular algae. Multicellular algae are better able to survive harsh conditions

C. They are multicellular when conditions are unfavorable, which favors sexual reproduction

D. Sexual reproduction produces larger numbers of offspring in less time

E. None of the above

35. Many cacti have large stems and no leaves. Which of these leaves' functions is assumed by the stems of these cacti?

A. absorb H_2O from the soil   **C.** absorb nutrients from the soil

B. transport materials throughout the plant   **D.** produce food by photosynthesis

     E. store excess H_2O

36. Which type of vascular tissue carries solutions of nutrients and carbohydrates produced by photosynthesis?

A. lignin   **B.** tracheids   **C.** phloem   **D.** xylem   **E.** rhizoid

37. Unlike roots, stems:

A. may be involved in photosynthesis   **C.** have ground tissue

B. are protected by epidermal cells   **D.** transport water   **E.** transport sugars

38. During primary growth, a stem:

A. increases in width   **C.** produces wood

B. fills with O_2 **D.** produces flowers

     E. increases in length

39. Hormones that stimulate cell elongation and are produced in the rapidly growing region near the tip of the plant's root or stem are:

A. cytokinins   **C.** auxins

B. gibberellins   **D.** ethylenes   **E.** sepals

40. What might a thin tree ring indicate?

A. decreased production of phloem   **C.** increased production of xylem

B. a year of drought   **D.** xylem production in winter

     E. decreased production of xylem

41. What is the phenomenon that is affecting the growth of the bean shoot in the figure below?

A. thigmotropism

B. rapid movement

C. gravitropism

D. phototropism

E. gibberellin

42. Bryophytes must live in moist areas because they lack vascular tissue. For what additional reason do they need to live in moist areas?

A. Carbon dioxide exchange is more efficient in wet areas

B. Without moisture, rhizoids cannot anchor the plants

C. Bryophytes need the extra water for photosynthesis

D. Oxygen exchange is more efficient in wet areas

E. The sperm of bryophytes need water to swim to an egg

43. Which structure includes a plant embryo, a food supply and a protective covering?

    A. Seed   **C.** Pollen grain

    B. Gametophyte   **D.** Spore   **E.** Xylem

44. When positioned horizontally, a plant's response to gravity is that:

    A. all parts of the plant bend upward, away from the force of gravity

    B. auxins in the shoot cause the growth of tendrils to pull the shoot upwards

    C. auxins in the lower sides of stems cause cell elongation that bends the stem upright

    D. the shoot bends to grow downward, toward the force of gravity

    E. none of the above

45. What is the source of ethylene gas in a plant?

    A. stems   **C.** roots

    B. fruit tissues   **D.** leaves   **E.** all of the above

46. The stomata of leaves are usually open:

    A. at night, if a plant has enough water

    B. during the day, if a plant has enough CO_2

    C. during the day, if a plant has enough water

    D. during the day, if a plant has too little water

    E. at night, if a plant has too little water

47. The yellow, orange and red colors of leaves in the fall are the result of the:

    A. decrease of auxins in the meristems

    B. production of anthocyanin and the breakdown of chlorophyll

    C. new synthesis of orange and yellow pigments in leaves

    D. movement of chlorophyll from the stems to the leaves

    E. increase of phytochrome in the leaves

48. O_2 and CO_2 move in and out of a leaf through:

    A. stomata   **C.** guard cells

    B. phloem   **D.** palisade mesophyll   **E.** xylem

49. The growth of plant seedlings is usually:

    A. gravitropic and influenced by ethylene   **C.** thigmotropic and gravitropic

    B. phototropic and influenced by ethylene   **D.** phototropic and gravitropic

                                           E. phototropic only

50. What is one reason that vascular plants are larger than nonvascular plants?

A. Nonvascular plants do not reproduce as quickly as vascular plants do, so they do not grow as large

B. Vascular tissue carries water and nutrients much more efficiently than can be carried by diffusion

C. Vascular plants live in areas where they can get more water than nonvascular plants

D. Nonvascular plants cannot live in as many areas as vascular plants, so they stay small

E. The rate of gas exchange is lower in nonvascular plants, which accounts for their diminished size

51. Most of the photosynthesis in plants takes place in:

A. xylem
B. stomata
C. mesophyll tissue
D. guard cells
E. phloem

52. Stomata are located only on the lower surface of the leaf in many plants. The most likely explanation for this fact is that:

A. water loss would be less on the shaded lower surface than in direct sun

B. gravity plays a role in gas exchange

C. stomata are closer to vascular bundles that bring water into the leaf

D. photosynthesis only occurs in the spongy mesophyll near the bottom of the leaf

E. photosynthesis only occurs in the regions surrounding the stomata

53. The growth of ivy tendrils that wrap around objects is an example of:

A. photoperiodism
B. chemotropism
C. phototropism
D. gravitropism
E. thigmotropism

54. Seed-bearing plants differ from all other plants in that:

A. they have true roots, stems, and leaves

B. their gametes do not require water for fertilization

C. they have a gametophyte generation

D. they have only xylem and no phloem

E. they have neither xylem nor phloem

55. Xylem tissue is important because it:

A. carries out photosynthesis

B. carries carbohydrates to all parts of the plant

C. allows water to diffuse from the roots

D. can conduct water over long distances

E. carries minerals to the roots

56. Corn, broccoli and cabbage were all developed by:

    A. auxins   **C.** plant propagation

    B. selective breeding   **D.** germination   **E.** pollination

57. In the figure below, the water pressure is:

    A. high in the guard cells

    B. low in the guard cells

    C. low in the stoma

    D. high in the stoma

    E. more than one of the above

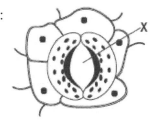

Underside of a leaf

58. Through which plant cells does water move by capillary action?

    A. mesophyll cells   **C.** phloem cells

    B. albuminous cells   **D.** guard cells   **E.** xylem cells

59. Which of the following gymnosperm reproductive structures is haploid?

    A. seed   **C.** pollen grains

    B. cone scale   **D.** seed cone   **E.** embryo

60. Which of the following BEST describes how plants are used by people?

    A. All medicines come from plants

    B. Plants break down decomposing organisms and wastes

    C. People use only plants for food

    D. Plants provide food, raw materials and several medicines

    E. Plants release carbon dioxide into the environment

61. In the figure shown, which letter indicates the structure where the embryo develops?

    A. A

    B. B

    C. C

    D. D

    E. F

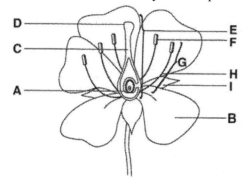

62. The movement of sugars in a plant can be explained by:

A. root pressure

B. the pressure-flow hypothesis

C. transpirational pull

D. diffusion gradient   **E.** capillary action

63. According to the pressure-flow hypothesis, which of the following statements is NOT true?

A. The movement of H_2O into a nutrient-rich region of the phloem decreases the pressure in that region

B. Phloem is able to move sugars in either direction to meet the nutritional needs of the plant

C. H_2O is necessary for sugars to move through phloem

D. H_2O moves from the xylem to the phloem of a plant

E. H_2O moves within the plant via capillary action

64. Angiosperms produce seeds inside protective structures called:

A. petals   **B.** cones   **C.** ovaries   **D.** pollen grains   **E.** stomata

65. Which of the following correctly relates an angiosperm structure and what it develops into?

A. embryo → fruit

B. ovule → fruit

C. pollen → grain seed

D. fruit → ovule

E. ovary → fruit

66. A pollen grain landing on a stigma of a flower of the same species produces:

A. flower

B. stamen

C. gametophyte

D. pollen tube   **E.** root hairs

67. Flowering plants that grow, flower, produce seeds and die in one year are called:

A. biennials

B. perennials

C. dicots

D. annuals   **E.** monocots

68. Ground tissue is found in a plant's:

A. roots and stems only

B. roots, stems, and leaves

C. stems only

D. stems and leaves only

E. roots only

69. A seed plant is held in the ground by its:

A. phloem   **B.** epidermis   **C.** stems   **D.** xylem   **E.** roots

Answer Key

1: E	11: C	21: A	31: D	41: D	51: C	61: A
2: A	12: B	22: B	32: E	42: E	52: A	62: B
3: B	13: E	23: E	33: B	43: A	53: E	63: A
4: A	14: D	24: D	34: A	44: C	54: B	64: C
5: E	15: C	25: C	35: D	45: E	55: D	65: E
6: C	16: A	26: E	36: C	46: C	56: B	66: D
7: D	17: D	27: A	37: A	47: B	57: A	67: D
8: C	18: E	28: C	38: E	48: A	58: E	68: B
9: B	19: A	29: B	39: C	49: D	59: C	69: E
10: D	20: B	30: D	40: B	50: B	60: D	

Chapter 4.2: Endocrine System

1. HMG-CoA reductase is the key enzyme in cholesterol biosynthesis. If a patient takes Mevacor, a potent inhibitor of this enzyme, production of which hormone will not decrease?

A. insulin
B. cortisol

C. testosterone
D. aldosterone

E. progesterone

2. Which of these hormones is NOT produced by the anterior pituitary gland?

A. growth hormone
B. prolactin

C. luteinizing hormone
D. thyroxine

E. leptin

3. Molecular signals that travel to distant cells are known as:

A. paracrine signals
B. parasitic signals

C. autocrine signals
D. hormones

E. responders

4. The class of hormones affecting cellular targets by starting or ending transcription is:

A. steroids
B. eicosanoids

C. peptides
D. amino acids

E. neurotransmitters

5. Which of these is expected to be observed in a patient with acromegaly, which is a condition from oversecretion of growth hormone, considering that growth hormone decreases cellular receptors' sensitivity to insulin?

A. decreased urine volume
B. decreased cardiac output

C. low blood glucose concentration
D. high blood glucose concentration
E. decreased osmolarity of urine

6. Which endocrine gland releases vasopressin, a hormone involved in water balance?

A. posterior pituitary
B. hypothalamus

C. thyroid
D. adrenal cortex

E. adrenal medulla

7. Which of the following is TRUE about hormones?

A. Most hormones operate by activation of cyclic cAMP
B. The circulating level is held constant through a series of positive feedback loops
C. Both lipid-soluble and water-soluble hormones bind to intracellular protein receptors
D. The ducts of endocrine organs release their contents into the bloodstream
E. They regulate cellular functions and are generally controlled via negative feedback

8. Which of these regulates temperature?

A. cerebrum   **C.** medulla oblongata

B. hypothalamus   **D.** pons   **E.** pineal gland

9. In the last mile of a long-distance run, all of these physiological and hormonal effects of stress would be observed in a runner, EXCEPT:

A. decreased ACTH secretion   **C.** increased glucagon secretion

B. decreased blood flow to the small intestine   **D.** increased heart rate

     E. pupil dilation

10. In general, cell signaling causes:

A. increased expression of genes   **C.** protein kinase activity

B. an influx of ions   **D.** G protein activation

     E. a change in receptor conformation

11. What kind of messenger is produced by all endocrine cells?

A. intracellular second messenger

B. extracellular messenger carried by the lymph

C. extracellular neurotransmitter released from nerve endings

D. extra- or intra-cellular messenger carried by the blood

E. a messenger molecule secreted into a duct

12. Given that parathyroid hormone plays an important role in the control of blood Ca^{2+} ion levels, it is an important hormone for:

     I. bone density

     II. renal calcium reabsorption

     III. blood calcium concentration

A. I only   **C.** I and III only

B. I and II only   **D.** II and III only   **E.** I, II and III

13. All of the following statements are characteristic of peptide hormone activity, EXCEPT:

A. hormone is transmitted via blood circulation

B. the target organ is at a distant site from the release of the hormone

C. cellular effects within cells often require the activity of a protein kinase

D. hormones pass into the target cell's membrane and enter the nucleus

E. cellular effects within cells often are mediated by second messengers

14. A hormone released by the posterior pituitary is:

A. TSH   **B.** prolactin   **C.** oxytocin   **D.** progesterone   **E.** calcitonin

15. Deficiency of which hormone causes female infertility due to the ovulation of immature ova?

    A. oxytocin   **B.** estrogen   **C.** FSH   **D.** LH   **E.** prolactin

16. The receptor of estrogen is:

    A. an ion channel receptor   **C.** a G protein

    B. a protein kinase receptor   **D.** located on the extracellular side of the membrane

                                     E. located within the cytoplasm

17. Which endocrine gland synthesizes ACTH?

    A. hypothalamus   **C.** anterior pituitary

    B. thalamus   **D.** medulla   **E.** posterior pituitary

18. All of the following hormones utilize a second messenger system, EXCEPT:

    A. TSH   **B.** estrogen   **C.** cAMP   **D.** insulin   **E.** epinephrine

19. Why is it that some cells respond differently to the same peptide hormones?

    A. Different target cells have different genes

    B. Each cell knows how it fits into the body's master plan

    C. A target cell's response is determined by a signal transduction pathway

    D. The circulatory system regulates responses to hormones by routing them to specific targets

    E. The hormone is chemically altered in different ways as it travels in the bloodstream

20. All of the following are secreted by the thyroid, EXCEPT:

    A. triiodothyronine   **C.** thyroxine

    B. TSH   **D.** all of the above   **E.** none of the above

21. Which of the following clinical manifestations would be expected of a mouse with a partial deletion of both IGF-1 genes?

    A. Decreased Ca^{2+} levels and brittle bones   **D.** Deficiency in growth

    B. Anemia   **C.** Deficiency in the digestion of lipids

                             E. Increased heart rate and no thyroid hormone release

22. Vitamin A is a relatively small, lipid-soluble molecule that can behave as a hormone. Which of the following would you predict about its receptor?

    A. It would be an ion channel receptor

    B. It would be a protein kinase receptor

    C. It would involve a G protein

    D. It would not exist; vitamin A would not have a receptor

    E. It would not be connected to the plasma membrane

23. GH, PRL and ACTH hormones are all secreted by:

A. hypothalamus
B. anterior pituitary

C. posterior pituitary
D. adrenal glands

E. medulla

24. Serum auto-antibody binding to TSH receptors of the thyroid gland causes a patient to have:

A. low thyroid hormone levels, because auto-antibodies suppress the thyroid gland
B. high thyroid hormone levels, because the thyroid is over-stimulated by the auto-antibodies
C. low thyroid hormone levels, because auto-antibodies block TSH from binding to its receptor
D. asymptomatic condition, because negative feedback regulates thyroid hormone levels
E. an unchanged basal level of thyroid hormone

25. If a hormone administered to mice intravenously accumulates rapidly inside renal cells without endocytosis, this hormone is most likely a:

A. neurotransmitter
B. second messenger

C. steroid
D. polypeptide

E. amine

26. A gland that produces both exocrine and endocrine secretions is:

A. adrenal   **B.** pancreas   **C.** parathyroid   **D.** pituitary   **E.** parotid

27. Female infertility caused by a failure to ovulate is most likely the result of:

A. low levels of LH
B. a dilation of the cervix

C. high levels of FSH
D. high levels of LH
E. high levels of LH and low levels of FSH

28. From start to finish, the order of the basic steps of a signal transduction pathway is:

A. signal → responder → receptor → effects
B. receptor → signal → responder → effects

C. signal → receptor → responder → effects
D. signal → receiver → responder → effects
E. signal → effects → receiver → responder

29. During stress, the adrenal cortex responds by secreting the following hormone:

A. adrenaline   **B.** norepinephrine   **C.** ACTH   **D.** acetylcholine   **E.** cortisol

30. A neuroendocrine tumor of the adrenal glands releases epinephrine at abnormally high levels. Which of the following symptoms would be observed in a patient with such a tumor?

A. pupil constriction
B. abnormally low heart rate

C. reduced blood pressure
D. elevated blood pressure
E. decreased blood flow to skeletal muscles

31. The hypothalamus controls the anterior pituitary by means of:

    A. cytokines

    B. second messengers

    C. releasing hormones

    D. antibodies

    E. bidirectional nervous inputs

32. The concentration of blood Ca^{2+} is raised by:

    A. calcitonin

    B. parathyroid hormone

    C. aldosterone

    D. glucagon

    E. antidiuretic hormone

33. Which of the following is an example of antagonistic endocrine relationships that maintain homeostasis?

    A. ACTH — TSH

    B. oxytocin — prolactin

    C. vitamin D — parathyroid hormone

    D. insulin — glucagon

    E. estrogen — insulin

34. Which of the following signals do cells receive?

    A. light   **B.** sound   **C.** hormones   **D.** odorants   **E.** all of the above

35. Which endocrine gland synthesizes PTH?

    A. adrenal   **B.** anterior pituitary   **C.** kidneys   **D.** parathyroid   **E.** thyroid

36. What would be the expected immediate response by the adrenal gland to modulate elevated levels of K^+ in the blood?

    A. block the secretion of aldosterone and increase release of corticotrophins

    B. stimulate the secretion of aldosterone from the adrenal gland

    C. stimulate the secretion of aldosterone from the anterior pituitary gland

    D. no response, because aldosterone only increases Na^+ reabsorption in the nephron

    E. decrease blood Na^+ levels

37. What is the classification of a feedback system where a response enhances the original stimulus?

    A. enhancing

    B. responsive

    C. negative

    D. intermittent

    E. positive

38. Which of these hormones directly affects blood sugar?

    A. calcitonin   **B.** estrogen   **C.** glucagon   **D.** oxytocin   **E.** thyrotropin

39. Which hormone acts as an inhibitor for the *hormone-sensitive lipase* cascade that releases fatty acids stored in adipose tissue?

A. insulin   **C.** estrogen

B. epinephrine   **D.** glucagon   **E.** norepinephrine

40. The molecular signals that bind to receptors of the same cell that made them are referred to as:

A. paracrine signals   **C.** autocrine signals

B. responders   **D.** hormones   **E.** second messengers

41. Which endocrine gland synthesizes melatonin?

A. pineal   **C.** anterior pituitary

B. hypothalamus   **D.** posterior pituitary   **E.** thalamus

42. Which paracrine hormone is secreted by the pancreatic islets of Langerhans to inhibit the release of insulin and glucagon (also secreted by the islets)?

A. cortisol   **C.** somatostatin

B. trypsin   **D.** pepsin   **E.** epinephrine

43. During a workout, a person starts to perspire profusely. Sweat glands are which component of the feedback loop?

A. controlled condition   **C.** stimulus

B. receptors   **D.** effectors   **E.** control center

44. Which statement is true for epinephrine?

A. It causes bronchial constriction

B. It is released during parasympathetic stimulation

C. It is a steroid hormone

D. It is synthesized by the adrenal cortex

E. It is released by the adrenal medulla

45. Which gland produces growth-hormone-releasing hormone (GHRH) that stimulates the transcription of the growth hormone gene?

A. parathyroid   **C.** hypothalamus

B. anterior pituitary   **D.** liver   **E.** adrenal

46. In what way do ligand-receptor interactions differ from enzyme-substrate reactions?

    A. The ligand signal is not metabolized into useful products
    B. The enzyme-substrate reactions and the ligand-receptor interactions are similar
    C. Inhibitors never bind to the ligand-binding site
    D. Reversibility does not occur in the ligand-receptor interactions
    E. Receptor-ligand interactions do not obey the laws of mass action

47. Which of the following is associated with steroid hormone activity?

    A. enzyme activity   **C.** neural activity
    B. gene expression   **D.** extracellular receptors   **E.** second messenger

48. Which of the following basic categories is the correct chemical classification for hormones?

    A. female and male hormones   **C.** carbohydrates, proteins and steroids
    B. peptides and steroids   **D.** steroid, peptide and amines
                                       E. stimulator and receptor hormones

49. One of the most common second messengers is:

    A. ATP   **C.** acetylcholine
    B. growth hormone   **D.** adrenaline   **E.** cyclic AMP

50. Which of the following is NOT a pancreatic exocrine secretion?

    A. glucagon   **C.** lipase
    B. amylase   **D.** protease   **E.** bicarbonate ions

51. Insulin and aldosterone are secreted by these endocrine organs, respectively:

    A. pancreas and adrenal cortex   **C.** liver and pancreas
    B. spleen and red bone marrow   **D.** adrenal gland and kidney
                                         E. liver and adrenal glands

52. Which of the following is/are TRUE?

    I. For most ligand–receptor complexes, binding is favored
    II. Ligand–receptor interactions are reversible
    III. Most drugs that alter human behavior bind to specific receptors in the brain

    A. I and II only   **C.** I and III only
    B. II and III only   **D.** I, II and III   **E.** I only

53. The ability of a tissue or organ to respond to the presence of a hormone depends on:

  A. nothing; all hormones stimulate all cell types, because hormones are powerful and nonspecific

  B. the location of the tissue or organ with respect to the circulatory path

  C. the membrane potential of the cells of the target organ

  D. the presence of the appropriate receptors on/in the target tissue or the organ cell

  E. neuronal connectivity

54. The parathyroid gland is one of the two glands that control blood calcium levels, and is anatomically located:

  A. superior to the kidney   **C.** posterior in the neck, near the larynx

  B. connected to the hypothalamus   **D.** within the duodenum of the small intestine

                                                   E. inferior to the kidneys

55. A patient who has gained 35 lbs in the past 3 months visits her physician and complains of fatigue. She is diagnosed with a goiter and decreased metabolic rate. Based on this, which hormone is most likely to be deficient in this patient?

  A. estrogen   **C.** progesterone

  B. cortisol   **D.** aldosterone   **E.** thyroxine

56. During periods of dehydration, the main activating factor of aldosterone secretion is:

  A. ACTH   **C.** sympathetic nervous system

  B. renin   **D.** spontaneous adrenal release

                        E. parasympathetic nervous system

57. Several hormones are synthesized by the hypothalamus and transported to the anterior pituitary gland. The mechanism of transport from hypothalamus to anterior pituitary gland takes place in the:

  A. general circulatory system   **C.** feedback loop

  B. hypophyseal portal system   **D.** hepatic portal system   **E.** adrenal axis

58. A radio-labeled hormone was introduced to a culture of liver cells. After 5 hours of incubation, the cells were separated and the radioactivity was observed primarily in the nucleus. Which of these conclusions about the hormone is the most consistent with the observations?

  A. It is a steroid, because it functions as a transcriptional activator by binding to DNA

  B. It is a steroid, because it contains hydrophilic regions that allow crossing the nuclear membrane

  C. It is a peptide, because it functions as a transcriptional activator by binding to DNA

  D. It is a peptide, because it contains hydrophilic amino acids that allow crossing the nuclear membrane

  E. It is a peptide, because it contains hydrophobic amino acids that allow crossing of the plasma membrane

Answer Key

1: A	11: D	21: D	31: C	41: A	51: A
2: D	12: E	22: E	32: B	42: C	52: D
3: D	13: D	23: B	33: D	43: D	53: D
4: A	14: C	24: B	34: E	44: E	54: C
5: D	15: C	25: C	35: D	45: C	55: E
6: A	16: E	26: B	36: B	46: A	56: B
7: E	17: C	27: A	37: E	47: B	57: B
8: B	18: B	28: C	38: C	48: D	58: A
9: A	19: C	29: E	39: A	49: E	
10: E	20: B	30: D	40: C	50: A	

Chapter 4.3: Nervous System

1. Which of the following is NOT a function of astrocytes?

   A. Guiding migration of developing neurons

   B. Controlling the chemical environment around neurons

   C. Assisting in maturation of Schwann cells

   D. Anchoring neurons to blood vessels

   E. Providing nutrients to the neurons

2. All of the following are correct regarding the occurrence of a sensation, EXCEPT:

   A. stimulus energy must be converted into threshold energy

   B. generator potential in the associated sensory neuron must reach threshold

   C. stimulus energy must match the specificity of the receptor

   D. stimulus energy must occur within the receptor's receptive field

   E. all are true statements

3. Which of these effects would result from the stimulation of the parasympathetic nervous system?

   A. Relaxation of the bronchi   **C.** Increased gut motility

   B. Dilation of the pupils   **D.** Increased heart rate

                                   E. Increased respiratory volume

4. Autonomic nervous system is comprised of:

   A. sensory neurons that supply the digestive tract

   B. sensory neurons that convey information from somatic receptors for special senses of vision, hearing, taste and smell

   C. CNS motor fibers that conduct nerve impulses from the CNS to skeletal muscles

   D. motor fibers that conduct nerve impulses from CNS to smooth muscle, cardiac muscle and glands

   E. motor fibers that supply the digestive tract

5. The specific molecule in the eye that has to absorb photons for the perception of vision to occur is the:

   A. opsin   **B.** retinal   **C.** cGMP   **D.** G-protein   **E.** cAMP

6. The central nervous system determines the strength of a stimulus by the:

   A. amplitude of action potentials   **C.** type of stimulus receptor

   B. wavelength of action potential   **D.** origin of the stimulus

                                   E. frequency of action potentials

7. In a myopic eye, the inverted image formed by the lens falls:

  A. on the optic nerve   **C.** behind the retina

  B. in front of the retina   **D.** on the retina   **E.** on the optic disc

8. Bipolar neurons are:

  A. not found in ganglia   **C.** motor neurons

  B. found in the organ of Corti   **D.** called neuroglial cells

                                 E. found in the retina of the eye

9. The smallest distance that can be resolved with the unaided eye is:

  A. 200 mm   **B.** 0.2 mm   **C.** 20 mm   **D.** 0.2 nm   **E.** 2 micrometers

10. The blood-brain barrier is effective against:

  A. metabolic waste such as urea   **C.** alcohol

  B. nutrients such as glucose   **D.** anesthetics   **E.** psychotropics

11. Which structure of the brain controls the breathing rate?

  A. cerebrum   **C.** medulla oblongata

  B. cerebellum   **D.** hypothalamus   **E.** pituitary gland

12. An excitatory neurotransmitter secreted by motor neurons innervating skeletal muscle is:

  A. gamma amine   **C.** norepinephrine

  B. acetylcholine   **D.** epinephrine   **E.** alpha peptide

13. The partial decussation (i.e. crossing) of axons in the optic chiasm results in:

  A. both sides of the visual field being represented together in each hemisphere of the primary visual cortex

  B. representation of the left side of the visual field remaining ipsilateral

  C. both visual fields having the same representation

  D. the left side of the visual field being represented on the left hemisphere of the primary visual cortex

  E. the left side of the visual field being represented on the right hemisphere of the primary visual cortex

14. All of the following is correct about neurons, EXCEPT that they:

  A. are mitotic   **C.** conduct impulses

  B. have high metabolic rates   **D.** have extreme longevity

                                E. have a greater concentration of Na^+ outside

15. The autonomic nervous system's motor pathway contains:

    A. three neurons   **C.** a single long neuron

    B. many neurons   **D.** two neurons   **E.** a single short neuron

16. Which ion channel opens in response to a change in membrane potential and is involved in the generation and conduction of action potentials?

    A. leakage channel   **C.** mechanically gated channel

    B. ligand-gated channel   **D.** voltage-gated channel

    E. all of the above

17. Which of the following is most characteristic of an accident victim who sustained isolated damage to the cerebellum?

    A. loss of coordination of smooth muscle contractions

    B. loss of speech

    C. loss of voluntary muscle contraction

    D. loss of sensation in the extremities

    E. loss of muscular coordination

18. The somatic nervous system:

    A. innervates skeletal muscles   **C.** innervates cardiac muscle

    B. innervates glands   **D.** innervates smooth muscle of the digestive tract

    E. innervates peristalsis of the gastrointestinal tract

19. The fovea has a very high concentration of:

    A. cones   **C.** rods

    B. axons leaving the retina for the optic nerve   **D.** blood vessels   **E.** ganglions

20. All of the following are a chemical class of neurotransmitters, EXCEPT:

    A. ATP and other purines   **C.** amino acids

    B. biogenic amines   **D.** nucleic acids   **E.** acetylcholine

21. Somatic sensory nerve cell bodies are located in the:

    A. brain   **C.** dorsal root ganglion

    B. ventral horn   **D.** spinal cord   **E.** nerve plexus

22. Which of the following is the most correct statement?

A. Ganglia are collections of neuron cell bodies in the spinal cord that are associated with efferent fibers

B. Ganglia are associated with afferent nerve fibers containing cell bodies of sensory neurons

C. The dorsal root ganglion is only a motor neuron structure

D. The cell bodies of afferent ganglia are located in the spinal cord

E. The dorsal root ganglion contains both sensory and motor neurons

23. All of these matches of brain structure to its function are correct, EXCEPT:

A. reticular activating system – sensory processing

B. hypothalamus – appetite

C. cerebellum – motor coordination

D. cerebral cortex – higher intellectual function

E. medulla oblongata – basic emotional drives

24. Which of the following is a FALSE statement?

    I. An excitatory postsynaptic potential occurs if the excitatory effect is greater than the inhibitory effect but less than threshold

    II. A nerve impulse occurs if the inhibitory and excitatory stimuli are equal

    III. An inhibitory postsynaptic potential occurs if the inhibitory effect is greater than the excitatory that causes hyperpolarization of the membrane

A. I only   **B.** II only   **C.** III only   **D.** I and II only   **E.** I and III only

25. The phenomenon of being able to make out objects 5-10 min after lights in a bright room are turned off is due to:

A. sudden bleaching of cones

B. sudden bleaching of rods

C. sudden bleaching of rhodopsin

D. slow unbleaching of rods

E. fast adaptation from the bleaching of rods

26. Which statement about synapses is correct?

A. The synaptic cleft prevents an impulse from being transmitted directly between neurons

B. Neurotransmitter receptors are located on the axon terminal of the cell

C. Release of neurotransmitter molecules gives cells the property of being electrically coupled

D. Cells with gap junctions use chemical synapses

E. Calcium is not required for the release of the vesicles containing the neurotransmitter

27. Which of the following results from parasympathetic stimulation?

   A. Piloerection of the hair cells of the skin

   B. Contraction of the abdominal muscles during exercise

   C. Vasodilation of the afferent arterioles to the kidneys

   D. Increased heart rate

   E. Dilation of the pupils

28. A graded potential:

   A. is voltage stimulus to initiate action potential

   B. is voltage regulated repolarization

   C. is long distance signaling

   D. is amplitude of various sizes

   E. does not use summation

29. Which of the following would occur following an ingestion of the insecticide Diazinon, known to block acetylcholinesterase function?

   A. decrease in postsynaptic receptors

   B. decrease in ACh concentration in synapses

   C. decrease in postsynaptic depolarization

   D. all synaptic nervous transmission ceases

   E. increase in ACh concentration in synapses

30. Which cells are functionally similar to Schwann cells?

   A. oligodendrocytes

   B. astrocytes

   C. ependymal cells

   D. microglia

   E. glutamate transporters

31. Most refraction (bending of light in the eye) is accomplished by:

   A. the pupil

   B. photoreceptors

   C. the vitreous humor

   D. the cornea

   E. rod cells

32. Which of the following is correct regarding the movement of ions across excitable membranes?

   A. Ions move from an area of higher concentration to an area of lower concentration

   B. Sodium gates in the membrane can open in response to electrical potential changes

   C. Ions move passively across membranes

   D. Ions move actively across membranes through leakage channels

   E. The Na^+ ion is concentrated inside the cell during a resting potential

33. Release of which hormone most closely resembles a response to sympathetic stimulation?

   A. aldosterone

   B. dopamine

   C. acetylcholine

   D. insulin

   E. epinephrine

34. Which of the following is NOT a location where white matter would be found?

   A. pyramidal tracts   **C.** corpus callosum

   B. outer portion of the spinal cord   **D.** cerebral cortex   **E.** all of the above

35. Which function would be significantly impaired by destroying the cerebellum?

   A. thermoregulation   **C.** sense of smell

   B. coordinated movement   **D.** urine formation   **E.** blood pressure

36. A second nerve impulse cannot be generated until:

   A. proteins have been resynthesized   **C.** the membrane potential has been reestablished

   B. all sodium gates are closed   **D.** Na^+ ions are pumped back into the cell

                                       E. K^+ ions are pumped back out of the cell

37. Which of the following choices is the *correct pathway* for a beam of light entering the eye?

   A. lens → pupil → vitreous humor → retina

   B. lens → pupil → retinal ganglion cells → photoreceptor outer segments

   C. pupil → vitreous humor → bipolar cells → retinal ganglion cells

   D. lens → retinal ganglion cells → photoreceptor outer segments → bipolar cells

   E. cornea → aqueous humor → lens → vitreous humor

38. Compared to the external surface of a cell membrane for a resting neuron, the interior surface is:

   A. positively charged and contains more sodium

   B. negatively charged and contains more sodium

   C. negatively charged and contains less sodium

   D. positively charged and contains less sodium

   E. positively charged and contains more potassium

39. Detection of sound involves air pressure waves that are converted into neural signals in the:

   A. semicircular canals   **C.** tympanic membrane

   B. cochlea   **D.** retina   **E.** sclera

40. If an electrode is placed at the midpoint along the length of the axon, the impulse would:

   A. move bidirectionally

   B. move to the axon terminal, and muscle contraction would not occur

   C. not move to the axon terminal, but muscle contraction would not occur

   D. not move to the axon terminal, but muscle contraction would occur

   E. not move because of the phase change in amplitudes

41. Which of these systems, when stimulated, results in increased heart rate, blood pressure and blood glucose levels?

A. central nervous system

B. somatic nervous system

C. sympathetic nervous system

D. parasympathetic nervous system

E. motor nervous system

42. Which of the following neurotransmitters inhibits pain in a manner similar to morphine?

A. acetylcholine

B. norepinephrine

C. serotonin

D. nitric oxide

E. endorphin

43. When light strikes the eye, photoreceptors:

A. depolarize

B. enter refractory

C. hyperpolarize

D. release neurotransmitter

E. contract

44. The brain stem consists of:

A. midbrain

B. pons, medulla, midbrain and cerebellum

C. pons, medulla and midbrain

D. cerebrum, pons and medulla

E. medulla, cerebellum, and midbrain

45. Which portion of the brain do nerve cells that control thermoregulation concentrate in?

A. medulla

B. hypothalamus

C. cerebellum

D. cerebrum

E. substantia nigra

46. All of the following are found in normal cerebrospinal fluid, EXCEPT:

A. potassium

B. protein

C. glucose

D. red blood cells

E. sodium

47. Which condition is NOT an eye disorder?

A. glaucoma   B. amblyopia   C. myopia   D. hyperopia   E. otitis

48. Which area of the brain is where the primary auditory cortex is located?

A. temporal lobe

B. parietal lobe

C. prefrontal lobe

D. frontal lobe

E. occipital lobe

49. Which of the following statements is correct about a reflex arc?

   A. Responses do not involve the central nervous system

   B. It involves inhibition along with excitation of muscles

   C. It involves motor neurons exiting the dorsal side of the spinal cord

   D. Cerebral cortex provides fine motor control for the muscle responses

   E. Sensory inputs are processed by the brain prior to a standard response

50. Cell bodies of the sensory neurons of the spinal nerves are located in the:

   A. sympathetic ganglia   **C.** ventral root ganglia of the spinal cord

   B. thalamus   **D.** dorsal root ganglia of the spinal cord   **E.** nerve plexus

51. Which of the following statements is true regarding a reflex arc?

   A. There are two types of reflex arc: autonomic and somatic

   B. The motor response occurs without synaptic delay

   C. A minimum of three neurons must participate

   D. Sensory and motor neurons can synapse outside of the spinal cord

   E. Reflex arc neurons synapse in the brain

52. The white matter of the spinal cord contains:

   A. myelinated nerve fibers

   B. neither myelinated nor unmyelinated nerve fibers

   C. myelinated and unmyelinated nerve fibers

   D. unmyelinated nerve fibers

   E. unmyelinated nerve fibers and glial cells

53. The one place in the body where a physician can directly observe blood vessels is the:

   A. eardrum   **C.** skin of the eyelid

   B. pointer finger   **D.** retina   **E.** lumen of the digestive system

54. Which statement is true regarding the Broca's area?

   A. It is considered a motor speech area

   B. It serves for the recognition of complex objects

   C. It is located in the right hemisphere

   D. It serves for auditory acuity

   E. It serves for visual acuity

55. Which of the following structures contain(s) hair cells that detect motion?

       I. the skin   II. the organ of Corti   III. the semicircular canals

A. I only   **B.** III only   **C.** II only   **D.** I, II and III   **E.** II and III only

56. The part of the cerebral cortex involved in cognition, personality, intellect and recall is:

A. the limbic association area
B. combined primary somatosensory cortex and somatosensory association
C. the prefrontal cortex
D. the cortex posterior association area
E. Broca's area

57. Which of these processes is controlled by the sympathetic nervous system?

A. increased gastric secretions   **C.** constriction of pupils
B. increased respiration   **D.** decreased heart rate
                                      E. decreased adrenaline secretions

58. Nerves that carry impulses towards the central nervous system are:

A. motor nerves   **C.** afferent nerves
B. mixed nerves   **D.** efferent nerves   **E.** sensory nerves

59. Which of the following is true regarding regeneration within the central nervous system?

A. It is prevented due to growth-inhibiting proteins of oligodendrocytes
B. It is promoted by growth inhibitors and glial scars
C. It is more successful than with the PNS
D. It is promoted by glial cells
E. It is facilitated via mitosis by astrocytes

60. In the nervous system, transduction is a conversion of:

A. electrical energy into mechanical energy
B. receptor energy into stimulus energy
C. stimulus energy into a change in the electrical potential of a membrane
D. presynaptic nerve impulses into postsynaptic nerve impulses
E. afferent impulses into efferent impulses

Answer Key

1: C	11: C	21: C	31: D	41: C	51: A
2: A	12: B	22: B	32: B	42: E	52: A
3: C	13: E	23: E	33: E	43: C	53: D
4: D	14: A	24: B	34: D	44: C	54: A
5: B	15: D	25: D	35: B	45: B	55: E
6: E	16: D	26: A	36: C	46: D	56: C
7: B	17: E	27: C	37: E	47: E	57: B
8: E	18: A	28: D	38: C	48: A	58: C
9: B	19: A	29: E	39: B	49: B	59: A
10: A	20: D	30: A	40: A	50: D	60: C

Chapter 4.4: Circulatory System

1. What cell is the precursor of all elements of blood?

A. polymorphonuclear cell   **C.** normoblast

B. hemocytoblast   **D.** megakaryocyte   **E.** platelet

2. Which of the following is correct about the hemoglobin molecule of the red blood cell?

A. At physiological conditions, it exists in the more oxidized form of Fe(II) rather than Fe(III)

B. It has a higher O_2 affinity in adults compared with fetal hemoglobin

C. It does not bind CO

D. It exhibits positive cooperative binding for O_2

E. does not bind CO_2

3. The osmotic pressure at the arterial end of a capillary bed:

    I. results in a net outflow of fluid

    II. is less than the hydrostatic pressure

    III. is greater than the hydrostatic pressure

A. I only   **C.** III only   **E.** I and II only

B. II only   **D.** I and III only

4. Comparing two liquids with a pH of 2 and a pH of 7, which liquid has the highest $[H^+]$?

A. Both liquids have the same $[H^+]$

B. A pH 2 liquid has a greater $[H^+]$ than a pH 7 liquid

C. A pH 7 liquid has a greater $[H^+]$ than a pH 2 liquid

D. Neither liquid contains $[H^+]$

E. None of the above

5. All of the following are functions of blood, EXCEPT:

A. transport of salts to maintain blood volume   **C.** transport of metabolic wastes from cells

B. transport of hormones to their target organs   **D.** delivery of O_2 to body cells

    E. transport of CO_2 to the lungs

6. Hill coefficient measures cooperativity, where Hill coefficient greater than 1 signifies positive cooperativity, less than 1 signifies negative cooperativity, and equal to 1 signifies the absence of cooperativity. What is the Hill coefficient of hemoglobin?

A. 2.3   **B.** 1   **C.** 0   **D.** -1.5   **E.** between −1 and 1

7. All of these statements about the circulatory system are true, EXCEPT:

    A. the ventricles are the pumping chambers of the heart

    B. oxygenated blood is typically transported in arteries

    C. mammals have a four chambered heart

    D. the thoracic duct returns lymphatic fluid to the circulatory system

    E. veins have a strong pulse

8. Which of the following is the *correct* path of blood flow through the heart?

    A. Superior/inferior vena cava → left atrium → left ventricle → pulmonary artery → lungs → pulmonary vein → right atrium → right ventricle → aorta

    B. Superior/inferior vena cava → right atrium → left ventricle → pulmonary artery → lungs → pulmonary vein → left atrium → right ventricle → aorta

    C. Superior/inferior vena cava → right atrium → right ventricle → pulmonary vein → lungs → pulmonary artery → left atrium → left ventricle → aorta

    D. Superior/inferior vena cava → right atrium → right ventricle → pulmonary artery → lungs → pulmonary vein → left atrium → left ventricle → aorta

    E. Superior/inferior vena cava → right atrium → left atrium → pulmonary vein → lungs → pulmonary artery → right ventricle → left ventricle → aorta

9. Which of the following is a protective role of blood?

    A. maintenance of pH in body tissue

    B. maintenance of body temperature

    C. prevention of blood loss

    D. maintenance of adequate fluid volume

    E. maintenance of osmolarity

10. Which of these statements is correct for myoglobin, considering that myoglobin accepts O_2 from hemoglobin and releases it to the cytochrome c oxidase system?

    A. myoglobin has a lower affinity for O_2 than hemoglobin

    B. myoglobin has a higher affinity for O_2 than hemoglobin

    C. cytochrome oxidase c system has a lower affinity for O_2 than hemoglobin

    D. cytochrome oxidase c system has a lower affinity for O_2 than myoglobin

    E. hemoglobin has a higher affinity for O_2 than myoglobin

11. Which blood component is involved in clot formation?

    A. erythrocytes   **C.** B cells   **E.** platelets

    B. macrophages   **D.** T cells

12. A patient with AB type blood is considered the *universal recipient* because this patient's blood:

    A. contains neither antibody

    B. contains both antibodies

    C. is most common

    D. is least common

    E. is not able to agglutinate

13. All of the following statements are true regarding blood, EXCEPT:

    A. It contains albumins to regulate osmolarity
    B. Its pH is normally between 7.34 and 7.45
    C. It varies from bright red to a dark red color
    D. It is denser and more viscous than water
    E. It carries body cells to injured areas for repair

14. What is the protein structure of the functional human myoglobin that is composed of 154 amino acid residues in a single chain?

    A. 1°   **B.** 2°   **C.** 3°   **D.** 4°   **E.** 5°

15. When inhaled, carbon monoxide can be lethal because it:

    A. irritates the pleura
    B. blocks electron transport within the cytochrome system
    C. binds with strong affinity to hemoglobin
    D. is insoluble in the bloodstream
    E. inhibits the Na^+/K^+ pump

16. Which statement correctly describes the relationship between osmotic and hydrostatic pressure differentials necessary for the exchange of fluid between the blood in the capillaries and the surrounding tissues?

    A. Osmotic pressure causes solutes and fluid to move out of the capillaries and into the tissues
    B. Hydrostatic pressure causes fluid to move out of the capillaries and into the tissues
    C. Osmotic pressure forces fluid to move out of the capillaries and into the tissues
    D. Hydrostatic pressure forces fluid to move out of the tissue and into the capillaries
    E. Both osmotic and hydrostatic pressure cause fluid to move out of the capillaries and into the tissues

17. Which statement is NOT true regarding the formation of blood cells?

    A. Platelets are formed from myeloblasts
    B. Lymphocytes are formed from lymphoblasts
    C. Eosinophils are formed from myeloblasts
    D. Erythrocytes are formed from erythroblasts
    E. None of the above

18. T and B-cell abnormalities can be caused by severe combined immunodeficiency (SCID), which develops due to a genetic disease where adenosine deaminase is deficient. Which organs would be underdeveloped in an individual suffering from SCID?

A. bone marrow only
B. bone marrow and thymus
C. thymus only
D. bone marrow and spleen
E. liver and spleen

19. Which of the following is characteristic of a capillary?

I. hydrostatic pressure is higher at the arteriole end than at the venule end
II. osmotic pressure is higher in the blood plasma than in the interstitial fluid
III. hydrostatic pressure results from heart contractions

A. I only
B. II only
C. II and III only
D. I, II and III
E. I and III only

20. What is the numerical difference in [H^+] between a liquid at pH 4 and 6?

A. 2 times   **B.** 10 times   **C.** 20 times   **D.** 100 times   **E.** 50% increase

21. Which plasma protein is the major contributor to osmotic pressure?

A. fibrinogen
B. albumin
C. alpha globulin
D. gamma globulin
E. hemoglobin

22. What is the primary function of the angiotensin converting enzyme (ACE) produced in the lungs?

A. to increase levels of angiotensinogen
B. to decrease levels of angiotensin II
C. to increase parathyroid hormone
D. to increase calcitonin synthesis
E. to increase levels of angiotensin II

23. All of the below statements about blood are correct, EXCEPT:

A. mature erythrocytes lack a nucleus
B. blood platelets are involved in the clotting process
C. erythrocytes develop in the adult spleen
D. leukocytes undergo phagocytosis of foreign matter
E. new red blood cells are constantly developing in the bone marrow

24. A patient has a blood clot forming in one of the large veins of his leg as a result of infection. If the clot dislodges and moves, what would be the most likely initial problem for this patient?

A. Cerebral stroke, because the clot has moved to the brain

B. Coronary thrombosis, because the clot has moved to the coronary circulation

C. Pulmonary embolism, because the clot has moved to the pulmonary capillaries

D. Renal shutdown, because the clot has moved to the kidney

E. Liver ischemia, because the clot has moved to the hepatic portal circulation

25. Which cell has no visible cytoplasmic granules?

A. monocyte   C. eosinophil

B. basophil   D. neutrophil   E. none of the above

26. Given that surfactant is produced by pneumocytes and is fully functional when it forms *micelles*, which of the following properties most likely describes a surfactant?

A. acidic molecule   C. neutral molecule

B. hydrophilic molecule   D. basic molecule

E. hydrophobic molecule

27. Which of the following is a normal blood flow pathway?

A. inferior vena cava to left atrium   C. pulmonary veins to left ventricle

B. right ventricle to aorta   D. pulmonary veins to left atrium

E. left ventricle to pulmonary artery

28. A patient with O type blood is considered the *universal donor* because this patient's blood:

A. contains neither antigen   C. is the least common

B. contains both antigens   D. is the most common

E. is not able to physically agglutinate

29. Which property is shared by all leukocytes?

A. They are the most numerous of the formed elements in blood

B. They are phagocytic   D. They are nucleated

C. They have cytoplasmic granules   E. They lack a nucleus

30. Interfacial tension between liquids is reduced by a soluble compound, surfactant, found in the:

A. larynx   B. alveoli   C. trachea   D. epiglottis   E. nasopharynx

31. The site of O_2 absorption in the capillaries of the lungs is:

A. alveoli   B. pleura   C. bronchi   D. bronchioles   E. trachea

---

32. In an infant born with a congenital heart defect, which of the following is most likely to result from the mixing of blood between the right and left ventricles?

A. recurrent fever

B. hemoglobin deficiency

C. myoglobin deficiency

D. low blood pressure

E. poor oxygenation of the tissues

33. Which statement is accurate regarding blood plasma?

A. It is over 90% water

B. It contains about 20 dissolved components

C. It is the same as serum but without the clotting proteins

D. The main protein component is hemoglobin

E. Erythrocytes lack a nucleus

34. Which organelle must be well developed within plasma cells for effective antibody synthesis?

A. storage vacuole

B. rough ER

C. smooth ER

D. mitochondria

E. lysosome

35. Which of these vessels has the highest partial pressure of oxygen in a healthy person?

A. aorta

B. coronary veins

C. superior vena cava

D. pulmonary arteries

E. inferior vena cava

36. The pH of cellular fluids (as well as blood fluids) is approximately:

A. 5   **B.** 6   **C.** 7   **D.** 8   **E.** 10

37. Which sequence of events is correct?

A. prothrombin → thrombin; formation of thromboplastin; fibrinogen → fibrin; clot retraction

B. formation of thromboplastin; clot retraction, fibrinogen → fibrin; prothrombin → thrombin

C. formation of thromboplastin; prothrombin → thrombin; fibrinogen → fibrin; clot retraction

D. fibrinogen → fibrin; clot retraction; formation of thromboplastin; prothrombin → thrombin

E. prothrombin → thrombin; fibrinogen → fibrin; clot retraction → formation of thromboplastin

38. Class II Major Histocompatibility Complex (MHC) proteins are found on the surface of leukocyte cells. Which type of cells does MHC II cell-mediated immunity require?

A. neutrophils

B. antibody-producing cells

C. erythrocytes

D. macrophages

E. T-helper cells

39. Which structure would be reached LAST by a tracer substance injected into the superior vena cava?

    A. left atrium   **C.** left ventricle

    B. pulmonary veins   **D.** right ventricle   **E.** tricuspid valve

40. In the mammalian heart, semi-lunar valves are found:

    A. where blood goes from atria to ventricles

    B. on the right side of the heart only

    C. where the pulmonary veins attach to the heart

    D. where blood leaves via the aorta and pulmonary arteries

    E. at the locations where the anterior and posterior venae cavae enter

41. Erythrocyte production is regulated by the:

    A. liver   **B.** pancreas   **C.** kidney   **D.** brain   **E.** spleen

42. Which statement is correct about the atrioventricular node?

    A. It regulates contraction rhythm of cardiac cells

    B. It delays the contraction of the heart ventricles

    C. It is the parasympathetic ganglion located in the left atrium of the heart

    D. It conducts action potentials from the vagus nerve to the heart

    E. It is the parasympathetic ganglion located in the right atrium of the heart

43. Stenosis is a condition whereby the valve's opening is narrowed and results in decreased blood flow through the valve. If diagnosed with stenosis of the mitral valve, where would a patient experience the greatest blood pressure increase?

    A. right atrium   **C.** left ventricle

    B. aorta   **D.** left atrium   **E.** right ventricle

44. A difference of 1 pH unit (e.g. pH 6 *vs.* 7) corresponds to what change in the $[H^+]$?

    A. 1   **B.** 2   **C.** 10   **D.** doubling   **E.** 100

45. Which of the following would result from a large loss of blood due to a hemorrhage?

    A. No change in blood pressure but a slower heart rate

    B. No change in blood pressure but a change in respiration

    C. Vasodilation

    D. Rise in blood pressure due to change in cardiac output

    E. Lowering of blood pressure due to change in cardiac output

46. What is the correct order for the cardiac conduction pathway?

  A. Purkinje fibers $\rightarrow$ SA node $\rightarrow$ AV node $\rightarrow$ bundle of His
  B. SA node $\rightarrow$ bundle of His $\rightarrow$ Purkinje fibers $\rightarrow$ AV node
  C. SA node $\rightarrow$ AV node $\rightarrow$ bundle of His $\rightarrow$ Purkinje fibers
  D. AV node $\rightarrow$ SA node $\rightarrow$ bundle of His $\rightarrow$ Purkinje fibers
  E. AV node $\rightarrow$ SA node $\rightarrow$ Purkinje fibers $\rightarrow$ bundle of His

47. Which of these heart structures has blood with the greatest O_2 content?

  A. right atrium   **C.** pulmonary artery
  B. left ventricle   **D.** thoracic duct   **E.** superior vena cava

48. When a swimmer holds their breath, which of the following blood gases changes first and creates the urge to breathe?

  A. rising O_2 **C.** rising CO_2
  B. falling O_2 **D.** falling CO_2 **E.** rising CO_2 and falling O_2

49. The left ventricular wall of the heart is thicker than the right wall, because it:

  A. pumps blood through a smaller valve   **C.** expands the thoracic cage during diastole
  B. pumps blood with greater pressure   **D.** accommodates a greater volume of blood
                                                   E. undergoes isotonic contractions

50. At high altitude, H_2O vapor pressure in the lungs remains constant while O_2 and CO_2 pressures fall. All of the following physiological changes assist a rock climber during low O_2 levels, EXCEPT:

  A. increased diffusion capacity of the lungs   **C.** increased blood flow to the tissues
  B. increased pulmonary ventilation   **D.** increased number of erythrocytes
                                                       E. decreased BPG synthesis

51. Which of these heart structures has blood with the lowest O_2 content?

  A. thoracic duct   **C.** pulmonary artery
  B. left ventricle   **D.** inferior vena cava   **E.** right atrium

52. In which of the following vessels would you expect the highest protein concentration?

  A. proximal tubule   **C.** inferior vena cava
  B. renal artery   **D.** vasa recta   **E.** afferent arteriole

53. Which statement regarding the heart valves is accurate?

    A. Aortic valves control the flow of blood into the heart

    B. AV valves are supported by chordae tendineae to prevent regurgitation of blood into the atria during ventricular contraction

    C. Mitral valve separates the right atrium from the right ventricle

    D. Tricuspid valve divides the left atrium from the left ventricle

    E. Pulmonary valves control the flow of blood into the heart

54. Which of these thermoregulation mechanisms is NOT involved in heat conservation?

    A. blood vessel constriction   **C.** perspiration

    B. shivering   **D.** piloerection   **E.** none of the above

55. Which blood cell is responsible for coagulation?

    A. leukocyte   **C.** lymphocyte

    B. erythrocyte   **D.** platelet   **E.** macrophage

56. Which of the following is NOT correct about the hepatic portal system?

    A. It branches off of the inferior vena cava

    B. It consists of a vein connecting two capillary beds together

    C. Its major vessels are the superior mesenteric, inferior mesenteric, and splenic veins

    D. It carries nutrients to the liver for processing

    E. It carries toxins and microorganisms to the liver for processing

57. When directed in right-to-left circulation in a fetus, blood travels directly from the:

    A. right atrium to the aorta   **C.** right lung to the left lung

    B. superior vena cava to the aorta   **D.** pulmonary vein to the pulmonary artery

    E. right atrium to the left atrium

58. How is most of the CO_2 transported by the blood?

    A. as bicarbonate ions in the plasma   **C.** as carbonic acid in the erythrocytes

    B. attached to hemoglobin   **D.** dissolved in the blood plasma

    E. as bicarbonate ions attached to hemoglobin

59. Which of the following is true according to Starling's hypothesis of capillary physiology?

    A. Filtration occurs where the hydrostatic pressure is less than the osmotic pressure

    B. Ultrafiltrate is returned to the bloodstream by the lymphatic system

    C. Hydrostatic pressure is lower at the arteriolar end

    D. Osmotic pressure drives fluid out of the capillary

    E. None of the above

Answer Key

1: B	11: E	21: B	31: A	41: C	51: C
2: D	12: A	22: E	32: E	42: B	52: E
3: E	13: E	23: C	33: A	43: D	53: B
4: B	14: C	24: C	34: B	44: C	54: C
5: A	15: C	25: A	35: A	45: E	55: D
6: A	16: B	26: E	36: C	46: C	56: A
7: E	17: A	27: D	37: C	47: B	57: E
8: D	18: B	28: A	38: E	48: C	58: A
9: C	19: D	29: D	39: C	49: B	59: B
10: B	20: D	30: B	40: D	50: E	

Chapter 4.5: Lymphatic and Immune Systems

1. All of the following statements regarding the thymus are correct, EXCEPT:

    A. its stroma consists of epithelial tissue
    B. it is smaller and less active in children than in adults
    C. it functions strictly in T lymphocyte maturation
    D. it does not fight antigens
    E. all of the above are true

2. Which of the following cell lines might be elevated in a patient exposed to prions?

    A. plasma cells
    B. macrophages
    C. T-cells
    D. monocytes
    E. erythrocytes

3. All of the following are involved in lymph transport, EXCEPT:

    A. lymph capillary minivalve action
    B. smooth muscle contraction in the lymph capillary walls
    C. thoracic pressure changes during breathing
    D. milking action of active muscle fibers
    E. lymph fluid rejoins the circulation at the superior vena cava

4. A person lacking gamma globulins would have:

    A. diabetes
    B. hemophilia
    C. severe allergies
    D. low resistance to infection
    E. none of the above

5. Which organ of the digestive system is the site of absorption of amino acids, fatty acids and sugars?

    A. large intestine
    B. small intestine
    C. gallbladder
    D. stomach
    E. colon

6. The thymus has levels of high activity during:

    I. neonatal development   II. pre-adolescence   III. middle age

    A. I only   **B.** II only   **C.** III only   **D.** I and III only   **E.** I and II only

7. Humoral immunity depends on the function of:

    A. erythrocytes
    B. cytotoxic T cells
    C. immunoglobulins
    D. albumin
    E. fibrin

8. The lymphatic capillaries are:

A. completely impermeable

B. as permeable as blood capillaries

C. less permeable than blood capillaries

D. more permeable than blood capillaries

E. smaller in diameter than blood capillaries

9. Which cell type produces antibodies?

A. neurons   **B.** T cells   **C.** B cells   **D.** macrophages   **E.** natural killer cells

10. Antibodies that function against a particular foreign substance are released by:

A. lymph nodes

B. medullary cords

C. T lymphocytes

D. plasma cells

E. natural killer cells

11. Through which process, similar to phagocytosis, do cells engulf droplets of liquid?

A. apoptosis   **B.** necrosis   **C.** exocytosis   **D.** endocytosis   **E.** pinocytosis

12. Lymph leaves a lymph node through:

A. the subcapsular sinus

B. the cortical sinus

C. afferent lymphatic vessels

D. efferent lymphatic vessels

E. capillaries

13. Based on the diagram showing the relationship between lymph flow and interstitial (extracellular) fluid pressure, would an increase in interstitial fluid protein result in increased lymph flow?

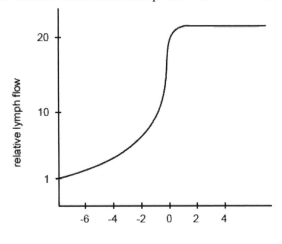

A. No, because fluid movement into the capillaries decreases interstitial fluid pressure

B. No, because the subsequent increase in interstitial fluid pressure reduces lymph flow

C. Yes, because fluid movement out of the capillaries increases interstitial fluid pressure

D. Yes, because the increased interstitial fluid protein reduces interstitial fluid volume

E. Yes, because fluid movement out of the capillaries decreases interstitial fluid pressure

14. Inflammatory responses may include which of the following?

    A. clotting proteins migrating away from the site of infection
    B. increased activity of phagocytes in an inflamed area
    C. reduced permeability of blood vessels to conserve plasma
    D. release of substances that decrease the blood supply to an inflamed area
    E. inhibiting the release of white blood cells from bone marrow

15. Which of the following statements about lymph transport is correct?

    A. It depends on the movement of adjacent tissues such as skeletal muscles
    B. It only occurs when illness causes tissue swelling
    C. It is faster than transport via veins
    D. Lymph vessels are high-pressure conduits
    E. It is rapid and continuous

16. The capsule of *B. vulgatus* bacteria is a major factor of its virulence, because it inhibits phagocytosis by preventing the activity of:

    A. monocytes   **B.** platelets   **C.** T-cells   **D.** B-cells   **E.** fibrin

17. Which statement about lymphocytes is correct?

    A. T cells are the precursors of B cells
    B. T cells are the only form of lymphocytes found in lymphoid tissue
    C. The two main types are T cells and macrophages
    D. B cells differentiate into plasma cells that secrete antibodies into the blood
    E. B cells are the precursors of T cells

18. Which of the following will most likely result in erythroblastosis fetalis?

    A. Rh⁻ mother and Rh⁻ fetus   **C.** Rh⁺ mother and Rh⁻ fetus
    B. Rh⁺ mother and Rh⁻ father   **D.** Rh⁺ mother and Rh⁺ fetus
                                            E. Rh⁻ mother and Rh⁺ fetus

19. Functions of the spleen include all of the following, EXCEPT:

    I. storage of iron   III. storage of blood platelets
    II. forming crypts that trap bacteria   IV. removal of old blood cells from the blood

    A. I only   **C.** III and IV only
    B. II only   **D.** I and II only   **E.** II and III only

20. The thymus gland is primarily responsible for:

    A. lymphoid T cell maturation

    B. removal and replacement of erythrocytes

    C. production of humoral immunity cells

    D. immunity only after a child matures into adulthood

    E. production of erythrocytes

21. Which structure(s) is/are the location of Peyer's patches?

    I. large intestine

    II. jejunum of the small intestine

    III. ileum of the small intestine

    A. I only   **B.** II only   **C.** I and II only   **D.** II and III only   **E.** I, II and III

22. If erythrocytes from a person with type A blood are added to three test tubes each containing a different blood type (AB, B and O), which test tube(s) will NOT agglutinate?

    I. type AB blood   II. type B blood   III. type O blood

    A. I only   **C.** III only

    B. II only   **D.** I and III only   **E.** II and III only

23. Which of the following would NOT be classified as a lymphatic structure?

    I. Peyer's patches of the intestine   III. tonsils

    II. spleen   IV. pancreas

    A. I only   **C.** III only

    B. II only   **D.** IV only   **E.** I and IV only

24. Lymph capillaries are found in:

    I. digestive organs   II. bone marrow   III. bones   IV. CNS

    A. I only   **C.** III only

    B. I and IV only   **D.** I and II only   **E.** II and III only

25. Which innate immune response is exhibited by a patient initially exposed to a pathogen?

    A. T killer cells   **C.** T helper cells

    B. neutrophils   **D.** B lymphocytes   **E.** immunoglobulin proliferation

26. The thymus is the only lymphoid organ that does NOT:

    A. directly fight antigens   **C.** produce hormones

    B. have a cortex   **D.** have lymphocytes   **E.** have a medulla

27. Which blood type(s) could be inherited by a child of a father with type AB blood and a mother with type O blood?

       I. type A   II. type B   III. type AB   IV. type O

  A. III only   **B.** I and IV only   **C.** II and IV only   **D.** I and III only   **E.** I and II only

28. Which of the following is NOT a lymphatic tissue associated with the digestive tract?

  A. islets of Langerhans   **C.** palatine tonsils

  B. lingual tonsils   **D.** Peyer's patches   **E.** all of the above

29. Which of the following would be impaired as a result of the HIV virus infecting and killing helper CD4 T lymphocytes?

  A. humoral immunity   **C.** non-specific immunity

  B. cell-mediated immunity   **D.** the bone marrow   **E.** transport of O_2 to tissue

30. Which of the following is/are function(s) of lymphoid tissue?

     I. To be involved in lymphocyte and macrophage activities
     II. To store and provide a proliferation site for lymphocytes
     III. To store and provide a proliferation site for neutrophils

  A. I only   **B.** II only   **C.** III only   **D.** I and II only   **E.** I and III only

31. What is the function of lymph nodes?

  A. breaking down hemoglobin   **C.** facilitating the absorption of amino acids

  B. filtering of lymph   **D.** increasing glucose concentrations in blood

                             E. carrying oxygen

32. What are antigens?

  A. proteins found in the blood that cause foreign blood cells to clump

  B. proteins embedded in B cell membranes

  C. proteins that consist of two light and two heavy polypeptide chains

  D. foreign molecules that trigger the production of antibodies

  E. proteins released during an inflammatory response

33. All of the following are functions of lymph nodes, EXCEPT:

  A. producing lymph fluid and cerebrospinal fluid

  B. serving as antigen surveillance areas

  C. acting as lymph filters

  D. producing lymphoid cells and storing granular leukocytes

  E. activating the immune system

34. Which cells are responsible for secreting virus-neutralizing antibodies?

A. T cells   **C.** macrophages
B. erythrocytes   **D.** thymus cells   **E.** plasma cells

35. Which of the following statements is correct regarding antibodies?

A. There are three binding sites per antibody monomer
B. They are incapable of being transferred from one person to another
C. They contain a prominent carbohydrate structure
D. They are composed of heavy and light polypeptide chains
E. They are held together by hydrophobic interactions

36. Which of these is absorbed by lacteals?

A. salts   **B.** fatty acids   **C.** proteins   **D.** carbohydrates   **E.** sugars

37. All of the following are NOT associated with passive immunity, EXCEPT:

A. booster shot of vaccine
B. differentiation of plasma cells
C. infusion of weakened viruses
D. exposure to an antigen
E. passage of IgG antibodies from a pregnant mother to her fetus

38. Hyperactivity of which organ causes anemia?

A. adenoids   **B.** spleen   **C.** parathyroid   **D.** thymus   **E.** thyroid

39. All of the following are T cell type, EXCEPT:

A. memory   **B.** regulatory   **C.** cytotoxic   **D.** antigenic   **E.** helper

40. All of the following are functions of the lymphatic system, EXCEPT:

A. draining excess cerebrospinal fluid from the brain
B. destruction and removal of foreign particles
C. removal of proteins from interstitial spaces
D. absorption of lipids from the small intestine
E. transport of white blood cells to and from the lymph nodes into the bones

41. Which of the following is NOT a function of the liver?

A. synthesis of red blood cells   **C.** conversion of carbohydrates to fats
B. deamination of amino acids   **D.** storage of glycogen
                                            E. metabolism of alcohol

42. B lymphocytes develop immunocompetence in:

 A. lymph nodes   **C.** the spleen

 B. bone marrow   **D.** the thymus   **E.** the liver

43. Antibodies act by:

 A. aiding in phagocytosis of antigens

 B. binding to antigens via the variable portion

 C. inhibiting stem cell production in the bone marrow

 D. binding to plasma cells and marking them for destruction via phagocytosis

 E. converting fibrinogen into fibrin

44. Inflammatory response is involved in all of the following processes, EXCEPT:

 A. disposing of cellular debris and pathogens

 B. setting the stage for repair processes

 C. preventing the spread of the injurious agent to nearby tissue

 D. replacing injured tissues with connective tissue

 E. increasing blood flow to the site of injury

45. According to the theory of blood groups, all of the following statements are true, EXCEPT:

 A. agglutination is caused by antibodies

 B. type O blood does not undergo agglutination

 C. Rh factor is a type of antigen found in human blood

 D. type B blood contains antibodies for type A antigens

 E. agglutinins are antigens

46. Which type of bond holds together antibody molecules?

 A. hydrophobic   **C.** van der Waals

 B. ionic   **D.** hydrogen   **E.** disulfide

47. Fluid absorbed from the interstitial spaces by lymphatic vessels is carried to the:

 A. kidneys, where it is excreted as urine

 B. lungs, where the fluid is expired during exhalation

 C. lymphatic ducts, where it returns to the circulation

 D. large intestine, where it is absorbed and returned to the bloodstream

 E. aorta, where it returns to the circulation

48. During clonal selection of B cells, which substance determines what cells will eventually become cloned?

 A. complement   **C.** interferon

 B. antibody   **D.** antigen   **E.** major histocompatibility complex (MHC)

49. B cells are dormant in lymph nodes and other lymphoid tissues until activated by specific lymphocyte antigens. When activated, B cells produce:

A. cytotoxic T cells

B. lymphokines

C. antibodies

D. macrophages

E. monocytes

50. In which of the following situations will helper T cells be activated?

A. When natural killer (NK) cells come in contact with a tumor cell

B. When a cytotoxic T cell releases cytokines

C. In the bone marrow during the self-tolerance test

D. When B cells respond to T-independent antigens

E. When an antigen is displayed by a dendritic cell

51. The correct sequence of events in phagocytosis is:

A. ingestion → adherence → chemotaxis → digestion → killing

B. chemotaxis → adherence → ingestion → digestion → killing

C. chemotaxis → ingestion → digestion → adherence → killing

D. adherence → digestion → killing → ingestion → chemotaxis

E. adherence → chemotaxis → digestion → killing → ingestion

52. All of the following are examples of innate immunity, EXCEPT:

A. phagocytotic cells

B. digestive enzymes and stomach acid

C. memory B cells

D. skin as a physical barrier to antigens

E. neutrophils

53. The sites of chronic infections are predominated by:

A. macrophages

B. B cells

C. basophils

D. eosinophils

E. plasma cells

54. Herbivores have longer alimentary canals in relation to body size than carnivores, because herbivores need:

A. reduced oxygen consumption

B. more time for digestion

C. a more robust hepatic portal vein

D. more surface area for absorption of nutrients

E. microorganisms to metabolize nutrients

55. Which cells are involved in cell-mediated immunity and respond to class I MHC molecule-antigen complexes?

A. cytotoxic T cells

B. natural killer cells

C. helper T cells

D. macrophages

E. B cells

56. The reason for *swollen glands* in the neck of a patient with a cold is that:

    A. blood pools in the neck to keep it warm

    B. lymph nodes swell during infection as white blood cells proliferate within them

    C. inflammation is initiated by the infection that causes fluid to be drained from the area

    D. fever initiates a general expansion of the tissues of the head and neck

    E. there is an increased degradation of erythrocytes in the lymph nodes

57. Which statement is true regarding interferons?

    A. An interferon produced against one virus would not protect cells against another virus

    B. They act by increasing the rate of cell division

    C. They interfere with viral replication within cells

    D. They are used in nasal sprays for the common cold

    E. They are virus-specific

58. Which of the following is correct about CD4 and CD8 cells?

    A. They are antigen-presenting cells that secrete proteins

    B. They are natural killer (NK) cells with receptors on the surface

    C. They are T cell-independent antigens

    D. They are T cells with glycoprotein molecules on their surface that enhance cellular interactions

    E. They are antigen-presenting cells that inhibit B cell activity

59. Which statement is correct about type A⁻ blood?

    A. type A⁻ blood produces antibodies that bind type O antigens only

    B. type A⁻ blood produces antibodies that bind both type B and type O antigens

    C. type A⁻ blood produces antibodies that bind type A, but not type B antigens

    D. type A⁻ blood produces antibodies that bind neither Rh nor type A antigens

    E. type A⁻ blood produces antibodies that bind type B, but not type A antigens

60. White blood cells at the scene of a wound:

    A. produce the blood clot

    B. secrete scar tissue fibers

    C. stimulate the epidermal cells to divide

    D. engulf microbes and cell debris

    E. convert fibrinogen to fibrin

Answer Key

1: B	11: E	21: D	31: B	41: A	51: B
2: E	12: D	22: A	32: D	42: B	52: C
3: B	13: C	23: D	33: A	43: B	53: A
4: D	14: B	24: A	34: E	44: D	54: E
5: B	15: A	25: B	35: D	45: B	55: A
6: E	16: A	26: A	36: B	46: E	56: B
7: C	17: D	27: E	37: E	47: C	57: C
8: D	18: E	28: A	38: B	48: D	58: D
9: C	19: B	29: B	39: D	49: C	59: E
10: D	20: A	30: D	40: A	50: E	60: D

Chapter 4.6: Digestive System

1. All of the following processes take place in the liver, EXCEPT:

    A. glycogen storage

    B. detoxification of poisons

    C. synthesis of adult erythrocytes

    D. conversion of amino acids to urea

    E. regulation of blood sugar levels

2. Glycogenolysis is the breakdown of glycogen stored in the liver and skeletal muscles. Which of the following hormones inhibits glycogenolysis?

    A. aldosterone

    B. glucagon

    C. adrenaline

    D. cortisol

    E. insulin

3. Which of the following statements is true of mammals?

    A. All foods begin their enzymatic digestion in the mouth

    B. The epiglottis prevents food from entering the trachea

    C. After leaving the oral cavity, the bolus enters the larynx

    D. Enzyme production continues in the esophagus

    E. The trachea leads to the esophagus and then to the stomach

4. What cartilaginous structure prevents food from going down the trachea?

    A. tongue

    B. larynx

    C. glottis

    D. epiglottis

    E. esophageal sphincter

5. The chemical and mechanical receptors that control digestion are located in the:

    A. pons and medulla

    B. oral cavity

    C. glandular tissue that lines the organ lumen

    D. walls of the tract organs

    E. hypothalamus

6. Which system is involved in distributing the absorbed lipids to the peripheral tissue?

    A. digestive

    B. integumentary

    C. lymphatic

    D. nervous

    E. excretory

7. What structure does food enter as it is swallowed and leaves the mouth?

    A. glottis

    B. esophagus

    C. trachea

    D. larynx

    E. fundus

8. Which of the following is a function of the hepatic portal circulation?

   A. Returning glucose to the general circulation when blood sugar is low
   B. Distributing hormones throughout the body
   C. Carrying nutrients to the spleen
   D. Carrying toxins to the venous system for disposal through the urinary tract
   E. Collecting absorbed nutrients for metabolic processing or storage

9. Which organ is the main site of fatty acid synthesis?

   A. smooth muscle   **C.** kidney
   B. spleen   **D.** liver   **E.** adrenal glands

10. Which organ is the site of bile production?

   A. large intestine   **C.** liver
   B. small intestine   **D.** gallbladder   **E.** pancreas

11. Which of the following is the correct order (from the lumen) of the four basic layers making up walls of every organ in the alimentary canal?

   A. mucosa → submucosa → muscularis externa → serosa
   B. submucosa → serosa → muscularis externa → mucosa
   C. serosa → mucosa → submucosa → muscularis externa
   D. muscularis externa → serosa → mucosa → submucosa
   E. mucosa→ serosa → submucosa → muscularis externa

12. Which of the following would be observed by a physician about her patient with type 1 diabetes and low blood insulin concentration?

   A. decreased levels of blood glucose
   B. increased insulin levels from thyroid stimulation   **D.** presence of glucose in urine
   C. decreased levels of circulating erythrocytes   **E.** increased hematocrit

13. The primary site of water absorption is the:

   A. duodenum   **C.** jejunum
   B. large intestine   **D.** ileum   **E.** mouth

14. The hormone that increases output of enzyme-rich pancreatic juice and stimulates contraction of the gallbladder to release bile is:

   A. gastric inhibitor peptide   **C.** secretin
   B. trypsin   **D.** gastrin   **E.** cholecystokinin

15. Which of these molecules has the highest lipid density?

    A. very low-density lipoprotein (VLDL)   **C.** high-density lipoprotein (HDL)

    B. low-density lipoprotein (LDL)   **D.** chylomicron

    E. all have the same lipid density

16. Which of these processes does NOT occur in the mouth?

    A. moistening of food   **C.** bolus formation

    B. mechanical digestion   **D.** chemical digestion of proteins

    E. chemical digestion of starch

17. All of the following statements about bile are correct, EXCEPT that:

    A. it functions to carry bilirubin formed from breakdown of worn-out RBCs

    B. it contains enzymes for digestion

    C. it is both an excretory product and a digestive secretion

    D. it functions to emulsify fats

    E. it is synthesized in the liver and stored in the gallbladder

18. An electrolyte is a compound that dissociates well in water; an example of a poor electrolyte would be:

    A. H_2SO_4 **B.** NaBr   **C.** glucose   **D.** KCl   **E.** $MgCl_2$

19. The principal function of $NaHCO_3$ is to:

    A. inactivate bile   **C.** dissolve CO_2

    B. combines with CO_2 in alveoli   **D.** combines with O_2 when hemoglobin is saturated

    E. buffer a solution

20. Saliva includes the following solutes:

    A. mucin, lysozyme, electrolytes, salts and minerals

    B. electrolytes, digestive enzyme, mucin, lysozyme and IgA

    C. digestive enzyme and electrolytes only

    D. proteases and amylase only

    E. salts and minerals only

21. Structures found at the esophagus-stomach, stomach-duodenum and ileum-colon junction sites are:

    A. sphincters   **C.** junction points for enzyme release

    B. regions of dense villi   **D.** sites for peristalsis   **E.** Peyer's patches

22. Which of these statements is NOT true about the digestive system?

  A. Peristalsis is a wave of smooth muscle contractions that proceed along the digestive tract
  B. Digestive enzymes from the pancreas are released via a duct into the duodenum
  C. Low pH of the stomach is essential for the functioning of carbohydrate digestive enzymes
  D. Villi in the small intestine absorb nutrients into both the lymphatic and circulatory systems
  E. Release of bile from the gallbladder is triggered by the hormone cholecsytokinin

23. In addition to mechanical breakdown and storage of food, the stomach:

  A. initiates protein digestion and denatures proteins
  B. is the first site where absorption takes place
  C. is the site of lipid digestion
  D. is the site of carbohydrate and lipid digestion
  E. is the site of carbohydrate digestion

24. Which of the following would be expected in a patient after the administration of Tagamet, a drug that is an antagonist of gastric parietal cell's H_2-receptors?

  A. increased $[H^+]$ in the stomach
  B. increased pH in the stomach
  C. increased ATP consumption by gastric parietal cells
  D. decreased pepsin release
  E. increased levels of pepsin

25. All of the following match nutrients to the specific digestive enzyme correctly, EXCEPT:

  A. fats – pancreatic lipase
  B. proteins – chymotrypsin
  C. proteins – carboxypeptidase
  D. carbohydrates – pancreatic amylase
  E. proteins – ptyalin (salivary amylase)

26. The acid secretions in the stomach are stimulated by the presence of:

  A. lipids and fatty acids
  B. simple carbohydrates and alcohols
  C. protein and peptide fragments
  D. starches and complex carbohydrates
  E. carbohydrates and lipids

27. For laboratory use, cellulose is a neutral polymer of glucose that can become either positive or negative by attaching cationic or anionic groups. What type of plasma compound is most likely to be filtered from blood during dialysis with anionic groups on the cellulose?

  A. negatively charged compounds
  B. positively charged compounds
  C. all compounds are filtered uniformly
  D. neutral charged compounds
  E. no compounds are filtered with efficiency

28. Which enzyme functions to convert disaccharides into monosaccharides?

  A. kinase   **B.** lactase   **C.** lipase   **D.** zymogen   **E.** phosphatase

29. The digestive enzyme pepsinogen is secreted by:

  A. Brunner's glands   **C.** foveolar cells of the stomach
  B. goblet cells of the duodenum   **D.** parietal cells of the stomach
                                  E. chief cells of the stomach

30. The gastrointestinal tract is made up of three layers: mucosa, submucosa and muscularis mucosae. The latter is composed of which type of muscle?

  A. voluntary skeletal   **C.** involuntary smooth
  B. voluntary smooth   **D.** involuntary skeletal
                                  E. voluntary smooth and skeletal

31. Which of these choices is the correct statement about lipids?

  A. They are composed of elements C, O, N & H
  B. Their secondary structure is composed of α helices and β pleated sheets
  C. They are molecules used for long term energy storage in animals
  D. Elements of C:H:O are in the ratio of 1:2:1
  E. Maltose is an example of a lipid

32. All of the following are functions of hepatocytes, EXCEPT:

  A. producing digestive enzymes   **C.** storing fat-soluble vitamins
  B. processing nutrients   **D.** detoxifying chemicals
                                  E. synthesizing cholesterol

33. Which is an example of an enzyme that is secreted as an inactive precursor and is converted into its active form in the lumen of the small intestine?

  A. protease   **C.** lipase
  B. salivary amylase   **D.** bicarbonate   **E.** trypsinogen

34. What is the site of amino acid absorption?

  A. stomach   **C.** large intestine
  B. gallbladder   **D.** small intestine   **E.** rectum

35. Which vitamin requires intrinsic factor for absorption?

  A. A   **B.** C   **C.** B_{12} **D.** K   **E.** E

36. What is the function of bicarbonate within the mucous secreted into the gastrointestinal tract by the epithelium?

       I. Digestion of proteins
       II. Functioning as a buffer for the contents within the gastrointestinal tract
       III. Preventing the gastrointestinal tract from becoming acidic

    A. I only
    B. I and II only
    C. I and III only
    D. I, II and III
    E. II and III only

37. Which nutrients yield 4 calories per gram?

    A. glucose and proteins
    B. fats and glucose
    C. proteins and lipids
    D. lipids and sugars
    E. glucose, proteins and fats

38. Which of the following is a function of goblet cells?

    A. Protection against invading disease-causing organisms, such as bacteria, that enter the digestive tract in food
    B. Secretion of buffers to keep the pH of the digestive tract close to neutral
    C. Producing mucus to protect parts of the digestive organs from the effects of protease enzymes needed for food digestion
    D. Absorption of nutrients from digested food
    E. Storing nutrients for future use

39. Which of the following is NOT an end product of digestion?

    A. amino acids   **B.** lactose   **C.** fructose   **D.** fatty acids   **E.** glucose

40. Which of these vitamins is needed for hepatic production of prothrombin?

    A. vitamin A   **B.** vitamin B_{12} **C.** vitamin D   **D.** vitamin K   **E.** vitamin E

41. Bacteria of the large intestine play an essential role in:

    A. synthesizing vitamin K and some B vitamins
    B. synthesizing vitamin C
    C. producing gas
    D. absorption of bilirubin
    E. synthesizing vitamin D

42. Which of the following enzymes is secreted by both saliva and the pancreas?

    A. chymotrypsin   **B.** lipase   **C.** pepsin   **D.** trypsin   **E.** secretin

43. Which of these controls the flow of material from the esophagus into the stomach?

    A. gallbladder
    B. pyloric sphincter
    C. epiglottis
    D. cardiac sphincter
    E. appendix

44. For normal hemoglobin production in RBCs, the necessary stomach secretion is:

- **A.** gastric lipase
- **B.** intrinsic factor
- **C.** pepsinogen
- **D.** HCl
- **E.** vitamin K

45. Cholecystokinin (CCK) is released by I cells of the small intestine in response to dietary fat and protein. What substance would NOT be released in response to CCK?

- **A.** trypsin
- **B.** pancreatic lipase
- **C.** pancreatic chymotrypsin
- **D.** bile
- **E.** salivary amylase

46. Which of these enzymes would be affected the most in a patient with a peptic ulcer after an overdose of antacid?

- **A.** procarboxypeptidase
- **B.** trypsin
- **C.** pepsin
- **D.** lipase
- **E.** amylase

47. Most nutrients are absorbed through the mucosa of the intestinal villi by:

- **A.** active transport
- **B.** bulk flow
- **C.** simple diffusion
- **D.** facilitated diffusion
- **E.** osmosis

48. Which organ secretes intrinsic factor and absorbs caffeine?

- **A.** liver
- **B.** pancreas
- **C.** duodenum
- **D.** ileum
- **E.** stomach

49. Lactic acid buildup results from continuous muscle contractions, because it is a:

- **A.** breakdown product of fatty acid degradation
- **B.** metabolic product of anaerobic metabolism
- **C.** product of phosphocreatine degradation
- **D.** breakdown product of ADP hydrolysis
- **E.** contraction induced by the lactic acid released from actin-myosin cross bridges

50. In the small intestine, chemical digestion needs:

- **A.** secretions from the spleen containing enzymes necessary for complete digestion
- **B.** bile salts that emulsify carbohydrates, so that they can be easily digested by enzymatic action
- **C.** a significant amount of enzyme secretion by the intestinal mucosa
- **D.** cholecystokinin (CCK), an intestinal hormone that stimulates gallbladder contraction
- **E.** pancreatic lipase that digests proteins

51. Which of the following stomach cells secretes intrinsic factor?

- **A.** parietal cell
- **B.** G cell
- **C.** chief cell
- **D.** mucous cell
- **E.** Peyer's patches

52. Which statement is true for bile?

A. It is a protease

B. It is a protein

C. It is an enzyme

D. It is a hormone

E. It is an emulsifying agent

53. As chyme passes from the stomach into the small intestine, the catalytic activity of pepsin:

A. decreases, because of increased pH

B. decreases, because pancreatic amylase degrades pepsin

C. decreases, because the pH decreases

D. decreases, because pepsinogen is converted into pepsin

E. increases, because of increased pH

54. Which of the following statements is correct about hormonal regulation of digestion?

A. Secretin targets the pancreas to release bile

B. Secretin targets the gallbladder to release digestive enzymes

C. Cholecystokinin (CCK) targets the pancreas to release digestive enzymes

D. Cholecystokinin (CCK) targets the pancreas to release sodium bicarbonate

E. Gastrin stimulates the gallbladder to contract

55. Most digestion of food in humans takes place in the:

A. liver   **B.** small intestine   **C.** mouth   **D.** stomach   **E.** pancreas

56. Which organ detoxifies the body like peroxisomes detoxify the cell?

A. the spleen   **B.** the liver   **C.** kidneys   **D.** the stomach   **E.** the colon

57. Which of the following is likely to occur within several hours after the pancreatic duct is obstructed?

I. diabetic crisis   II. acromegaly   III. impaired digestion

A. III only   **B.** I and II only   **C.** I only   **D.** II only   **E.** II and III only

58. During the process of digestion, fats are broken down when fatty acids are detached from glycerol, while proteins are digested into amino acids. What do these two processes have in common?

A. Both require ATP as an energy source

B. Both occur as intracellular processes in most organisms

C. Both are catalyzed by the same enzyme

D. Both require the presence of hydrochloric acid to lower the pH

E. Both involve the addition of H_2O to break bonds (hydrolysis)

59. Which nutrient would be the most difficult to digest if the duct leading from the gallbladder was blocked by a gallstone?

    A. fats   **B.** starch   **C.** amino acids   **D.** proteins   **E.** glycogen

60. In the symbiotic relationship between the human body and *Escherichia coli* bacteria (in the large intestine), all of the following are benefits to either organism, EXCEPT that the bacteria:

    A. produce toxins that inhibit the growth of non-symbiotic bacteria

    B. invade other tissues of the host

    C. metabolize organic material present in the large intestine

    D. synthesize vitamin K, which is absorbed in the intestine

    E. catabolize cellulose present in the large intestine

61. Pepsin is a hydrolytic enzyme produced by cells lining the stomach. Which of the following statements about pepsin is correct?

    A. Enzymes do not destroy their precursor cells

    B. Pepsin concentration in the stomach is so low, that it cannot destroy stomach cells

    C. Stomach is coated by a substance that inhibits the actions of hydrolytic enzymes

    D. Pepsin cleaves carbohydrates, and therefore the lining of the stomach is unaffected

    E. Pepsin remains as a zymogen until cleaved by HCl

Answer Key

1: C	11: A	21: A	31: C	41: A	51: A	61: E
2: E	12: D	22: C	32: A	42: B	52: E	
3: B	13: B	23: A	33: E	43: D	53: A	
4: D	14: E	24: B	34: D	44: B	54: C	
5: D	15: D	25: E	35: C	45: E	55: B	
6: C	16: D	26: C	36: E	46: C	56: B	
7: B	17: B	27: B	37: A	47: A	57: A	
8: E	18: C	28: B	38: C	48: E	58: E	
9: D	19: E	29: E	39: B	49: B	59: A	
10: C	20: B	30: C	40: D	50: D	60: B	

Chapter 4.7: Excretory System

1. The kidneys are stimulated to produce renin when:

A. blood pressure decreases
B. specific gravity of urine rises above 1.05

C. peritubular capillaries are dilated
D. pH of the urine decreases
E. aldosterone levels are high

2. A structure originating from the renal pelvis and extending to the urinary bladder is:

A. urethra
B. major calyx

C. ureter
D. vas deferen

E. collecting duct

3. All of these statements about aldosterone are correct, EXCEPT that it:

A. stimulates reabsorption of Na^+
B. stimulates secretion of K^+

C. results in the production of concentrated urine
D. is produced by the adrenal cortex
E. stimulates secretion of Na^+

4. Permeability properties of which structure allow the kidney to establish the medullary osmotic gradient?

A. collecting duct
B. distal convoluted tubule

C. loop of Henle
D. glomerular filtration membrane
E. proximal convoluted tubule

5. Which of the following results from increased hydrostatic pressure in Bowman's capsule due to the excretory system blockage by a calcified deposit in the urethra?

A. Increased filtration of blood toxins
B. Decreased filtration of blood toxins
C. Vasoconstriction of renal blood vessels
D. Increased filtrate production
E. No physiological, effect because humans have two kidneys

6. If a concentrated NaCl solution is infused directly into the renal tubules of a healthy person, what is the most likely effect?

A. Urine volume decreases because of decreased filtrate osmolarity
B. Urine volume increases because of decreased filtrate osmolarity
C. Urine volume remains unchanged
D. Urine volume decreases because of increased filtrate osmolarity
E. Urine volume increases because of increased filtrate volume

7. In the collecting tubule, the permeability of the cells to water increases because of a(n):

A. increase in the plasma levels of aldosterone

B. decrease in the osmolarity of blood plasma

C. decrease in the plasma levels of ADH

D. increase in the plasma levels of ADH

E. decrease in the plasma levels of cortisol

8. Which of the following is NOT reabsorbed by secondary active transport in the proximal convoluted tubule?

A. phosphate   **B.** urea   **C.** glucose   **D.** amino acids   **E.** all are reabsorbed

9. Certain mammals have unusually long loops of Henle to maintain a steep osmotic gradient, which allows the animal to excrete:

A. insoluble nitrogenous wastes

B. isotonic urine

C. less Na^+ across the membrane

D. hypertonic urine

E. hypotonic urine

10. Which of the choices below is NOT a function of the urinary system?

A. Eliminates solid undigested wastes and salts, and excretes CO_2 and H_2O

B. Maintains blood osmolarity

C. Regulates blood glucose levels and produces hormones

D. Helps maintain homeostasis by controlling blood volume

E. Helps maintain homeostasis by controlling blood pressure

11. Bowman's capsule functions to filter urea from the blood by:

A. ATP-hydrolysis for pumping urea into the filtrate via active transport

B. conversion of urea into amino acids

C. diffusion of urea into the filtrate via hydrostatic pressure

D. antiport exchange of urea for glucose

E. antiport exchange of urea for water

12. Which statement is true about the descending limb of the nephron loop?

A. It draws water by osmosis into the lumen of the tubule

B. It contains filtrate that becomes more concentrated as it moves down into the medulla

C. It is not permeable to water

D. It is permeable to sodium

E. It is permeable to urea

13. Which of the following is NOT normally contained in the blood that is filtered through the glomerulus in the kidney?

    A. sodium ions   **B.** potassium ions   **C.** amino acids   **D.** glucose   **E.** blood cells

14. Which of the following statements about ureters is correct?

    A. Ureters are innervated by only parasympathetic nerves
    B. Ureters are capable of peristalsis
    C. The epithelium is stratified squamous like the skin, which allows distensibility
    D. Ureters contain sphincters at the entrance to the bladder to prevent the backflow of urine
    E. The detrusor muscle is a layer of the urinary bladder wall made of skeletal muscle

15. A presence of glucose in a patient's urine indicates that:

    A. the proximal tubule is impervious to glucose
    B. the collecting ducts are defective
    C. glucose transporters in the loop of Henle are defective
    D. no clinical significance, because glucose in urine is normal
    E. glucose is entering the filtrate at a higher rate than it is being reabsorbed

16. What is the major force that moves solutes and water out of the blood across the filtration membrane?

    A. Glomerular hydrostatic pressure
    B. Size of the pores in the basement membrane of the capillaries
    C. Ionic electrochemical gradient
    D. Protein-regulated diffusion
    E. Osmotic pressure

17. Fish that live in seawater drink large amounts of salt water and use cells in their gills to pump excess salt out of the body in response to:

    A. loss of salt into their surroundings
    B. loss of H_2O by active transport to their hypertonic surroundings
    C. the need to maintain their tissues in a hypotonic state
    D. the influx of H_2O by osmosis into their tissues
    E. the need to maintain their tissues in a hypertonic state

18. Water reabsorption by the descending limb of loop of Henle takes place through:

    A. cotransport with sodium ions   **C.** osmosis
    B. filtration   **D.** active transport   **E.** facilitated transport

19. Gout is manifested by decreased excretion of uric acid. Which organ is mainly responsible for the elimination of uric acid?

A. spleen   **B.** large intestine   **C.** kidney   **D.** liver   **E.** duodenum

20. Most electrolyte reabsorption by the renal tubules is:

A. in the proximal convoluted tubule
B. in the descending loop of Henle
C. not limited by a transport maximum
D. in the distal convoluted tubule
E. in the ascending loop of Henle

21. The lowest solute concentration can be found in this region of the kidney:

A. cortex   **B.** nephron   **C.** pelvis   **D.** medulla   **E.** epithelia

22. The fluid in the Bowman's (glomerular) capsule is similar to plasma, but it does NOT contain a significant amount of:

A. urea   **B.** electrolytes   **C.** hormones   **D.** glucose   **E.** plasma protein

23. Which of the following is true of urea?

A. It is insoluble in water
B. It is more toxic to human cells than ammonia
C. It is the primary nitrogenous waste product of humans
D. It is the primary nitrogenous waste product of most birds
E. It is the primary nitrogenous waste product of most aquatic invertebrates

24. Alcohol acts as a diuretic because it:

A. increases secretion of ADH
B. inhibits the release of ADH
C. is not reabsorbed by the tubule cells
D. increases the rate of glomerular filtration
E. decreases the rate of glomerular filtration

25. Which area of the kidney is the site of passive diffusion of Na^+?

A. thick segment of ascending limb
B. distal convoluted tubule
C. loop of Henle
D. proximal convoluted tubule
E. collecting ducts

26. Angiotensin II functions to:

A. decrease water absorption
B. decrease arterial blood pressure
C. decrease aldosterone production
D. constrict arterioles and increase blood pressure
E. decrease plasma pH

27. The main mechanism of removing Na$^+$ ions from urine is the nephron sodium-hydrogen exchange carrier (an anti-port carrier). What is the approximate pH of the urine when the carrier is functioning?

A. 5.8   **B.** 8.2   **C.** 7.6   **D.** 11.1   **E.** 2

28. Inadequate secretion of ADH causes symptoms of polyuria in a disease known as:

A. diabetic acidosis   **C.** diabetes mellitus
B. nephrogenic diabetes insipidus   **D.** diabetes insipidus   **E.** diabetic alkalosis

29. Hydrolysis of which of the following results in the production of nitrogenous waste in the form of urea, uric acid or ammonia?

A. vitamins   **B.** fats   **C.** sugars   **D.** carbohydrates   **E.** proteins

30. The above normal hydrostatic pressure in Bowman's capsule would result in:

A. filtration increase in proportion to the increase in Bowman's capsule pressure
B. osmotic pressure compensating, so that filtration does not change
C. net filtration increase above normal
D. net filtration decrease below normal
E. no effect on filtration rate

31. Concentration of which of the following compounds is controlled by the respiratory system to maintain the pH of body fluids?

A. oxygen   **B.** bicarbonate   **C.** carbon dioxide   **D.** HCl   **E.** pepsin

32. All of the following statements regarding tubular reabsorption are true, EXCEPT that:

A. it involves hormonal signals in the collecting ducts
B. it is only a passive transport process
C. it occurs via transcellular or paracellular routes
D. it is a reclamation process
E. the transport mechanism of the components depends on their location within the nephron

33. Which of the following is part of both the excretory and reproductive systems of male mammals?

A. urethra   **B.** ureter   **C.** prostate   **D.** vas deferens   **E.** epididymis

34. High levels of glucose and amino acids are reabsorbed in the filtrate through:

A. counter-transport   **C.** facilitated diffusion
B. primary active transport   **D.** passive transport   **E.** secondary active transport

35. Which statement most accurately describes the expected levels of aldosterone and vasopressin in the blood of a dehydrated patient compared to a healthy individual?

A. Aldosterone levels are higher, and vasopressin levels are lower

B. Aldosterone levels are lower, and vasopressin levels are higher

C. Aldosterone and vasopressin levels are higher

D. Aldosterone and vasopressin levels are lower

E. Aldosterone and vasopressin levels are not related

36. Compared to the human kidney, the kidney of an animal living in an arid environment is capable of producing more concentrated urine, because it:

A. produces urine that is isotonic compared to the blood

B. maintains a greater osmolarity gradient in the medulla

C. maintains a greater hydrostatic pressure for filtration at the glomerulus

D. increases the rate of filtration

E. maintains a lower osmolarity gradient in the medulla

37. Which of the following is a function of the loop of Henle?

A. Absorption of water into the filtrate

B. Absorption of electrolytes by active transport and water by osmosis in the same segments

C. Forming a small volume of very concentrated urine

D. Forming a large volume of very concentrated urine

E. Absorption of electrolytes into the filtrate

38. Which of the following nephron structures is/are the site(s) of removing H_2O to concentrate the urine?

    I. ascending loop of Henle
    II. descending loop of Henle
    III. proximal convoluted tubule

A. III only   **B.** I and III only   **C.** II only   **D.** I, II and III   **E.** II and III only

39. If a particular amino acid has a T_m of 140 mg/100 ml and its concentration in the blood is 230 mg/100 ml, this amino acid will:

A. be completely reabsorbed by secondary active transport

B. appear in the urine

C. be completely reabsorbed into the filtrate by facilitated transport

D. be completely reabsorbed by primary active transport

E. be completely reabsorbed into the filtrate by diffusion

40. Which of the following would be the most likely result of increased plasma osmolarity?

A. dehydration

B. increased ADH secretion

C. decreased H_2O permeability in the nephron

D. excretion of dilute urine

E. increased calcitonin release

41. If the renal clearance value of substance X is zero, it means that:

A. normally, all substance X is reabsorbed

B. the value is relatively high in a healthy adult

C. the substance X molecule is too large to be filtered via the kidneys

D. most of substance X is filtered via the kidneys and is not reabsorbed in the convoluted tubules

E. substance X is mostly excreted in the urine

42. Which of the following sequences correctly represents the pathway of anatomical structures passed by a probe being inserted into the urethra of a female patient?

A. urethra → bladder → opening to the ureter → ureter → prostate → renal pelvis

B. ureter → opening to the ureter → prostate → vas deferens → epididymis

C. urethra → bladder → opening to the ureter → ureter → renal pelvis

D. kidney → ureter → opening to the bladder → bladder → urethra

E. opening to the ureter → bladder → ureter → prostate → renal pelvis

43. Which of the following is observed in a patient affected by chronic adrenal insufficiency caused by decreased activity of the adrenal cortex?

A. high sex hormone concentrations

B. high plasma control

C. increased resistance to stress

D. high urinary output

E. decreased rate of filtration

44. Which of the following sequences correctly represents the pathway of anatomical structures passed by a probe being inserted into the urethra of a male patient?

A. urethra → vas deferens → prostate → ejaculatory duct → seminiferous tubules → epididymis

B. urethra → prostate → vas deferens → ejaculatory duct → seminiferous tubules → epididymis

C. urethra → ejaculatory duct → prostate → vas deferens → epididymis → seminiferous tubules

D. urethra → ejaculatory duct → vas deferens → epididymis → seminiferous tubules

E. urethra → prostate → ejaculatory duct → vas deferens → epididymis → seminiferous tubules

45. Renin is a polypeptide hormone that:

    A. is produced in response to increased blood volume
    B. is produced in response to decreased blood pressure
    C. acts on the pituitary gland
    D. is produced in response to concentrated urine
    E. is secreted in the proximal convoluted tubules of the kidney

46. What is unique about transport epithelial cells in the ascending loop of Henle in humans?

    A. They are the largest epithelial cells in the body
    B. They are not in contact with interstitial fluid
    C. They are membranes that are impermeable to water
    D. 50% of their cell mass is comprised of smooth endoplasmic reticulum
    E. They are not affected by high levels of nitrogenous wastes

47. Which symptom is characteristic of a patient with *Nephrogenic diabetes insipidus*, whereby kidneys are unresponsive to ADH hormone?

    A. dilute urine   **C.** detection of glucose in urine
    B. decreased levels of plasma ADH   **D.** elevated plasma glucose levels
                                      E. concentrated urine

48. Excretion of dilute urine requires the:

    A. presence of vasopressin
    B. transport of sodium ions out of the descending nephron loop
    C. impermeability of the collecting tubule to water
    D. relative permeability of the distal tubule to water
    E. transport of chloride ions out of the descending nephron loop

49. All of these are reabsorbed from the glomerular filtrate in the ascending loop of Henle, EXCEPT:

    A. K^+ **B.** Cl^- **C.** Na^+ **D.** glucose   **E.** amino acids

50. The epithelial cells of the proximal convoluted tubule contain a brush border the primary purpose of which is to:

    A. slow the rate of movement of the filtrate through the nephron
    B. move the filtrate via cilia action through the nephron
    C. increase the volume of urine produced
    D. increase the amount of filtrate entering the loop of Henle
    E. increase the absorptive surface area

51. Which of the following structures within the nephron is the site of aldosterone function?

   A. Bowman's capsule   **C.** ascending loop of Henle

   B. distal convoluted tubule   **D.** descending loop of Henle   **E.** glomerulus

52. Which of the following would NOT be observed in a child with acute inflammation of the glomerulus, causing its failure to filter adequate quantities of fluid?

   A. high urine [Na^+]   **C.** high plasma [urea]

   B. decreased urine output   **D.** excess interstitial fluid   **E.** low urine [Na^+]

53. In the ascending limb of the nephron loop, the:

   A. thin segment is not permeable to chloride

   B. thin segment is not permeable to sodium

   C. thick segment is permeable to water

   D. thin segment is freely permeable to water

   E. thick segment moves ions out into interstitial spaces for reabsorption

54. What is the clinical result from the administration of a pharmaceutical agent that selectively binds and inactivates renin?

   A. Decreased Na^+ reabsorption by the distal tubule

   B. Increase in the amount of filtrate entering Bowman's capsule

   C. Increased Na^+ reabsorption by the distal tubule

   D. Increase of blood pressure

   E. Excretion of platelets in the urine

55. An example of a properly functioning homeostatic control system is seen when:

   A. core body temperature of a runner rises gradually from 37° to 45°C

   B. kidneys excrete salt into the urine when dietary salt levels rise

   C. blood cell shrinks when placed in a solution of salt and water

   D. blood pressure increases in response to an increase in blood volume

   E. the level of glucose in the blood is abnormally high whether or not a meal has been eaten

56. Within the kidney:

   A. filtration of blood begins at the glomerulus and ends at the loop of Henle

   B. glucose and H_2O are actively reabsorbed from the glomerular filtrate

   C. ammonia is converted to urea

   D. nephrons are located only in the cortex of the kidney

   E. antidiuretic hormone causes the collecting tubule to reabsorb H_2O

57. What hormone(s) will most likely be elevated in the blood of a patient who has been in the desert without water or food all day?

    I. ADH   III. insulin

    II. aldosterone   IV. epinephrine

  A. I only   **B.** II only   **C.** I and II only   **D.** I, II and IV   **E.** I and IV only

58. In a patient with highly concentrated urine, the filtrate entering the loop of Henle likely has:

  A. less volume than the filtrate leaving the loop of Henle

  B. more volume than the filtrate leaving the loop of Henle

  C. lower osmolarity than the filtrate leaving the loop of Henle

  D. higher osmolarity than the filtrate leaving the loop of Henle

  E. same osmolarity as the filtrate leaving the loop of Henle

59. Which of the following would most likely occur if a diabetic patient accidentally overdoses with insulin?

  A. increased urine excretion leading to dehydration

  B. increased conversion of glycogen to glucose

  C. decreased plasma glucose levels leading to convulsions

  D. increased glucose concentration in urine

  E. decreased urine excretion

Answer Key

1: A	11: C	21: A	31: C	41: A	51: B
2: C	12: B	22: E	32: B	42: C	52: A
3: E	13: E	23: C	33: A	43: D	53: E
4: C	14: B	24: B	34: E	44: E	54: A
5: B	15: E	25: C	35: C	45: B	55: B
6: E	16: A	26: D	36: B	46: C	56: E
7: D	17: C	27: A	37: C	47: A	57: C
8: B	18: C	28: B	38: E	48: C	58: D
9: D	19: C	29: E	39: B	49: D	59: C
10: A	20: A	30: D	40: B	50: E	

Chapter 4.8: Muscle System

1. What part of the sarcolemma contains acetylcholine receptors?

   A. the part contiguous to another muscle cell
   B. entire sarcolemma
   C. motor end plate
   D. distal end of the muscle fiber
   E. presynaptic membrane

2. How is cardiac muscle different from both smooth and skeletal muscle?

   A. It lacks sarcoplasmic reticulum
   B. It lacks actin and myosin fibers
   C. It is under involuntarily control
   D. It has striated appearance
   E. It has T tubules

3. The energy for muscle contraction is supplied by:

   A. lactose
   B. ADP
   C. lactic acid
   D. phosphocreatine and ATP
   E. cAMP and ATP

4. The role of tropomyosin in skeletal muscles is to:

   A. serve as a contraction inhibitor by blocking the actin binding sites on the myosin molecules
   B. be the receptor for the motor neuron neurotransmitter
   C. be the chemical that activates the myosin heads
   D. provide the energy for myosin heads to dissociate from the actin filaments
   E. serve as a contraction inhibitor by blocking myosin binding sites on the actin molecules

5. Which of the following takes place during muscular contraction?

   A. Neither the thin nor the thick filament contracts
   B. Both the thin and thick filaments contract
   C. The thin filament contracts, while the thick filament remains constant
   D. The thin filament remains constant, while the thick filament contracts
   E. Both the thin and thick filaments undergo depolymerization

6. At the neuromuscular junction, the depression on the sarcolemma is:

   A. motor end plate
   B. junctional fold
   C. synaptic knob
   D. synaptic vesicle
   E. boundary of the sarcomere

7. Most skeletal muscles consist of:

   A. predominantly slow oxidative fibers
   B. predominantly fast oxidative fibers
   C. muscle fibers of the same type
   D. a mixture of fiber types
   E. predominantly fast glycolytic fibers

8. Irreversible sequestering of Ca^{2+} in the sarcoplasmic reticulum most likely results in:

    A. depolymerization of actin filaments within the sarcomere

    B. permanent contraction of the muscle filaments analogous to *rigor mortis*

    C. increased bone density as Ca^{2+} is resorbed from bones to replace sequestered Ca^{2+}

    D. prevention of actin cross-bridges to the myosin thin filament

    E. prevention of myosin cross-bridges to the actin thin filament

9. Which of the following does NOT happen during muscle contraction?

    A. H band shortens or disappears

    B. I band shortens or disappears

    C. A band shortens or disappears

    D. sarcomere shortens

    E. Z disks approach each other

10. The strongest muscle contractions are normally achieved by:

    A. increasing the stimulation up to the maximal stimulus

    B. recruiting small and medium muscle fibers

    C. increasing stimulus above the threshold

    D. increasing stimulus above the *treppe* stimulus

    E. all of the above

11. The SA node, innervated by the vagus nerve, is a collection of cardiac muscle cells that have the capacity for self-excitation. The effect of the vagus nerve on the frequency of the heartbeat for self-excitation is likely:

    A. slower, because innervation decreases heart rate

    B. faster, because innervation decreases heart rate

    C. slower, because innervation increases heart rate

    D. faster, because innervation increases heart rate

    E. no effect on heart rate, because cardiac cells maintain auto-rhythmicity

12. Continuous repetitive muscle contraction results in a buildup of lactic acid because it is the:

    A. degradation product of ADP

    B. degradation product of phosphocreatine

    C. organic product of anaerobic metabolism

    D. degradation product of fatty acid degradation

    E. releasing factor for hydrolysis actin-myosin crosslinking

13. The point of attachment of a nerve to a muscle fiber is the:

    A. contraction point

    B. synapse

    C. relaxation point

    D. neuromuscular junction

    E. Z disc

14. Which of the following is used later in muscle stimulation when contractile strength increases?

    A. Motor units with larger, less excitable neurons

    B. Large motor units with small, highly excitable neurons

    C. Many small motor units with the ability to stimulate other motor units

    D. Motor units with the longest muscle fibers

    E. Large motor units with small, less excitable neurons

15. The reason for sudden vasodilation, manifested by flush skin, before the onset of frostbite is:

    A. blood shunting via sphincters within the smooth muscle

    B. rapid contractions of skeletal muscles

    C. paralysis of skeletal muscle encapsulating the area

    D. onset of tachycardia from the increase in blood pressure

    E. paralysis of smooth muscle in the area

16. Which of the following is NOT correct regarding cardiac muscle?

    A. The ANS is required for stimulation

    B. Large mitochondria fill about 25% of the cell

    C. It is rich in glycogen and myoglobin

    D. It has one or possibly two nuclei

    E. It is repaired is primarily by fibrosis

17. In the muscle cell, the function of creatine phosphate is:

    A. to induce a conformational change in the myofilaments

    B. to store energy that will be transferred to ADP to synthesize ATP

    C. to form a temporary chemical compound with myosin

    D. to form a chemical compound with actin

    E. to bind to the troponin

18. Shivering functions to increase core body temperature because it:

    A. increases contractile activity of skeletal muscles

    B. signals the hypothalamus that body temperature is higher than actual

    C. causes bones to rub together to generate heat

    D. signals the body that the core temperature is low and encourages behavioral modification

    E. increases contractile activity of smooth muscles

19. Which is NOT a property of muscle tissue?

    A. extensibility   **C.** communication

    B. contractility   **D.** excitability   **E.** all of the above

20. The individual muscle cell is surrounded by:

    A. fascicle   **B.** epimysium   **C.** periosteum   **D.** perimysium   **E.** endomysium

21. Which of the following muscle types are involuntary muscles?

A. cardiac and smooth   **C.** skeletal and smooth

B. cardiac, smooth and skeletal   **D.** only smooth   **E.** cardiac and skeletal muscle

22. Myoglobin:

A. phosphorylates ADP directly   **C.** produces the end plate potential

B. stores oxygen in muscle cells   **D.** breaks down glycogen

                                          E. transports carbon dioxide in the blood

23. During an action potential, a cardiac muscle cell remains depolarized much longer than a neuron in order to:

A. prevent a neuron from depolarizing twice in rapid succession

B. permit adjacent cardiac muscle cells to contract at different times

C. prevent the initiation of another action potential during contraction of the heart

D. ensures that Na^+ voltage-gated channels remain open so Na^+ exits the cell

E. ensures that K^+ voltage-gated channels remain open so K^+ exits the cell

24. A characteristic of cardiac muscle that can be observed with the light microscope is:

A. somatic motor neurons   **C.** intercalated discs

B. no nuclei   **D.** single cells   **E.** none of the above

25. Rigor mortis (stiffness of the corpse's limbs after death) occurs because:

A. no ATP is available to release attached actin and myosin molecules

B. proteins are beginning to hydrolyze and break down

C. the cells are dead

D. sodium ions leak into the muscle causing continued contractions

E. calcium ions are prevented from flowing to their target

26. Muscles undergo movement at joints by:

A. increasing in length and pushing the origin and insertion of the muscle together

B. filling with blood and increasing the distance between the ends of a muscle

C. increasing in length and pushing the origin and insertion of the muscle apart

D. depolarizing neurons which initiate electrical twitches at the tendons

E. decreasing in length and moving the origin and insertion of the muscle closer

27. The plasma membrane of a muscle cell is:

A. sarcolemma   **B.** sarcoplasmic reticulum   **C.** titin   **D.** myofibril   **E.** periosteum

28. T tubules function to:

A. hold cross bridges in place in a resting muscle

B. synthesize ATP to provide energy for muscle contraction

C. stabilize the G and F actin

D. facilitate cellular communication during muscle contraction

E. store calcium when the muscle is at rest

29. What is the role of Ca^{2+} in muscle contraction?

A. Transmitting the action potential across the neuromuscular junction

B. Reestablishing the polarization of the plasma membrane after an action potential

C. Propagation of the action potential through the transverse tubules

D. Binding to troponin, which changes the conformation of tropomyosin and permits the myosin to bind to actin

E. Binding to troponin, which changes the conformation of tropomyosin and permits the actin heads to bind to the myosin thin filament

30. Wavelike contraction produced by smooth muscle is:

A. vasoconstriction **B.** myoblastosis **C.** vasodilation **D.** circulitus **E.** peristalsis

31. The striations of a skeletal muscle cell are due to:

A. T tubules

B. sarcoplasmic reticulum

C. arrangement of myofilaments

D. thickness of the sarcolemma

E. repetition of actin thick filaments

32. For the movement of a pair of antagonistic muscles, one of the muscles:

A. contracts in an isometric action

B. acts synergistically by contracting to stabilize the moving bone

C. contracts in an isotonic action

D. relaxes

E. establishes an insertion point to slide along the bone to allow a larger range of movement

33. Which skeletal muscle fiber type is most resistant to fatigue?

A. no-twitch

B. fast glycolytic

C. intermediate

D. fast-oxidative

E. slow-oxidative

34. During muscle contraction, the thick filament cross bridges attach to:

A. Z discs **B.** T tubules **C.** myosin filaments **D.** actin filaments **E.** M line

35. What structure connects the biceps muscle to the radius bone?

A. biceps cartilage   **C.** biceps muscle

B. biceps ligament   **D.** annular ligament of the radius   **E.** biceps tendon

36. Which of the following is the correct sequence of events for muscle contractions?

A. Motor neuron action potential → muscle cell action potential → neurotransmitter release → release of calcium ions from SR → sliding of myofilaments → ATP-driven power stroke

B. Neurotransmitter release → motor neuron action potential → muscle cell action potential → release of calcium ions from SR → ATP-driven power stroke

C. Motor neuron action potential → neurotransmitter release → muscle cell action potential → release of calcium ions from SR → ATP-driven power stroke → sliding of myofilaments

D. Neurotransmitter release → muscle cell action potential → motor neuron action potential → release of calcium ions from SR → sliding of myofilaments → ATP-driven power stroke

E. Muscle cell action potential → neurotransmitter release → ATP-driven power stroke → calcium ion release from SR → sliding of myofilaments

37. The release of which molecule from the sarcoplasmic reticulum is normally required for the contraction of a muscle?

A. calcium   **B.** potassium   **C.** water   **D.** sodium   **E.** magnesium

38. The autonomic nervous system controls which action of the muscle?

A. Contraction of the diaphragm

B. Peristalsis of the gastrointestinal tract

C. Conduction of cardiac muscle action potentials

D. The reflex arc of the knee jerk response when the patella is struck

E. Contraction of the bicep and the opposing relaxation of the triceps

39. Which molecule breaks the bonding between the two contractile proteins for the myosin-actin cross-bridges?

A. ATP   **B.** sodium   **C.** acetylcholine   **D.** calcium   **E.** potassium

40. The contraction of smooth muscle is different from skeletal muscle because in smooth muscle contraction:

A. ATP energizes the sliding process

B. the site of calcium regulation differs

C. the trigger for contraction is a rise in intracellular calcium

D. actin and myosin interact by the sliding filament mechanism

E. none of the above

41. Which of the following is NOT present in a skeletal muscle?

A. sarcoplasmic reticulum

B. multinucleated cells

C. individual innervations of each muscle fiber

D. intercellular conductivity of action potentials

E. regular array of molecular components

42. Excitation-contraction coupling requires which of the following substances?

     I. ATP   II. glucose   III. Ca^{2+}

A. I only   **B.** III only   **C.** I and II only   **D.** I and III only   **E.** I, II and III

43. All of the following statements are true for smooth muscle, EXCEPT:

A. Its contractions are involuntary

B. It does not require Ca^{2+} for contraction

C. Its contractions produce a chemical change near the smooth muscle

D. Its contractions are longer than the contractions in skeletal muscle

E. Its contractions are slower than the contractions in skeletal muscle

44. Which molecule is released by synaptic vesicles into the synaptic cleft for skeletal muscles?

A. water   **B.** sodium   **C.** acetylcholine   **D.** calcium   **E.** calcitonin

45. Which statement about smooth muscle is correct?

A. Smooth muscle cannot stretch as much as skeletal muscle

B. Smooth muscle has well-developed T tubules at the site of invagination

C. Smooth muscle stores calcium in the sarcoplasmic reticulum

D. Smooth muscle contains troponin as a calcium binding site

E. Certain smooth muscle cells can actually divide to increase their numbers

46. Which of the following muscles is under voluntary control?

A. smooth muscle in the gastrointestinal tract

B. iris of the eye

C. diaphragm

D. cardiac tissue

E. vasodilation

47. The thick filament of skeletal muscle fibers is composed of:

A. myosin   **B.** troponin   **C.** tropomyosin   **D.** actin   **E.** none of the above

48. Which of the following is NOT a characteristic of a smooth muscle?

   A. Noncontractile intermediate filaments attach to dense bodies within the cell

   B. There are no sarcomeres

   C. Calmodulin is a regulatory protein to regulate the level of calcium within the cell

   D. It lacks troponin

   E. There are more thick filaments than thin filaments

49. Which of the following statements describes the structure of smooth muscle?

   A. Multicellular units of muscle tissue are under voluntary control

   B. Peristalsis results from single-unit muscle cells within the gastrointestinal tract

   C. Ca^{2+} distribution occurs via an extensive network of T-tubules

   D. Ca^{2+} binds to troponin and changes the conformation of the tropomyosin

   E. Sarcomeres are visible as repeating motifs of actin and myosin

50. Which molecule, abundant in the sarcoplasm, provides stored energy for the muscle to use during exercise?

   A. glycogen   **B.** calcium   **C.** water   **D.** myosin   **E.** actin

51. Which of the following is a true statement about muscle?

   A. Striated muscle cells are long and cylindrical with many nuclei

   B. Cardiac muscle cells are found in the heart and large blood vessels

   C. Cardiac muscle cells have many nuclei

   D. Smooth muscle cells have T tubules

   E. Smooth muscles have extensive gap junctions for rapid communication

52. Which of the following statements is true for cardiac muscle?

   A. It is multi-nucleated

   B. It is not striated

   C. It does not require Ca^{2+}

   D. It is innervated by the somatic motor nervous system

   E. It is under involuntary control

53. A typical skeletal muscle:

   A. is innervated by the somatic nervous system

   B. is innervated only by the autonomic nervous system

   C. lines the walls of glands and organs

   D. has myosin and actin that lack a striated appearance

   E. is innervated only by the parasympathetic and sympathetic nervous system

54. A muscle type that has only one nucleus, no sarcomeres and rare gap junctions:

A. cardiac muscle

B. skeletal muscle

C. visceral smooth muscle

D. multiunit smooth muscle

E. cardiac and skeletal muscle

55. A protein found in muscle cells that binds oxygen is:

A. tropomyosin   **B.** glycogen   **C.** calmodulin   **D.** myosin   **E.** myoglobin

56. This muscle type is always characterized by multinucleated cells.

A. cardiac muscle

B. smooth muscle

C. skeletal muscle

D. A and C

E. B and C

57. In skeletal muscle cells, the structure that functions in calcium storage is:

A. intermediate filament network

B. myofibrillar network

C. calmodulin

D. sarcoplasmic reticulum

E. mitochondria

58. Which of the following statements is/are correct about cardiac muscle?

I. It acts as a functional syncytium

II. It is under the control of the autonomic nervous system

III. It is striated due to the arrangement of actin and myosin filaments

A. I only   **B.** III only   **C.** II and III only   **D.** I, II, and III   **E.** I and II only

59. The velocity and duration of muscle contraction depends on:

A. muscle length

B. load on the fiber

C. size of the muscle fibers stimulated

D. number of muscle fibers stimulated

E. size and number of the muscle fibers stimulated

60. All of these statements about acetylcholine (ACh) are correct, EXCEPT:

A. ACh is degraded by enzymes in the synaptic cleft

B. ACh is released at the neuromuscular junction

C. ACh binds to specific receptors on the postsynaptic membrane

D. ACh diffuses through the presynaptic membrane during release

E. ACh becomes depleted in a synapse from repeated action potentials

Answer Key

1: C	11: A	21: A	31: C	41: D	51: A
2: A	12: C	22: B	32: D	42: D	52: E
3: D	13: D	23: C	33: E	43: B	53: A
4: E	14: A	24: C	34: D	44: C	54: D
5: A	15: E	25: A	35: E	45: E	55: E
6: A	16: A	26: E	36: C	46: C	56: C
7: D	17: B	27: A	37: A	47: A	57: D
8: E	18: A	28: D	38: B	48: E	58: D
9: C	19: C	29: D	39: A	49: B	59: B
10: A	20: E	30: E	40: B	50: A	60: D

Chapter 4.9: Skeletal System

1. The vertebral curves function to:

    A. improve the cervical center of gravity
    B. accommodate the weight of the pelvic girdle
    C. provide resilience and flexibility
    D. accommodate muscle attachment
    E. provide a route for blood distribution

2. Which connective tissue is used for the attachment of muscle to bone?

    A. origin   B. ligament   C. aponeurosis   D. tendon   E. cartilage

3. The purpose of synovial fluid in a synovial joint is to:

    A. hydrate osteocyte cells
    B. create a rigid connection between opposing flat bones
    C. reduce friction between the ends of opposing bones
    D. create the structure for the necessary morphology of joints and bones
    E. create a physical barrier so cells do not migrate from the region

4. Which region of the human vertebral column bears most of the weight of the body and receives the most stress?

    A. thoracic region
    B. lumbar region
    C. cervical region
    D. sacral region
    E. cranial

5. During bone maturation, osteoblasts are trapped in their own matrix and then turn into which cells as they stop producing more matrix?

    A. matrixocytes
    B. trabeculocytes
    C. osteoclasts
    D. spiculoclasts
    E. osteocytes

6. Which of the following is characteristic of short bones?

    A. Consist mainly of hyaline cartilage
    B. Consist mostly of dense bone
    C. Consist of both spongy and dense bone
    D. Consist mostly of spongy bone
    E. Consist mostly of hyaline cartilage and dense bone

---

7. The major function of the axial skeleton is to:

    A. provide central support for the body and protect internal organs
    B. provide a space for the major digestive organs
    C. provide a conduit for the peripheral nerves
    D. provide an attachment point for muscles to allow movement
    E. give the body resilience

8. The skeletal system performs all of the following functions, EXCEPT:

    A. removal of toxins from the blood   **C.** protection of the viscera
    B. storage and release of minerals   **D.** production of the blood
                                        E. production of Vitamin C

9. Which tissue would result in the LEAST amount of pain when surgically cut?

    A. skin   **B.** smooth muscle   **C.** bone   **D.** cartilage   **E.** skeletal muscle

10. Which bone tissue is most adapted to support weight and withstand tension stress?

    A. compact bone   **C.** spongy bone
    B. trabecular bone   **D.** irregular bone   **E.** long bone

11. What disease in children results in soft and pliable bones due to the lack of vitamin D?

    A. Paget's disease   **C.** Rickets
    B. Brittle bones   **D.** osteoporosis   **E.** Turner's syndrome

12. All of the following structures are made of cartilage, EXCEPT:

    A. larynx   **B.** nose   **C.** middle ear   **D.** outer ear   **E.** skeletal joint

13. Yellow bone marrow contains a large percentage of:

    A. collagen fibers   **C.** blood-forming cells
    B. elastic tissue   **D.** fat   **E.** hydroxylapatite

14. The parathyroid hormone can most likely increase the production of:

    A. osteoblasts   **C.** osteocytes
    B. osteoclasts   **D.** osseous tissue   **E.** cartilage

15. The structural unit of compact bone is:

    A. lamellar bone   **C.** canaliculi
    B. spongy bone   **D.** osseous matrix   **E.** osteon

16. Which hormone maintains the appropriate levels of calcium in the blood, often at the expense of bone loss?

    A. vitamin D   **C.** androgens

    B. parathyroid hormone   **D.** calcitonin   **E.** estrogen

17. Which connective tissue type has a flexible and strong matrix?

    A. cartilage   **C.** reticular tissue

    B. dense regular connective tissue   **D.** areolar tissue

                                         E. collagenous tissue

18. During bone formation, a deficiency of growth hormone can cause:

    A. decreased proliferation of the epiphyseal plate cartilage

    B. increased osteoclast activity

    C. inadequate calcification of bone

    D. decreased osteoclast activity

    E. decreased formation of hydroxylapatite

19. Connective tissue holding the bones together in a synovial joint is:

    A. osteocyte   **B.** tendons   **C.** cartilage   **D.** periosteum   **E.** ligament

20. Which statement is correct regarding the ossification of the ends of long bones?

    A. It takes twice as long as diaphysis ossification

    B. It is produced by secondary ossification centers

    C. It involves medullary cavity formation

    D. It is a characteristic of intramembranous bone formation

    E. none of the above

21. The abnormal curvature of the thoracic spine that often comes from compression fractures of weakened vertebrae is:

    A. kyphosis   **B.** calluses   **C.** stooping   **D.** menopausal atrophy   **E.** scoliosis

22. Which of these substances is NOT involved in bone remodeling?

    A. calcitonin   **C.** thyroxine

    B. vitamin D   **D.** parathyroid hormone   **E.** all are involved

23. Until adolescence, the diaphysis of the bone is able to increase in length due to the:

A. epiphyseal plate   **C.** lacunae

B. epiphyseal line   **D.** Haversian system   **E.** canaliculi

24. Which of these is NOT the function of bone?

A. storage of lipids   **C.** structural support

B. facilitate muscle contraction   **D.** storage of minerals

     E. regulation of blood temperature

25. Which of the following is the single most important stimulus for epiphyseal plate activity during infancy and childhood?

A. cortisol   **C.** calcium

B. growth hormone   **D.** parathyroid hormone   **E.** thyroid hormone

26. The normal histological make up of bone is concentric lamellae called:

A. periosteum   **B.** endosteum   **C.** osteon   **D.** trabeculae   **E.** canaliculi

27. Parathyroid hormone elevates blood calcium levels by:

A. increasing the number and activity of osteoblasts

B. decreasing the number and activity of osteoblasts

C. decreasing the number and activity of osteoclasts

D. increasing the number and activity of osteoclasts

E. has no effect on bone cells

28. The inner osteogenic layer of periosteum consists primarily of:

A. hyaline and cartilage   **C.** cartilage and compact bone

B. chondrocytes and osteocytes   **D.** marrow and osteons   **E.** progenitor cells

29. Which of these is NOT found in compact bone?

A. yellow marrow   **C.** Haversian canals

B. canaliculi   **D.** Volkmann's canals   **E.** osteons

30. The Haversian canal runs through the core of each osteon and is the site of:

A. yellow marrow and spicules   **C.** cartilage and interstitial lamellae

B. blood vessels and nerve fibers   **D.** adipose tissue and nerve fibers

     E. adipose tissue and cartilage

31. The process when bone develops from hyaline cartilage is:

    A. periosteal ossification
    B. intramembranous ossification
    C. intermembranous ossification
    D. endochondral ossification
    E. none of the above

32. Connective tissue that connects bones to each other is:

    A. synovium
    B. osteoprogenitor cells
    C. ligaments
    D. sockets
    E. muscles

33. A failure of the lamina of the vertebrae to fuse properly in the lumbar region during development can produce an exposed spinal cord called:

    A. spondylosis
    B. spina bifida
    C. stenosis
    D. scoliosis
    E. kyphosis

34. The process of bones increasing in width is:

    A. concentric growth
    B. long bone growth
    C. epiphyseal plate opening
    D. closing of the epiphyseal plate
    E. appositional growth

35. Hydroxyapatite is the mineral portion of bone and does NOT contain:

    A. hydrogen
    B. calcium
    C. phosphate
    D. sulfur
    E. all are in bone

36. Osteoclast activity to release more calcium ions into the bloodstream is stimulated by:

    A. cortisol
    B. parathyroid hormone
    C. thyroxine
    D. calcitonin
    E. estrogen

37. Which type of tissue is bone classified as?

    A. epithelial   **B.** endothelial   **C.** muscle   **D.** skeletal   **E.** connective

38. What connective tissue stores calcium?

    A. muscle   **B.** tendon   **C.** ligament   **D.** bone   **E.** calyx of the kidneys

39. Bone growth and healing from a fracture are absolutely impossible without:

A. osteoclasts

B. osteoblasts

C. dietary intake of calcium and vitamin D

D. osteocytes

E. chondrocytes

40. The precursor of an osteoclast, the cell responsible for bone resorption, is:

A. erythrocyte   **B.** chondrocyte   **C.** osseous tissue   **D.** macrophage   **E.** canaliculi

41. Which of the following statements regarding interstitial growth is correct?

A. Unspecialized mesenchymal cells develop into chondrocytes that divide and form cartilage

B. Chondrocytes in the lacunae divide and secrete matrix that allows the cartilage to grow from within

C. Growth occurs in the lining of the long bones

D. Fibroblasts give rise to chondrocytes that differentiate and form cartilage

E. Chondrocytes in the lacunae divide and secrete matrix that allows the cartilage to grow from depositing new layers onto the surface of the bone

42. The shaft of long bones is:

A. cortex   **B.** medulla   **C.** epiphysis   **D.** diaphysis   **E.** endochondral ossification

43. The concentration of blood calcium is raised by which hormone?

A. insulin

B. glucagon

C. antidiuretic hormone

D. aldosterone

E. parathyroid hormone

44. In the epiphyseal plate, cartilage grows:

A. in a circular fashion

B. from the edges inward

C. by pushing the epiphysis away from the diaphysis

D. by pulling the diaphysis toward the epiphysis

E. by pulling the epiphysis toward the diaphysis

45. Which of these is the most important function of the spongy bone of the hips?

A. storage of fats

B. synthesis of lymph fluid

C. degradation of leukocytes

D. storage of erythrocytes

E. synthesis of erythrocytes

46. The structural unit of spongy bone is:

   A. trabeculae   **D.** lamellar bone

   B. osseous lamellae   **C.** osteons   **E.** Haversian canals

47. If a thyroid tumor secreted an excessive amount of calcitonin, this would result in:

   A. increase in blood calcium concentration

   B. reduction in the rate of endochondral ossification

   C. increase in the level of osteoblast activity

   D. increase in the level of osteoclast activity

   E. decrease in bone density

48. Which of the following can NOT be a factor that contributes to osteoporosis (decrease in bone mass and density)?

   A. menopause

   B. high blood levels of calcitonin hormone

   C. high sensitivity to endogenous parathyroid hormone

   D. high blood levels of parathyroid hormone

   E. impaired intestinal Ca^{2+} absorption

49. Connective tissue sacs lined with synovial membranes that act as cushions in places where friction develops are:

   A. tendons   **B.** ligaments   **C.** bursae   **D.** menisci   **E.** Haversian canal

50. The narrow space of hyaline cartilage that is the site of long bone growth until the end of puberty is:

   A. chondrocytic line   **C.** medullary cavity

   B. epiphyseal plate   **D.** mineralization

                           E. endochondral ossification

51. What is the structure, found on the ends of bones, that looks bluish-white and reduces friction:

   A. hyaline cartilage

   B. elastic cartilage

   C. areolar connective tissue

   D. fibrocartilage

   E. adipose tissue

52. Which of the following is one of the possible reasons why the epiphyseal plate of the long bones in some children closes too early?

  A. High levels of dietary vitamin D
  B. Overproduction of thyroid hormone
  C. Osteoblast activity exceeding osteoclast activity
  D. Elevated levels of sex hormones
  E. Abnormally low levels of osteoclast activity

53. Which of the following does NOT promote bone growth?

  A. estrogen and testosterone
  B. parathyroid hormone
  C. vitamin D
  D. calcium
  E. calcitonin

54. Tendon sheaths:

  A. are extensions of periosteum
  B. help anchor the tendon to the muscle
  C. act to reduce friction
  D. are lined with dense irregular connective tissue
  E. are junction points between tendons

55. Which type of bone cell produces the matrix of the bone?

  A. osteoclasts
  B. osteocytes
  C. endothelial
  D. osteoprogenitor cells
  E. osteoblasts

56. Where would the highest concentration of strontium in a child exposed to it be, considering that strontium is preferentially incorporated into growing long bone?

  A. near the epiphyseal plates of long bones
  B. evenly distributed throughout long bones
  C. in the center of long bones
  D. in the cartilage lining the joints
  E. in the ligaments

57. Synovial fluid within joint cavities of freely movable joints contains:

  A. hydrochloric acid
  B. hyaluronic acid
  C. lactic acid
  D. pyruvic acid
  E. uric acid

58. The inorganic matter found in the matrix of osseous tissue is primarily:

  A. calcium carbonate
  B. glycosaminoglycans
  C. collagen
  D. hydroxyapatite
  E. proteoglycans and glycoprotein

Answer Key

1: C	11: C	21: A	31: D	41: B	51: A
2: D	12: C	22: C	32: C	42: D	52: D
3: C	13: D	23: A	33: B	43: E	53: B
4: B	14: B	24: E	34: E	44: C	54: C
5: E	15: E	25: B	35: D	45: E	55: E
6: D	16: B	26: C	36: B	46: A	56: A
7: A	17: A	27: D	37: E	47: C	57: B
8: E	18: A	28: E	38: D	48: B	58: D
9: D	19: E	29: A	39: B	49: C	
10: A	20: B	30: B	40: D	50: B	

Chapter 4.10: Respiratory System

1. Which of the following is NOT a function of the respiratory system?

   A. Helping to expel abdominal contents during defecation and childbirth
   B. Helping to transport gases to tissues
   C. Allowing exchange of O_2 between blood and air
   D. Contributing to maintenance of the pH balance
   E. Allowing exchange of CO_2 between blood and air

2. The loudness of a person's voice depends on:

   A. force with which air rushes across the vocal folds
   B. strength of the intrinsic laryngeal muscles
   C. length of the vocal folds
   D. thickness of vestibular folds
   E. all of the above

3. Which of the following nervous systems controls the vasoconstriction of the arterioles involved in heat conduction to the skin?

   A. motor   **B.** somatic   **C.** sensory   **D.** autonomic   **E.** sympathetic

4. The function of type I cells of the alveoli is to:

   A. form the structure of the alveolar wall
   B. protect the lungs from bacterial invasion
   C. secrete surfactant
   D. trap dust and other debris
   E. replace mucus in the alveoli

5. In which portion of the brain are the inspiratory and expiratory centers located?

   A. cerebral hemispheres   **C.** pons
   B. substantia nigra   **D.** cerebellum   **E.** medulla oblongata

6. Air moves out of the lungs when the pressure inside the lungs is:

   I. greater than the intra-alveolar pressure
   II. greater than the pressure in the atmosphere
   III. less than the pressure in the atmosphere

   A. I only   **B.** II only   **C.** III only   **D.** I and II only   **E.** I and III only

7. Which two factors are necessary for expiration?

   A. Negative feedback of expansion fibers used during inspiration and the outward pull of surface tension due to surfactant

   B. Combined amount of CO_2 in the blood and air in the alveoli

   C. Recoil of elastic fibers that were stretched during inspiration and the inward pull of surface tension due to the film of alveolar fluid

   D. Expansion of respiratory muscles that were contracted during inspiration and the lack of surface tension on the alveolar wall

   E. Combined amount of O_2 in the blood and CO_2 in the alveoli

8. During breathing, inhalation results from:

   A. forcing air from the throat down into the lungs

   B. contracting the abdominal muscles

   C. relaxing the muscles of the rib cag

   D. muscles of the lungs expanding the alveoli

   E. contracting the diaphragm

9. In a panting animal that breathes in and out rapidly, a large portion of new air comes into contact with the upper regions of the respiratory passages in order to:

   A. decrease body heat via evaporation

   B. rapidly increase CO_2 expiration

   C. moisten the mucosa of the respiratory passages

   D. minimize the movement of respiratory muscles

   E. decrease CO_2 expiration

10. What maintains the openness (i.e. patency) of the trachea?

   A. relaxation of smooth muscle

   B. C-shaped cartilage rings

   C. surfactant production

   D. surface tension of water

   E. relaxation of skeletal muscle

11. Why are there rings of hyaline cartilage in the trachea?

   A. To prevent choking

   B. To keep the passageway open for the continuous flow of air

   C. To provide support for the mucociliary escalator

   D. To provide support for the passage of food through the esophagus

   E. To provide the surface for gas exchange

12. Intrapulmonary pressure is:

    A. negative pressure in the intrapleural space

    B. the difference between atmospheric pressure and respiratory pressure

    C. pressure within the pleural cavity

    D. pressure within the alveoli of the lungs

    E. difference between pressure in the pleural cavity and the atmospheric pressure

13. The gas exchange between the blood and tissues takes place in:

    A. arteries, arterioles and capillaries

    B. the entire systemic circulation

    C. pulmonary arteries only

    D. pulmonary veins only

    E. capillaries only

14. Surfactant helps to prevent the alveoli from collapsing by:

    A. protecting the surface of alveoli from dehydration

    B. disrupting the cohesiveness of H_2O thereby reducing the surface tension of alveolar fluid

    C. warming the air before it enters the lungs

    D. humidifying the air before it enters the lungs

    E. protecting the surface of alveoli from environmental variations

15. Through the Bohr effect, more oxygen is released to the tissue because:

    A. increase in pH weakens the hemoglobin-oxygen bond

    B. increase in pH strengthens the hemoglobin-oxygen bond

    C. the tissue-oxygen bond is stronger

    D. decrease in pH strengthens the hemoglobin-oxygen bond

    E. decrease in pH weakens the hemoglobin-oxygen bond

16. In a healthy person, the most powerful respiratory stimulus for breathing is:

    A. acidosis

    B. alkalosis

    C. a decrease of oxygen in tissues

    D. an increase of carbon dioxide in blood

    E. increase in oxygen in lungs

17. Which cells produce pulmonary surfactant?

    A. type II alveolar cells   **C.** alveolar macrophages

    B. goblet cells   **D.** type I alveolar cells   **E.** none of the above

18. The graph shows the relationship between the total alveolar ventilation (the rate at which air reaches the alveoli) and O_2 consumption during exercise.

What is the net effect of exercise on arterial partial pressure of oxygen (P_{O_2})?

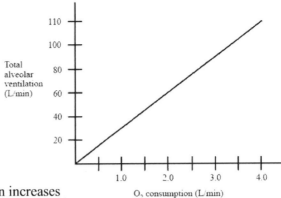

A. P_{O_2} decreases because O_2 consumption increases

B. P_{O_2} initially decreases and then rebounds because ventilation increases

C. P_{O_2} remains the same because ventilation increases as metabolism increases

D. P_{O_2} increases significantly because ventilation increases as metabolis decreases

E. P_{O_2} increases because O_2 consumption decreases

19. Ductus arteriosus connects the aorta and the pulmonary arteries during fetal development and closes at birth. *Patent ductus arteriosus*, the congenital heart defect where ductus arteriosus fails to close, would NOT lead to increased:

A. cardiac output by the left ventricle

B. cardiac output by the right ventricle

C. pulmonary plasma [O_2]

D. systemic plasma [O_2]

E. heart rate

20. After inhalation, lungs pull away from the thoracic wall during elastic recoil due to:

A. natural tendency for the lungs to recoil and transpulmonary pressure

B. compliance and the surface tension of the alveolar fluid

C. elastic fibers of the connective tissue and the surface tension of the alveolar fluid

D. compliance and transpulmonary pressure

E. none of the above

21. Which of the following would cause an increased breathing rate?

A. Low partial pressure of CO_2 in blood

B. High partial pressure of CO_2 in blood

C. High partial pressure of O_2 in blood

D. High pH of blood

E. High number of red blood cells

22. Lung compliance is determined by:

A. alveolar surface tension

B. muscles of inspiration

C. flexibility of the thoracic cage

D. airway opening

E. muscles of expiration

23. What would cause bronchodilation?

- **A.** inspiratory center of the medulla
- **B.** asthma
- **C.** vagus nerve
- **D.** sympathetic nerves
- **E.** parasympathetic nerves

24. Tidal volume is the amount of air:

- **A.** forcibly inhaled after normal inspiration
- **B.** forcibly expelled after normal expiration
- **C.** remaining in the lungs after forced expiration
- **D.** equal to total lung capacity
- **E.** exchanged during normal breathing

25. Working muscles produce lactic acid that lowers blood pH. For homeostasis, the lungs alter the blood ratio of CO_2 to HCO_3^- as shown in the following reaction:

$$CO_2 + H_2O \leftrightarrow H^+ + HCO_3^-$$

Which mechanism is utilized by the lungs as a compensatory action?

- **A.** increased production of O_2
- **B.** increased exhalation of CO_2
- **C.** decreased ventilation rate
- **D.** increased conversion of CO_2 to H^+ and CO_3^-
- **E.** decreased exhalation of CO_2

26. The direction of respiratory gas movement is determined by:

- **A.** molecular weight of the gas molecule
- **B.** average mean temperature
- **C.** partial pressure gradient
- **D.** density
- **E.** size of the gas molecule

27. Since fetal lungs are nonfunctional prior to birth, they are supplied with only enough blood to nourish the lung tissue itself. Obstruction of which of these structures would result in an increased blood supply to fetal lungs?

- **A.** ductus venosus
- **B.** pulmonary artery
- **C.** aorta
- **D.** ductus arteriosus
- **E.** pulmonary vein

28. Which is a possible cause of hypoxia?

- **A.** taking several rapid deep breaths
- **B.** very cold climate
- **C.** low atmospheric oxygen levels
- **D.** obstruction of the esophagus
- **E.** none of the above

29. Where does the lower respiratory tract begin?

- **A.** bronchioles
- **B.** primary bronchi
- **C.** choanae
- **D.** glottis
- **E.** trachea

30. All of the following are stimuli for breathing, EXCEPT:

    I. decrease in plasma pH   II. decreased O_2 levels
    III. elevated blood pressure   IV. elevated CO_2 levels

    A. I only   **B.** II only   **C.** III only   **D.** I and IV only   **E.** II and III only

31. Which of these structures would NOT be affected by a respiratory tract infection?

    A. alveoli   **B.** bronchi   **C.** esophagus   **D.** trachea   **E.** bronchioles

32. Respiratory control centers are located in the:

    A. pons and midbrain   **C.** midbrain and medulla
    B. upper spinal cord and medulla   **D.** medulla and pons
                                 E. upper spinal cord and pons

33. Which of the following sequences is correct for air movement during exhalation?

    A. alveoli → trachea → bronchi → bronchioles → larynx → pharynx
    B. alveoli → bronchi → bronchioles → trachea → pharynx
    C. alveoli → bronchi → lungs → bronchioles → trachea → pharynx
    D. alveoli → bronchi → trachea → bronchioles → pharynx → larynx
    E. alveoli → bronchioles → bronchi → trachea → larynx → pharynx

34. All of the following statements about CO_2 are correct, EXCEPT:

    A. more is carried by erythrocytes than is dissolved in blood plasma as bicarbonate
    B. its concentration is greater in venous blood than in arterial blood
    C. its concentration in the blood decreases by hyperventilation
    D. its accumulation in the blood is associated with increased plasma acidity
    E. it stimulates breathing

35. The projection from the inside of the nose into the breathing passage is called:

    A. meatus   **B.** concha   **C.** vibrissae   **D.** naris   **E.** nasopharynx

36. In the lungs and through all cell membranes, oxygen and carbon dioxide are exchanged through the process of:

    A. active transport   **C.** diffusion
    B. filtration   **D.** osmosis   **E.** facilitated transport

37. In a healthy individual, the highest blood pressure would most likely be found in the:

    A. superior vena cava   **C.** pulmonary capillaries
    B. aorta   **D.** systemic capillaries   **E.** inferior vena cava

38. The majority of CO_2 is transported in the blood:

    A. combined with the heme portion of hemoglobin

    B. as carbonic acid in the plasma

    C. as bicarbonate ions in the plasma after first entering the erythrocytes

    D. combined with the amino acids of hemoglobin as carbaminohemoglobin in the erythrocytes

    E. combined with albumin

39. Which physiological process is generally passive?

    I. gas exchange   II. inhalation   III. exhalation

    A. I only   **B.** I and II only   **C.** III only   **D.** I, II and III   **E.** I and III only

40. All of the following are possible mechanisms of CO_2 transport, EXCEPT:

    A. attached to the heme part of hemoglobin

    B. as bicarbonate ion in plasma

    C. about 10% of CO_2 is carried in the form of carbaminohemoglobin

    D. about 7-10% of CO_2 is dissolved directly into the plasma

    E. all are possible mechanisms of CO_2 transport

41. The point of division of the trachea into the right and left primary bronchi is the:

    A. trachea   **B.** esophagus   **C.** carina   **D.** glottis   **E.** alveoli

42. Oxygen binding to and dissociation from hemoglobin is promoted by all of the following factors, EXCEPT:

    A. partial pressure of CO_2 **C.** partial pressure of O_2

    B. number of erythrocytes   **D.** pH of the blood   **E.** BPG

43. Hypovolemic shock occurs when a patient's blood volume falls abruptly and is likely the result of:

    A. depleted Na^+ consumption   **C.** venous bleeding

    B. excessive Na^+ consumption   **D.** high levels of aldosterone

                                            E. arterial bleeding

44. When an individual goes from a low to a high altitude, the erythrocyte count increases after a few days because:

    A. concentration of O_2 and/or total atmospheric pressure is lower at high altitudes

    B. concentration of O_2 and/or total atmospheric pressure is higher at higher altitudes

    C. basal metabolic rate is higher at high altitudes

    D. temperature is lower at higher altitudes

    E. the concentration of CO_2 is higher at high altitudes

45. All of these statements regarding the respiratory system are true, EXCEPT:

   A. when the pulmonary pressure is less than atmospheric pressure, air flows out of the lungs
   B. thoracic cavity enlargement causes the pressure of air within the lungs to decrease
   C. contraction of the diaphragm enlarges the thoracic cavity
   D. ciliated nasal membranes warm, moisten and filter inspired air
   E. respiratory process is a cycle of repetitive inspirations and expirations

46. Most inhaled particles (e.g. dust) do not reach the lungs because of the:

   A. action of the epiglottis
   B. porous structure of turbinate bones
   C. abundant blood supply to nasal mucosa
   D. ciliated mucous lining in the nose
   E. ciliated lining of the alveoli

47. Which is NOT a part of the pharynx?

   A. laryngopharynx   **C.** nasopharynx
   B. oropharynx   **D.** mesopharynx   **E.** vertebropharynx

48. Which statement regarding the physical factors influencing pulmonary ventilation is correct?

   A. Surfactant helps increase alveolar surface tension
   B. As alveolar surface tension increases, additional muscle action is required
   C. A lung that is less elastic requires less muscle action to perform adequate ventilation
   D. A decrease in compliance causes an increase in ventilation
   E. None are correct

49. Oxygenated blood is pumped by:
   A. left ventricle and left atrium
   B. left ventricle and right atrium
   C. left and right atria
   D. left and right ventricles
   E. inferior vena cava

50. Which statement regarding oxygen transport in blood is correct?

   A. Increased BPG levels in erythrocytes enhance oxygen-carrying capacity
   B. A 50% O_2 saturation level of blood returning to the lungs might indicate an elevated activity level
   C. During normal activity, hemoglobin returning to the lungs carries one molecule of O_2
   D. During acidosis, hemoglobin carries O_2 more efficiently
   E. none of the above

51. Why is the air that enters the lungs of a patient, who has a tube inserted directly into the trachea, colder and dryer than normal and often causes lung crusting and infection?

A. Because it is not properly humidified by the larynx
B. Because it is not filtered as it enters the respiratory system
C. Because it does not flow past the mouth and tongue
D. Because it does not flow through the nasal passageways
E. Because it enters the respiratory system too rapidly

52. All of the following influence hemoglobin saturation, EXCEPT:

A. carbon dioxide
B. nitric oxide
C. temperature
D. BPG
E. pH

53. The airway from the nose to the larynx is the:

A. lower respiratory tract
B. posterior lung system
C. respiratory division
D. esophagus
E. upper respiratory tract

54. The factors responsible for holding the lungs to the thorax wall are:

A. surface tension from pleural fluid and negative pressure in the pleural cavity
B. visceral pleurae
C. diaphragm and the intercostal muscles
D. smooth muscles of the lung
E. changing volume of the lungs

55. The smallest passageways in the lungs to have ciliated epithelia are:

A. tertiary bronchi
B. alveolar ducts
C. terminal bronchioles
D. respiratory bronchioles
E. trachea

56. Which of the following factors influences the rate and depth of breathing?

A. stretch receptors in the alveoli
B. temperature of alveolar air
C. thalamic control
D. vascularity of the alveoli
E. voluntary cortical control

57. Which structure is most important in keeping food out of the trachea?

A. vocal folds
B. soft palate
C. epiglottis
D. glottis
E. nasopharynx

Answer Key

1: B	11: B	21: B	31: C	41: C	51: D
2: A	12: D	22: A	32: D	42: B	52: B
3: D	13: E	23: D	33: E	43: E	53: E
4: A	14: B	24: E	34: A	44: A	54: A
5: E	15: E	25: B	35: B	45: A	55: D
6: B	16: D	26: C	36: C	46: D	56: E
7: C	17: A	27: D	37: B	47: E	57: C
8: E	18: C	28: C	38: C	48: B	
9: A	19: D	29: E	39: E	49: A	
10: B	20: C	30: C	40: A	50: B	

Chapter 4.11: Skin System

1. The primary role of melanin in the skin is to:

   A. provide a waterproof layer for the skin
   B. shield the nucleus from damage by UV radiation
   C. be an integral component of collagen fibers
   D. keep the body cool via evaporation
   E. provide a "healthy tan look" for the person

2. Which layer is NOT part of the skin?

   A. hypodermis   C. dermis
   B. papillary   D. epidermis   E. keratinocytes

3. Most of the sensations of the skin are due to nerve endings in the:

   A. medulla   B. hypodermis   C. dermis   D. epidermis   E. subcutaneous layer

4. The touch sensors of the epidermis are:

   A. nociception   C. melanocytes
   B. dendritic cells   D. keratinocytes   E. tactile cells

5. Which of the following are the sites where the apocrine glands are found?

   A. Palms of the hands and soles of the feet
   B. Under arms and in the external genitalia areas
   C. Beneath the flexure lines in the body
   D. All body regions, buried deep in the dermis
   E. Buried deep in the dermis

6. Which layer of the skin is composed of dense, irregular connective tissue?

   A. reticular layer of the dermis   C. epidermis
   B. hypodermis   D. papillary layer of the dermis
                                             E. subcutaneous layer

7. The most important function of the eccrine sweat gland is:

   A. sebum production   C. milk production
   B. earwax production   D. stress-induced sweating
                                 E. body temperature regulation

8. What causes "goosebumps" (the hair standing on end)?

A. contraction of the epidermal papillae
B. contraction of the epidermal ridges
C. contraction of the arrector pili
D. contraction of the dermal papillae
E. none of the above

9. In addition to waterproofing and lubricating the skin, another important function of sebum is providing protection against:

A. abrasions or cuts to the skin
B. harmful bacteria
C. overheating
D. overexposure to UV light
E. changes in pH of the integumentary

10. Which part of the hair cells has pigment granules that are responsible for the color of hair?

A. cuticle   **B.** bulb   **C.** medulla   **D.** cortex   **E.** eccrine

11. Loss of skin, such as with a severe burn, leads to an increased risk of:

I. dehydration
II. bacterial infection
III. inadequate body temperature maintenance

A. I only   **B.** II only   **C.** III only   **D.** I, II and III   **E.** I and II only

12. Which glands are numerous throughout most of the body and produce a watery sweat that cools the body?

A. merocrine (eccrine) glands
B. apocrine glands
C. ceruminous glands
D. sebaceous glands
E. holocrine

13. Which of the following contribute(s) to skin color?

I. carotene   II. melanin   III. keratin   IV. hemoglobin

A. II only
B. I and II only
C. II and IV only
D. I, II and IV only
E. I, II, III and IV

14. Which degree of burn results in the complete destruction of the epidermis and dermis and frequently requires a skin graft?

A. first-degree burn
B. second-degree burn
C. third-degree burn
D. fourth-degree burn
E. fifth-degree burn

15. Which layer of the skin has no blood vessels?

A. papillary   **B.** hypodermis   **C.** epidermis   **D.** dermis   **E.** none of the above

16. Which of the following is NOT a function of the integument?

   A. thermoregulation   **C.** water retention
   B. vitamin E synthesis   **D.** infection resistance
                              E. barrier to UV energy

17. Which of the following is NOT associated with the dermis?

   A. elastin   **C.** blood vessels
   B. extrafibrillar matrix   **D.** collagen   **E.** keratin

18. The outside layer of the skin is the:

   A. epidermis   **C.** apocrine layer
   B. dermis   **D.** lamellar layer   **E.** hypodermis

19. Sweat is mostly composed of:

   A. electrolytes   **C.** sodium chloride
   B. metabolic wastes   **D.** antibodies   **E.** water

20. The inside layer of the two main layers of the skin is the:

   A. dermis   **B.** epidermis   **C.** lamellar layer   **D.** apocrine layer   **E.** hypodermis

21. Which skin glands play an important role in body temperature regulation?

   A. sebaceous   **C.** apocrine
   B. ceruminous   **D.** eccrine   **E.** all of the above

22. Which kind of cells comprises up to 30 or more layers of stratum corneum?

   A. keratinized   **C.** non-keratinized
   B. tactile "Merkel"   **D.** stem   **E.** melanocytes

23. Which skin gland is also a holocrine gland?

   A. eccrine   **C.** endocrine
   B. sudoriferous   **D.** ceruminous   **E.** sebaceous

24. The factor in the hue of the skin is the amount of:

   A. oxygen in the blood   **C.** keratin
   B. blood vessels   **D.** melanin   **E.** lamellar layer

25. All of the following are accessory glands of the skin, EXCEPT:

A. sebaceous
B. mammary

C. ceruminous
D. all are accessory glands of the skin
E. none are accessory glands of the skin

26. The distinctive fingerprints humans have are due to:

A. friction ridges
B. hypodermal channels

C. oily hair
D. sudoriferous glands

E. keratin

27. The principal tissue found in the dermal layer is:

A. dense regular connective tissue
B. dense irregular connective tissue

C. areolar connective tissue
D. stratified squamous epithelium
E. stratified irregular connective tissue

28. Which is NOT a derivative of the epidermis?

A. nails   **B.** glands   **C.** adipose tissue   **D.** hair   **E.** all of the above

29. Which protein provides the epidermis with protective properties?

A. elastin   **B.** carotene   **C.** melanin   **D.** collagen   **E.** keratin

30. Which of the following is NOT a zone of the hair along its length?

A. cuticle   **B.** shaft   **C.** bulb   **D.** root   **E.** all of the above

31. Which of the following is NOT associated with hair?

A. medulla   **B.** lunula   **C.** cuticle   **D.** keratin bundles   **E.** matrix

32. What layer of the integumentary system contains adipose tissue?

A. dermis
B. hypodermis

C. epidermis
D. muscular layer

E. reticular dermis

33. When people gain weight, they often accumulate fat in:

A. hypodermis
B. reticular dermis

C. papillary dermis
D. epidermis

E. dermis

34. The apocrine sweat glands are found in certain locations on the body in association with:

A. soles of the hand and feet
B. nails

C. hair
D. exposed areas to the sun
E. none of the above

35. Which statement regarding skin cancer is the most accurate?

   A. Basal cell carcinomas are the least common but most malignant

   B. Melanomas are rare but must be removed quickly to prevent them from metastasizing

   C. Most tumors that arise on the skin are malignant

   D. Squamous cell carcinomas arise from the stratum corneum

   E. Basal cell carcinomas often metastasize

36. Skin cancer is most often due to exposure to:

   A. UV light   **B.** chemicals   **C.** infrared light   **D.** x-rays   **E.** gamma rays

37. In which order would a needle pierce the epidermal layers of the skin?

   A. granulosum → basale → spinosum → corneum → lucidum

   B. corneum → lucidum → granulosum → spinosum → basale

   C. basale → spinosum → lucidum → granulosum → corneum

   D. basale → spinosum → granulosum → lucidum → corneum

   E. corneum → lucidum → basale → granulosum → spinosum

38. Which layer of the epidermis would be affected first by a drug that inhibits cell division (e.g. chemotherapy drug)?

   A. stratum spinosum   **C.** stratum corneum

   B. stratum granulosum   **D.** stratum lucidum   **E.** stratum basale

39. Mitosis occurs primarily in which stratum of the epidermis?

   A. spinosum   **B.** granulosum   **C.** corneum   **D.** basale   **E.** lucidum

40. Which layer of the skin is responsible for fingerprints?

   A. hypodermis   **C.** reticular dermis

   B. papillary dermis   **D.** epidermis   **E.** stratum corneum

41. Which cutaneous receptor is utilized for the reception of touch or light pressure?

   A. Ruffinian endings   **C.** Meissner's corpuscles

   B. end-bulbs of Krause   **D.** Pacinian corpuscles   **E.** free nerve endings

42. The aging of the integumentary system involves changes in the:

   I. hair   II. sebaceous glands   III. blood vessels

   A. I only   **B.** II only   **C.** I and III only   **D.** I, II and III   **E.** I and II only

43. Which of the following statements is correct about integument function?

    A. Epidermal blood vessels serve as a blood reservoir
    B. Body cools by increasing the action of sebaceous glands during high-temperature conditions
    C. Dermis provides the major mechanical barrier to chemicals and other external substances
    D. Dermis provides the major mechanical barrier to water
    E. Resident macrophage-like cells ingest antigenic invaders and present them to the immune system

44. Glands only found in the auditory canal are:

    A. mammary glands   **C.** sebaceous glands
    B. apocrine glands   **D.** cerumin glands   **E.** eccrine glands

45. All of the following are the major regions of a hair shaft, EXCEPT:

    A. medulla   **C.** external root sheath
    B. cortex   **D.** cuticle   **E.** all of the above

46. Oil glands that are normally associated with hair shafts and provide lubrication for hair and skin are:

    A. apocrine glands   **C.** cerumen glands
    B. mammary glands   **D.** sebaceous glands   **E.** eccrine glands

47. Which glands produce ear wax?

    A. ceruminous glands   **C.** merocrine glands
    B. eccrine glands   **D.** apocrine glands   **E.** sudoriferous

48. What is normally associated with the hair shaft for tactile sensations?

    A. sebaceous glands   **C.** oil glands
    B. keratin   **D.** muscle   **E.** free nerve endings

49. The hypodermis acts as a shock absorber because:

    A. cells that make up the hypodermis secrete a protective mucus
    B. major part of its makeup is adipose tissue
    C. the basement membrane can absorb shock
    D. it is located just below the epidermis and protects the dermis from shock
    E. it has no delicate nerve endings and can therefore absorb more shock

50. Which of these is generally NOT involved in non-specific immunity?

    A. tears   **B.** macrophages   **C.** T cells   **D.** neutrophils   **E.** skin

51. Keratinocytes protect the skin from UV damage by:

    A. maintaining the appropriate pH in order for the melanocyte to synthesize melanin granules

    B. maintaining the appropriate temperature so melanocyte proteins are not denatured

    C. providing the melanocyte with nutrients necessary for melanin synthesis

    D. accumulating melanin granules on their superficial portion

    E. maintaining the appropriate pH so melanocyte proteins are not denatured

52. Animals that migrate annually to warmer climates dissipate metabolic heat through:

    A. increased metabolic rate   **C.** shivering

    B. increased contraction of muscles   **D.** vasoconstriction

                                            E. vasodilation

53. Which of the following layers is responsible for cell division and replacement?

    A. stratum basale   **C.** stratum granulosum

    B. stratum spinosum   **D.** stratum corneum   **E.** stratum lucidum

54. The epidermis is composed of:

    A. stratified squamous epithelium

    B. keratinized nonstratified squamous epithelium

    C. stratified columnar epithelium

    D. keratinized stratified squamous epithelium

    E. keratinized stratified columnar epithelium

55. A dendritic or Langerhan cell is a specialized:

    A. melanocyte   **C.** phagocytic cell

    B. nerve cell   **D.** squamous epithelial cell

                                   E. keratinocyte

56. The composition of the eccrine glands secretions is:

    A. fatty substances, proteins, antibodies

    B. metabolic wastes

    C. primarily uric acid

    D. primarily minerals and vitamins

    E. primarily water and sodium chloride with trace amounts of wastes

Answer Key

1: B	11: D	21: D	31: B	41: C	51: D
2: A	12: A	22: A	32: E	42: D	52: E
3: C	13: D	23: E	33: A	43: E	53: A
4: E	14: C	24: D	34: C	44: D	54: D
5: B	15: C	25: D	35: B	45: C	55: C
6: A	16: B	26: A	36: A	46: D	56: E
7: E	17: E	27: B	37: B	47: A	
8: C	18: A	28: C	38: E	48: E	
9: B	19: E	29: E	39: D	49: B	
10: D	20: A	30: A	40: B	50: C	

Chapter 4.12: Reproductive System

1. Which of the following events is/are essential for the menstrual cycle?

    I.   adrenal medulla releases norepinephrine
    II. ovarian follicle development is stimulated by FSH
    III. progesterone stimulates the formation of the endometrial lining

    A. II only   **B.** III only   **C.** I and II only   **D.** II and III only   **E.** I and III only

2. What is the number of double stranded DNA molecules in a single mouse chromosome immediately after the gametes are formed?

    A. 0   **B.** 1   **C.** 2   **D.** 4   **E.** 8

3. All of the following about human gamete production is true, EXCEPT:

    A. meiosis in females produces four egg cells
    B. sperm develop in the seminiferous tubules within the testes
    C. eggs develop in the ovarian follicles within the ovaries
    D. FSH stimulates gamete production in both males and females
    E. gametes arise via meiosis

4. Common red-green color blindness is an X-linked trait. When a woman, whose father is color blind, has a son with a non-afflicted man, what is the probability that their son will be color blind?

    A. 0   **B.** 1/4   **C.** 1/2   **D.** 3/4   **E.** 1

5. At birth, a woman possesses a finite number of ova. In oogenesis, meiotic division is arrested at which stage until she reaches menarche?

    A. ovum   **C.** ova
    B. oogonium   **D.** secondary oocytes   **E.** primary oocytes

6. The most likely gamete cell to be produced from meiosis in the seminiferous tubules is:

    A. diploid 2° spermatocytes   **C.** haploid spermatids
    B. haploid 1° spermatocytes   **D.** diploid spermatids   **E.** none of the above

7. A human cell after the first meiotic division is:

    A. 2N and 2 chromatids   **C.** 1N and 2 chromatids
    B. 2N and 4 chromatids   **D.** 1N and 1 chromatid   **E.** None of the above

8. All of these statements regarding the menstrual cycle are true, EXCEPT:

    A. Graafian follicle, under the influence of LH, undergoes ovulation

    B. follicle secretes estrogen as it develops

    C. corpus luteum develops from the remains of the post-ovulatory Graafian follicle

    D. FSH causes the development of the primary follicle

    E. FSH and LH are both secreted by the posterior pituitary

9. Which of the following is the initial site of spermatogenesis?

    A. seminiferous tubules   **C.** vas deferens

    B. seminal vesicles   **D.** epididymis   **E.** prostate

10. What distinguishes meiosis from mitosis?

    I. Genetic recombination

    II. Failure to synthesize DNA between successive cell divisions

    III. Separation of homologous chromosomes into distinct cells

    A. I only   **B.** II only   **C.** I and III only   **D.** II and III only   **E.** I, II and III

11. The development of oocytes in females until ovulation is arrested in:

    A. interphase   **C.** prophase I of meiosis

    B. prophase II of meiosis   **D.** prophase of mitosis   **E.** G1

12. Which of the following events occurs first?

    A. formation of corpus luteum   **C.** secretion of estrogen

    B. rupture of the Graafian follicle   **D.** release of progesterone

                                          E. decrease in FSH release by pituitary

13. SRY gene encoding for the testis-determining factor is the master sex-determining gene that resides on:

    A. pseudoautosomal region of the Y chromosome

    B. short arm of the Y chromosome, but not in the pseudoautosomal region

    C. X chromosome

    D. pseudoautosomal region of the X chromosome

    E. autosomes

14. The number of chromosomes contained in the human primary spermatocyte is:

    A. 23   **B.** 23,X/23,Y   **C.** 92   **D.** 184   **E.** 46

15. Which of the following processes does NOT contribute to genetic variation?

    A. Random segregation of homologous chromosomes during meiosis
    B. Random segregation of chromatids during mitosis
    C. Recombination
    D. Mutation
    E. All of the above

16. Which statement correctly describes the role of LH in the menstrual cycle?

    A. stimulates the ovary to increase LH secretions
    B. inhibits secretions of GnRH
    C. stimulates development of the endometrium for implantation of the zygote
    D. induces the corpus luteum to secrete estrogen and progesterone
    E. stimulates milk production after birth

17. What is the most likely cause of a condition in which the testes do not fully descend into the scrotum due to abnormal testicular development?

    A. cortisol deficiency   **C.** excess LH
    B. testosterone deficiency   **D.** excess estrogen   **E.** excess FSH

18. Testosterone is synthesized primarily by:

    A. sperm cells
    B. hypothalamus
    C. Leydig cells
    D. anterior pituitary gland
    E. seminiferous tubules

19. In human females, secondary oocytes do not complete meiosis II until:

    A. menarche   **B.** menstruation   **C.** puberty   **D.** menopause   **E.** fertilization

20. 47, XXY is a condition known as:

    A. Turner syndrome   **C.** trisomy X syndrome
    B. double Y syndrome   **D.** Klinefelter syndrome   **E.** fragile X syndrome

21. Which of these conditions is the LEAST likely cause of male infertility?

    A. acrosomal enzymes denaturation   **C.** abnormal mitochondria
    B. immotility of cilia   **D.** testosterone deficiency
                                       E. abnormal flagellum

22. The surgical removal of the seminiferous tubules would likely cause:

    A. sterility, because sperm would not be produced
    B. sterility, because sperm would not be able to exit the body
    C. reduced volume of semen
    D. enhanced fertilization potency of sperm
    E. testes to migrate back into the abdominal cavity

23. All of these cell types contain the diploid (2N) number of chromosomes EXCEPT:

    A. primary oocyte   **B.** spermatogonium   **C.** spermatid   **D.** zygote   **E.** oogonium

24. The probability that all children in a four-children family will be males is:

    A. 1/2   **B.** 1/4   **C.** 1/8   **D.** 1/16   **E.** 1/64

25. Which of the following can result from scars formed in the reproductive system of women with high risk of infections (e.g. chlamydia)?

    A. elevated levels of estrogen   **C.** infertility
    B. decreased ovulation   **D.** reduced gamete production
    E. decreased levels of estrogen

26. The surgical removal of the seminal vesicles would likely cause:

    A. sterility, because sperm would not be produced
    B. sterility, because sperm would not be able to exit the body
    C. testes to migrate back into the abdominal cavity
    D. enhanced fertilization potency of sperm
    E. reduced volume of semen

27. Unequal cytoplasm division is characteristic of:

    A. binary fission of bacteria   **C.** production of a sperm
    B. mitosis of a kidney cell   **D.** production of an ovum   **E.** none of the above

28. All of the following are clinicial manifestations of Kartagener's syndrome which results from defective dynein that causes paralysis of cilia and flagella, EXCEPT:

    A. chronic respiratory disorders
    B. cessation of ovulation
    C. male infertility
    D. ectopic pregnancy
    E. decreased absorption within the duodenum

29. The primary difference between estrous and menstrual cycles is that:

    A. in estrous cycle, endometrium is shed and reabsorbed by the uterus, while in menstrual cycle the shed endometrium is excreted from the body

    B. behavioral changes during estrous cycles are much less apparent than those of menstrual cycles

    C. season and climate have less pronounced effects on estrous cycle than they do on menstrual cycles

    D. copulation normally occurs across the estrous cycle, whereas in menstrual cycles copulation only occurs during the period surrounding ovulation

    E. most estrous cycles are much longer in duration compared to menstrual cycles

30. Translation, transcription and replication take place in this phase of the cell cycle:

    A. G1   **B.** G2   **C.** metaphase   **D.** anaphase   **E.** S

31. Progesterone is primarily secreted by the:

    A. primary oocyte   **C.** corpus luteum

    B. hypothalamus   **D.** anterior pituitary gland   **E.** endometrial lining

32. All of the following statements are true for a normal human gamete, EXCEPT:

    A. it originates via meiosis from a somatic cell

    B. it contains genetic material that has undergone recombination

    C. it contains a haploid number of genes

    D. it always contains an X or Y chromosome

    E. it forms from meiotic process

33. All of the following statements are correct regarding cleavage in human embryos, EXCEPT:

    A. blastomeres are genetically identical to the zygote

    B. holoblastic cleavage occurs in one portion of the egg

    C. morula is a solid mass of cells produced via cleavage of the zygote

    D. size of the embryo remains constant throughout cleavage of the zygote

    E. none of the above

34. One of the functions of estrogen is to:

    A. induce the ruptured follicle to develop into the corpus luteum

    B. stimulate testosterone synthesis in males

    C. maintain female secondary sex characteristics

    D. promote development and release of the follicle

    E. lower blood glucose

35. Which type of inheritance has the pattern where an affected male has daughters that are all affected, but no affected sons?

   A. X-linked recessive   **C.** Autosomal dominant

   B. Y-linked   **D.** Autosomal recessive   **E.** X-linked dominant

36. Increasing levels of estrogen during the female menstrual cycle trigger which feedback mechanism?

   A. positive, which stimulates LH secretion by the anterior pituitary

   B. positive, which stimulates FSH secretion by the anterior pituitary

   C. negative, which stimulates the uterine lining to be shed

   D. negative, which inhibits progesterone secretion by the anterior pituitary

   E. none of the above

37. Which cell division process results in four genetically different daughter cells that contain one haploid set of chromosomes?

   A. interphase   **C.** cell division

   B. somatic cell regeneration   **D.** mitosis

                               E. meiosis

38. Crossing over occurs during:

   A. anaphase I   **B.** telophase I   **C.** prophase I   **D.** metaphase I   **E.** interkinesis

39. The epididymis functions to:

   A. synthesize and release testosterone

   B. store sperm until they are released during ejaculation

   C. initiate the menstrual cycle by secreting FSH and LH

   D. provide a conduit for the ovum as it moves from the ovary into the uterus

   E. provide the majority of the fluid that comprises semen

40. A genetically important event of crossing over occurs during:

   A. metaphase II   **C.** anaphase I

   B. telophase I   **D.** prophase I   **E.** telophase II

41. Which of the following does NOT describe events occurring in prophase I of meiosis?

   A. chromosomal migration   **C.** formation of a chiasma

   B. genetic recombination   **D.** spindle apparatus formation

                               E. tetrad formation

42. The difference between spermatogenesis and oogenesis is:

    A. spermatogenesis produces haploid cells, while oogenesis produces diploid cells
    B. spermatogenesis produces gametes, while oogenesis does not produce gametes
    C. oogenesis is a mitotic process, while spermatogenesis is a meiotic process
    D. spermatogenesis is a mitotic process, while oogenesis is a meiotic processs
    E. spermatogenesis produces 4 1N sperms, while oogenesis produces 1 egg cell and polar bodies

43. What is the purpose of the cilia that covers the inner linings of the Fallopian tubes?

    A. Preventing polyspermy by immobilizing additional incoming sperm after fusion
    B. Facilitating movement of the ovum towards the uterus
    C. Removing particulate matter that becomes trapped in the mucus layer
    D. Protecting the ovum from pH fluctuations
    E. Protecting the ovum from temperature fluctuations

44. How many Barr bodies are present in white blood cells of a 48, XXYY individual?

    A. 0   **B.** 1   **C.** 2   **D.** 3   **E.** 4

45. Polar bodies are the products of:

    A. meiosis in females
    B. meiosis in males
    C. mitosis in females
    D. mitosis in males
    E. two of the above

46. Which of the following endocrine glands initiates the production of testosterone?

    A. adrenal medulla   **C.** pancreas
    B. hypothalamus   **D.** anterior pituitary   **E.** posterior pituitary

47. Turner syndrome results from:

    A. extra chromosome number 13
    B. absence of Y chromosome
    C. presence of an extra Y chromosome
    D. trisomy of X chromosome
    E. monosomy of X chromosome

48. Decreasing progesterone during the luteal phase of the menstrual cycle results in:

    A. increased secretion of estrogen in the follicle followed by the menstruation phase

    B. degeneration of the corpus luteum in the ovary

    C. increased secretion of LH, which produces the luteal surge and onset of ovulation

    D. thickening of the endometrial lining in preparation for implantation of the zygote

    E. none of the above

49. Vasectomy procedure prevents the:

    A. movement of sperm along the vas deferens

    B. transmission of sexually transmitted disease

    C. production of sperm via spermatogenesis

    D. production of semen in the seminal vesicles

    E. synthesis of seminal fluid

Answer Key

1: D	11: C	21: B	31: C	41: A
2: B	12: C	22: A	32: A	42: E
3: A	13: B	23: C	33: B	43: B
4: C	14: E	24: D	34: C	44: B
5: E	15: B	25: C	35: E	45: A
6: C	16: D	26: E	36: A	46: B
7: C	17: B	27: D	37: E	47: E
8: E	18: C	28: B	38: C	48: B
9: A	19: E	29: A	39: B	49: A
10: D	20: D	30: E	40: D	

Chapter 4.13: Development

1. Which of the following statements is true regarding respiratory exchange during fetal life?

    A. Respiratory exchanges are made through the placenta
    B. Since lungs develop later in gestation, a fetus doesn't need a mechanism for respiratory exchange
    C. Respiratory exchange is made through the ductus arteriosus
    D. Respiratory exchange is not necessary
    E. Respiratory exchange is made through the ductus venosus

2. When each cell in a very early stage of embryonic development still has the ability to develop into a complete organism, it is known as:

    A. gastrulation   **C.** determinate cleavage
    B. blastulation   **D.** indeterminate cleavage   **E.** none of the above

3. Mutations in *Drosophila*, which result in the transformation of one body segment into a different one, are mutations in:

    A. maternal-effect genes   **C.** execution genes
    B. homeotic genes   **D.** segmentation genes   **E.** gastrulation genes

4. Which of the following structures is developed from embryonic ectoderm?

    A. connective tissue   **C.** rib cartilage
    B. bone   **D.** epithelium of the digestive system   **E.** hair

5. Placenta is a vitally important metabolic organ made from a contribution by both the mother and the fetus. The portion of the placenta contributed by the fetus is:

    A. amnion   **B.** yolk sac   **C.** umbilicus   **D.** chorion   **E.** none of the above

6. Which of the following occurs when an embryo lacks synthesis of human chorionic gonadotropin (hCG)?

    A. The embryo does not support the maintenance of the corpus luteum
    B. The embryo increases the production of progesterone
    C. The embryo develops immunotolerance
    D. The placenta forms prematurely
    E. Conception does not occur

7. Which germ layer gives rise to smooth muscle?

    A. ectoderm   **B.** epidermis   **C.** mesoderm   **D.** endoderm   **E.** hypodermis

8. In early embryogenesis, the most critical morphological change in the development of cellular layers is:

- **A.** epigenesist
- **B.** gastrulation
- **C.** conversion of morula to blastula
- **D.** acrosomal reaction
- **E.** fertilization membrane

9. During the first 8 weeks of development, all of the following events occur, EXCEPT:

- **A.** myelination of the spinal cord
- **B.** presence of all body systems
- **C.** formation of a functional cardiovascular system
- **D.** beginning of ossification
- **E.** limb buds appear

10. Which structures derive from the ectoderm germ layers?

- **A.** blood vessels, tooth enamel and epidermis
- **B.** heart, kidneys and blood vessels
- **C.** nails, epidermis and blood vessels
- **D.** epidermis and adrenal cortex
- **E.** epidermis and neurons

11. A diagram of the blastoderm that identifies regions from which certain adult structures are derived is a:

- **A.** phylogenetic tree
- **B.** pedigree diagram
- **C.** fate map
- **D.** linkage map
- **E.** lineage diagram

12. Which tissue is the precursor of long bones in the embryo?

- **A.** hyaline cartilage
- **B.** fibrocartilage
- **C.** dense fibrous connective tissue
- **D.** elastic cartilage
- **E.** costal cartilage

13. What changes are observed in the developing cells as development proceeds?

- **A.** Cytoskeletal elements involved in forming the mitotic spindle
- **B.** Energy requirements of each cell
- **C.** Genetic information that was duplicated with each round of cell division
- **D.** Composition of polypeptides within the cytoplasm
- **E.** Composition of nucleotides within the nucleus

14. During fertilization, what is the earliest event to occur in the process?

- **A.** sperm contacts the cortical granules around the egg
- **B.** sperm nucleic acid enters the egg's cytoplasm to form a pronucleus
- **C.** acrosome releases hydrolytic enzymes
- **D.** sperm contacts the vitelline membrane around the egg
- **E.** hyaluronic acid forms a barrier to prevent polyspermy

15. The embryonic ectoderm layer gives rise to all of the following, EXCEPT:

  A. eyes   **C.** integument

  B. fingernails   **D.** nervous system   **E.** blood vessels

16. All the following are correct matches of a fetal structure with what it becomes at birth, EXCEPT:

  A. ductus venosus—ligamentum venosum

  B. umbilical arteries—medial umbilical ligament

  C. foramen ovale—fossa ovalis

  D. ductus arteriosus—ligamentum teres

  E. all choices are correct

17. All of the following statements regarding gastrulation are true, EXCEPT:

  A. the primitive gut that results from gastrulation is called the archenteron

  B. for amphibians, gastrulation is initiated at the gray crescent

  C. after gastrulation, the embryo consists of two germ layers of endoderm and ectoderm

  D. for amphibians, blastula cells migrate during gastrulation through the invagination region known as the blastopore

  E. the mesoderm develops as the third primary germ layer during gastrulation

18. A homeobox is a:

  A. protein involved in the control of meiosis

  B. transcriptional activator of other genes

  C. DNA binding site

  D. sequence responsible for post-transcriptional modifications

  E. sequence that encodes for a DNA binding motif

19. After the gastrulation's completion, the embryo undergoes:

  A. cleavage   **C.** blastocoel formation

  B. blastulation   **D.** neurulation   **E.** dedifferentiation

20. Shortly after implantation:

  A. trophoblast forms two distinct layers

  B. embryo undergoes gastrulation within 3 days

  C. maternal blood sinuses bathe the inner cell mass

  D. myometrical cells cover and seal off the blastocyst

  E. the umbilical cord forms within 48 hours

21. What is the name of the anatomical connection between the placenta and embryo?

A. chorion   **C.** endometrium

B. umbilical cord   **D.** corpus luteum   **E.** vas deferens

22. Which statement correctly illustrates the principle of induction during vertebrate development?

A. Neural tube develops into the brain, the spinal cord and the nervous system

B. Secretion of TSH stimulates the release thyroxine hormone

C. Ectoderm develops into the nervous system

D. Neurons synapse with other neurons via neurotransmitters

E. Presence of a notochord beneath the ectoderm results in the formation of a neural tube

23. The acrosomal reaction by the sperm is:

A. a transient voltage change across the vitelline envelope

B. the jelly coat blocking penetration by multiple sperm

C. the consumption of yolk protein

D. hydrolytic enzymes degrading the plasma membrane

E. the inactivation of the sperm acrosome

24. Which primary germ layer gives rise to the cardiovascular system, bones and skeletal muscles?

A. endoderm   **C.** mesoderm

B. blastula   **D.** ectoderm   **E.** gastrula

25. The dorsal surface cells of the inner cell mass form:

A. notochord   **C.** one of the fetal membranes

B. placenta   **D.** structure called the embryonic disc   **E.** primitive streak

26. All of the following occurs in a newborn immediately after birth, EXCEPT:

A. the infant completely stops producing fetal hemoglobin

B. resistance in the pulmonary arteries decreases

C. pressure in the left atrium increases

D. pressure in both the inferior vena cava and the right atrium increase

E. ductus arteriosus constricts

27. Homeobox sequences are present in:

A. introns   **C.** 3'-untranslated regions

B. exons   **D.** 5'-untranslated regions   **E.** exon-intron boundaries

28. After 30 hours of incubation, the ectoderm tissue within the gastrula differentiates into different specific tissues, which supports the conclusion that:

       I. cells become either endoderm or mesoderm
       II. cells contain a different genome from their parental cells
       III. gene expression is altered

A. III only   **C.** I only
B. I and III only   **D.** I and II only   **E.** II and III only

29. What is the function of the yolk sac in humans?

    A. forms into the placenta
    B. secretes progesterone in the fetus
    C. gives rise to blood cells and gamete-forming cells
    D. stores embryonic wastes
    E. stores nutrients for the embryo

30. The trophoblast is mostly responsible for forming:

A. placental tissue   **C.** allantois
B. lining of the endometrium   **D.** archenteron   **E.** chorion

31. Certain lesions to the mesodermal embryonic primary germ layer may simulate the development of a human condition *spina bifida*, a congenital fissure in the lower vertebrae. What other structures, besides the spinal column, would most likely be affected by such lesions?

    I. intestinal epithelium   III. blood vessels
    II. skin and hair   IV. muscles

A. I and II only   **C.** III and IV only
B. II and III only   **D.** I, III and IV only   **E.** IV only

32. The homeotic genes encode for:

    A. repressor proteins
    B. transcriptional activator proteins
    C. helicase proteins
    D. single-strand binding proteins
    E. restriction enzymes

33. Which of the following cell groups gives rise to muscles in a frog embryo?

A. neural tube   **B.** ectoderm   **C.** endoderm   **D.** mesoderm   **E.** notochord

34. Which of the following is NOT involved in the implantation of the blastocyst?

    A. Settling of the blastocyst onto the prepared uterine lining

    B. Adherence of the trophoblast cells to the uterine lining

    C. Phagocytosis by the trophoblast cells

    D. Inner cell mass giving rise to the primitive streak

    E. Endometrium proteolytic enzymes produced by the trophoblast cells

35. During labor, which hormone stimulates contractions of uterine smooth muscle?

    A. oxytocin   **B.** prolactin   **C.** luteinizing hormone   **D.** hCG   **E.** estrogen

36. In a chick embryo, certain ectoderm cells give rise to wing feathers, while others develop into thigh feathers or feet claws. The cells from an ectoderm area that normally develop into wing feathers were transplanted to an area that develops into claws of the feet, and it was observed that the transplanted cells developed into claws. Which of the following best explains the result of the experiment?

    A. ectoderm cells possess positional information

    B. destiny of the cells was already determined

    C. ectoderm cells can develop into any type of tissue

    D. cells were induced by underlying mesoderm

    E. ectoderm released growth factors

37. The *slow block* to polyspermy is due to:

    A. transient voltage change across the membrane

    B. jelly coat blocking sperm penetration

    C. consumption of yolk protein

    D. formation of the fertilization envelope

    E. inactivation of the sperm acrosome

38. Polyspermy in humans results in:

    A. mitotic insufficiency   **C.** nonfunctional zygote

    B. interruption of meiosis   **D.** multiple births

    E. formation of multiple placentas

39. Mesoderm gives rise to:

    A. intestinal mucosa   **C.** skin

    B. nerves   **D.** lung epithelium   **E.** heart

40. Genomic imprinting is:

 A. inactivation of a gene by interruption of its coding sequence
 B. organization of molecules in the cytoplasm to provide positional information
 C. DNA modification in gametogenesis that affects gene expression in the zygote
 D. suppression of a mutant phenotype because of a mutation in a different gene
 E. mechanism by which enhancers distant from the promoter can still regulate transcription

41. Comparing a developing frog embryo and an adult organism, cells of which organism have a greater rate of translation?

 A. Adult, because ribosomal production is more efficient in a mature organism
 B. Adult, because a mature organism has more complex metabolic requirements
 C. Embryo, because a developing organism requires more protein production than an adult
 D. Embryo, because ribosomal production is not yet under regulatory control by DNA
 E. Embryo, because the proteins must undergo more extensive post-translation processing

42. Which structure is the first to form during fertilization in humans?

 A. blastula   **B.** morula   **C.** neural tube   **D.** ectoderm   **E.** zygote

43. It is impossible for sperm to be functional (i.e. able to fertilize the egg) until after:

I. they undergo capacitation	III. they have been in the uterus for several days
II. the tail disappears	IV. they become spermatids

 A. I only
 B. II and IV only
 C. I and IV only
 D. I and III only
 E. IV only

44. Where is the most likely deformity if a teratogen affects the development of the endoderm shortly after gastrulation?

 A. nervous system   **C.** liver
 B. lens of the eye   **D.** skeleton   **E.** connective tissue

45. Mutations that cause cells to undergo developmental fates of other cell types are:

 A. heterochronic mutations   **C.** transection mutations
 B. loss-of-function mutants   **D.** execution mutations   **E.** homeotic mutations

46. Each of the following is true for the cells of an early gastrula's eye field, EXCEPT that they are:

 A. terminally differentiated   **C.** derived from the ectoderm layer
 B. competent   **D.** capable of becoming other ectoderm structures
                        E. undifferentiated ectoderm

47. What is the role of proteases and acrosin enzymes in reproduction?

    A. They degrade the nucleus of the egg and allow the sperm to enter

    B. They degrade the protective barriers around the egg and allow the sperm to penetrate

    C. They direct the sperm to the egg through chemotaxis messengers

    D. They neutralize the mucous secretions of the uterine mucosa

    E. They degrade the protective barriers around the sperm and allow the egg to penetrate

48. When does the human blastocyst implant in the uterine wall?

    A. about a week past fertilization   **C.** a few hours past fertilization

    B. at blastulation   **D.** at primary germ formation

    E. at primitive streak formation

49. Which primary layer(s) develop(s) into the retina of the eye?

    I. endoderm   II. ectoderm   III. mesoderm

    A. I only   **B.** II only   **C.** III only   **D.** II and III   **E.** I and III

50. What changes must take place in the cardiovascular system of a newborn after the infant takes its first breath?

    A. Ductus arteriosus constricts and is converted to the ligamentum arteriosum

    B. Urinary system is activated at birth

    C. Ductus venosus is disconnected at the severing of the umbilical cord and all visceral blood goes into the vena cava

    D. Foramen ovale between the atria of the fetal heart closes at the moment of birth

    E. Foramen ovale becomes the medial umbilical ligament

51. Early activation of the mammalian zygote nucleus may be necessary, because:

    A. most developmental decisions are made under the influence of the paternal genome

    B. mammalian oocytes are too small to store molecules needed to support the cleavage divisions

    C. in gametogenesis, certain genes undergo imprinting

    D. mammals don't have maternal-effect genes

    E. none of the above

52. In reptiles, the aquatic environment necessary for embryonic development of amphibians is replaced by:

    A. use of lungs instead of gills   **C.** shells which prevent the escape of gas

    B. humid atmospheric conditions   **D.** intrauterine development

    E. amniotic fluid

53. What would be expected to form from a portion of cells destined to become the heart if it was excised from an early gastrula and placed in a culture medium?

A. undifferentiated mesoderm

B. undifferentiated ectoderm

C. differentiated endoderm

D. undifferentiated endoderm

E. differentiated ectoderm

54. When, during gestation, can the sex of the fetus be determined by ultrasound?

A. at the midpoint of the first trimester

B. about 18 weeks after fertilization

C. at the end of the first trimester

D. at the midpoint of the second trimester

E. at the midpoint of the third trimester

55. How long is the egg viable and able to be fertilized after ovulation?

A. 36-72 hours

B. a full week

C. up to 6 hours

D. 24-36 hours

E. 12-24 hours

56. Which stage of embryonic development is associated with a hollow ball of cells surrounding a fluid filled center?

A. 3-layered gastrula

B. blastula

C. morula

D. zygote

E. 2-layered gastrula

57. Which statement describes the acrosome of a sperm cell?

A. It contains the nucleic acid

B. It contains hydrolytic enzymes which are released when the sperm encounters the jelly coat of the egg

C. It fuses with the egg's cortical granules

D. It functions to prevent polyspermy

E. It is used for the motility of the sperm along the Fallopian tubes (oviducts)

58. Which statement is correct about fertilization?

A. Most sperm cells are protected and remain viable once inside the uterus

B. If estrogen is present, the pathway through the cervical opening is blocked from sperm entry

C. Millions of sperm cells are destroyed by the vagina's acidic environment

D. Spermatozoa remain viable for about 72 hours in the female reproductive tract

E. The ovulated secondary oocyte is viable for about 72 hours in the female reproductive tract

Answer Key

1: A	11: C	21: B	31: C	41: C	51: C
2: D	12: A	22: E	32: B	42: E	52: E
3: B	13: D	23: D	33: D	43: A	53: A
4: E	14: C	24: C	34: E	44: C	54: C
5: D	15: E	25: E	35: A	45: E	55: E
6: A	16: D	26: D	36: D	46: A	56: B
7: C	17: C	27: B	37: D	47: B	57: B
8: B	18: E	28: A	38: C	48: A	58: C
9: A	19: D	29: C	39: E	49: B	
10: E	20: A	30: A	40: C	50: A	

Chapter 4.14: Animal Behavior

1. Which of the following behaviors must be learned from another animal?

   A. potato washing in macaques

   B. suckling of newborn mammals

   C. web building in spiders

   D. nest building in birds

   E. location identity in pigeons

2. A cat hears a rustling in the leaves that sounds like a mouse digging for food. The cat crouches, becoming very still and focused on the leaves as it waits for the right moment to pounce. This combination of movements is an example of:

   A. courtship

   B. stimulus

   C. imprinting

   D. circadian rhythm

   E. behavior

3. For a behavior to evolve under the influence of natural selection, that behavior must be:

   A. acquired through learning

   B. influenced by genes

   C. related to predator avoidance

   D. neither adaptive nor harmful

   E. independent of reproductive success

4. If a dog that barks when indoors is always let outside immediately, it will learn to bark whenever it wants to go outside. This change in the dog's behavior is an example of:

   A. imprinting

   B. insight learning

   C. operant conditioning

   D. classical conditioning

   E. acquired behavior

5. A behavior is innate, rather than learned, when:

   A. all individuals perform the behavior the same way each time

   B. individuals become better at performing the behavior the more they practice it

   C. some individuals perform the behavior and some do not

   D. the behavior is different in individuals that have had different experiences

   E. different individuals perform the behavior at different stages of development

6. The terms "inborn behavior" and "instinct" have the same meaning as:

   A. courtship behavior

   B. imprinting

   C. learned behavior

   D. innate behavior

   E. operant conditioning

7. When disturbed, certain moths lift their front wings to expose eyelike markings on their hind wings. This behavior would be most effective against predators that hunt by:

   A. touch   **B.** sight   **C.** smell   **D.** sound   **E.** innate predation

8. The type of learning that occurs when a stimulus produces a particular response, because it is associated with a positive or negative experience, is:

A. trial-and-error learning

B. habituation

C. operant conditioning

D. classical conditioning

E. imprinting

9. Certain behaviors are innate in animals because they are essential for:

A. acquiring other behaviors

B. developing a circadian rhythm

C. survival immediately after birth

D. guarding territory

E. maintenance of homeostasis

10. Imprinting is a form of behavior that:

A. requires practice for the animal to become good at the activity

B. always involves the sense of sight

C. is restricted to birds

D. is often used in the training of adult animals

E. occurs during a specific time in young animals

11. When people first move into an apartment near railroad tracks, they are awakened at night each time they hear a train go by. Which of the following responses to the sound of the train would be the result of habituation?

A. People begin to sleep through the sound of the train over the next few nights

B. People learn that they can cover the sound of the train by sleeping with the radio on

C. People associate the sound of the train with the arrival of the morning newspaper

D. People learn they can sleep between the times that the train travels by their home

E. People learn they will be given a reward if they wake up when the train goes by

12. Aquarium fish often swim to the water's surface when a person approaches. Their behavior has probably formed through:

A. imprinting

B. insight learning

C. instinct

D. classical conditioning

E. habituation

13. What are specific chemical signals that insects use that affect the behavior or development of other individuals of the same species?

A. cilia

B. spiracles

C. pheromones

D. peptidoglycan

E. glycolipids

14. The ability of salmon to recognize their home stream at spawning time is an example of:

A. communication   **C.** competition
B. imprinting   **D.** insight learning   **E.** classical conditioning

15. Learning occurs whenever:

A. a stimulus has no effect on an animal the first time the animal encounters the stimulus
B. an animal leaves a chemical scent on its territory
C. an animal ignores the stimulus
D. an animal performs a task perfectly without prior experience
E. a stimulus causes an animal to change its behavior

16. The appearance of fireflies at dusk is an example of a circadian rhythm because it:

A. happens daily   **C.** is related to the phase of the moon
B. happens seasonally   **D.** is related to the temperature of the air
E. is related to the organism's reproductive cycle

17. During migration, animals:

A. search for new permanent habitats   **C.** repeat their daily cycle of behavior
B. enter a sleeplike state   **D.** conduct seasonal movement
E. change types of foods consumed

18. The theory that helping family members survive increases the chances that some of one's genes will be passed along to offspring is:

A. competition   **C.** kin selection
B. classical conditioning   **D.** imprinting   **E.** operant conditioning

19. In winter, bears settle into dens and enter a sleeplike state that lasts until spring. This state is:

A. imprinting   **C.** new permanent habitats
B. aggression   **D.** migration   **E.** hibernation

20. When a bird learns to press a button to get food, it has learned by:

A. insight learning   **C.** classical conditioning
B. operant conditioning   **D.** habituation   **E.** imprinting

21. To survive during winter, when resources are scarce, some animals enter a sleeplike state called:

A. reproductive rhythm   **C.** ritual
B. territoriality   **D.** circadian rhythm   **E.** dormancy

22. In some species of balloon flies, males spin balloons of silk and carry them while flying. If a female approaches one of the males and accepts his balloon, the two will fly off to mate. This type of behavior is an example of:

A. courtship   **C.** aggression

B. language   **D.** territorial defense   **E.** operant conditioning

23. Members of a society:

A. act independently for each individual's benefit

B. are usually unrelated to one another

C. belong to at least two species

D. belong to the same species and interact closely with each other

E. develop no direct relations with other members

24. Animals that use language are most likely those that have the greatest capacity for:

A. habituation   **C.** behavioral cycles

B. insight learning   **D.** innate behavior   **E.** imprinting

25. An animal can benefit the most by defending a territory if:

A. the animals it defends against do not use the same resources

B. there are more than enough resources in that territory for all competitors

C. that territory has more resources than surrounding areas

D. that territory has many predators

E. the territory is a temporary location for the members

26. An innate behavior:

A. requires reasoning

B. occurs with trial and error

C. requires habituation

D. appears in fully functional form the first time it is performed

E. improves in function with each additional practice

27. It is advantageous for grazing mammals to gather in groups because groups:

A. have to travel less distance to locate food compared to individuals

B. are more difficult for predators to locate than are individuals

C. can migrate more easily than individuals can

D. can make the available food resources last longer

E. offer greater protection from predation

28. Nocturnal animals that have a poorly developed sense of smell are most likely to communicate by:

A. pheromones   **C.** auditory signals

B. chemical signals   **D.** visual displays   **E.** olfactory signals

29. Many cat species mark their territory by rubbing glands on their faces against surfaces such as tree trunks. This form of communication relies on:

A. chemical messenger   **C.** visual signal

B. defensive display   **D.** sound signal   **E.** predatory behavior

30. When an animal associates a stimulus with a reward or a punishment, it has learned by:

A. imprinting   **C.** habituation

B. operant conditioning **D.** classical conditioning   **E.** insight

31. Cephalopods (i.e. squid and octopi) can communicate by changing skin colors and sometimes even skin patterns. This type of communication is an example of:

A. visual signal   **C.** sound signal

B. language   **D.** chemical signal   **E.** touch signal

32. Dolphins communicate with one another mainly through:

A. chemical signals   **C.** auditory signals

B. pheromones   **D.** visual displays   **E.** touch signals

Answer Key

1: A	11: A	21: E	31: A
2: E	12: D	22: A	32: C
3: B	13: C	23: D	
4: C	14: B	24: B	
5: A	15: E	25: C	
6: D	16: A	26: D	
7: B	17: D	27: E	
8: D	18: C	28: C	
9: C	19: E	29: A	
10: E	20: B	30: B	

UNIT 5. EVOLUTION AND DIVERSITY

Chapter 5.1: Evolution, Natural Selection, Classification, Diversity

1. During his trip to the Galapagos Islands, Darwin discovered that:

    A. local fossils were not related to contemporary organisms of that region
    B. all organisms were related
    C. organisms in tropical regions were closely related independent of the region
    D. several species of finches varied from island to island
    E. organisms had no relationship to others

2. The *Canis lupus* is a wolf that is in the Canidae family. Which of the following statements is accurate about its members?

    A. They may be classified as *lupus* but not *Canis*
    B. They may be classified as *Canis* but not Canidae
    C. More are classified as *lupus* than *Canis*
    D. More are classified as Canidae than *Canis*
    E. None of the above

3. In evolutionary terms, organisms belonging to which category would be the most similar?

    A. genus   **B.** family   **C.** order   **D.** kingdom   **E.** class

4. What is the original source of genetic variation that serves as the raw material for natural selection?

    A. random fertilization   **C.** gene flow
    B. sexual recombination   **D.** random mating   **E.** mutation

5. In modern evolutionary theory, chloroplasts probably descended from:

    A. free-living cyanobacteria   **C.** mitochondria
    B. red algae   **D.** aerobic prokaryote   **E.** eukaryotic cells

6. Speciation is the evolutionary creation of new, genetically distinct populations from a common ancestral stock due to:

    I. random mutation   II. geographic isolation   III. reduction of gene flow

    A. I and III only   **C.** I and II only
    B. II and III only   **D.** I, II, and III   **E.** III only

7. Darwin hypothesized that the mechanism of evolution involves:

A. selective pressure   **C.** natural selection

B. epigenetics   **D.** selective breeding   **E.** random selection

8. Two species of the same order must be members of the same:

A. class   **B.** habitat   **C.** genus   **D.** family   **E.** subfamily

9. The complexity of the organisms that exist on earth today is the result of:

I. multicellularity   II. photosynthesis   III. eukaryotic cell development

A. I only   **C.** I and II only

B. II only   **D.** II and III only   **E.** I, II and III

10. The Archaean Eon contains the oldest known fossil record dating to about:

A. 1.5 billion years   **C.** 7,000 years

B. 2.0 million years   **D.** 3.5 billion years   **E.** 7 billion years

11. Urey-Miller experiment demonstrates that:

A. life may have evolved from inorganic precursors

B. humans have evolved from photosynthetic cyanobacteria

C. life existed on prehistoric earth

D. small biological molecules cannot be synthesized from inorganic material

E. the early earth lacked oxygen

12. To ensure the survival of their species, animals that do not care for their young:

A. have protective coloring   **C.** lay eggs

B. produce many offspring   **D.** have the ability to live in water and on land

    E. have internal fertilization

13. Homology serves as the evidence of:

A. balancing selection   **C.** genetic mutation

B. biodiversity   **D.** sexual selection

    E. common ancestry

14. Which statement is correct about orthologous genes?

A. They are related through speciation   **C.** They are repetitive

B. They are house-keeping genes   **D.** They are related via gene duplication within species

    E. They are viral oncogenes

15. Which of the following best describes the relationship between the nitrogen-fixing bacteria that derive their nutrition from plants and the plants which in turn benefit from the nitrogen supplied by the bacteria?

    A. parasitism   **C.** mutualism

    B. commensalism   **D.** mimicry   **E.** amensalism

16. Which given pair represents homologous structures?

    A. Mouth of a fly and beak of a hummingbird

    B. Wings of a dragonfly and of a blue jay

    C. Forelimb of a human and of a dog

    D. Wings of a pigeon and of a bat

    E. Forelimb of a raccoon and the hind limb of a wolf

17. Humans belong to the order:

    A. hominidae   **B.** primate   **C.** chordata   **D.** vertebrata   **E.** sapiens

18. Which of the following selection types is most likely to lead to speciation?

    A. sexual selection   **C.** directional selection

    B. stabilizing selection   **D.** disruptive selection

                                 E. nondirectional selection

19. The similarity among the embryos of fish, amphibians, reptiles and humans is evidence of:

    A. genetic drift   **C.** analogous traits

    B. sexual selection   **D.** common ancestry   **E.** genetic equilibrium

20. Selective breeding of soybeans by humans has genetically altered the soybean, so that it could not survive in the wild without human intervention. Soybean population is controlled, and most of the soybean seeds are eaten or become spoiled. The relationship between humans and soybeans is best described as:

    A. commensalism, because both species benefit

    B. parasitism, because humans benefit and soybeans are harmed

    C. commensalism, because there is no benefit to either species

    D. commensalism, because humans benefit and soybeans are neither benefited nor harmed

    E. mutualism, because both species benefit

21. In a climax community, which of these would be a dominant species?

    A. shrubs   **C.** mosses

    B. annual grasses   **D.** deciduous trees   **E.** evergreens

22. Natural selection leads to:

   A. larger population

   B. population most adapted to its present environment

   C. phenotypic diversity

   D. broad genetic variation

   E. population well adapted to changes in the environment

23. From an evolutionary perspective, which property of a cell is of paramount importance?

   A. passing genetic information to progeny   **C.** containing mitochondria

   B. containing a nucleus   **D.** interacting with other cells

                                    E. extracting energy from the environment

24. Inherited traits determined by elements of heredity are:

   A. crossing over   **C.** locus

   B. homologous chromosomes   **D.** p elements   **E.** genes

25. Which statement is correct about the evolutionary process?

   A. Organisms develop traits that they need for survival

   B. Mutations decrease the rate of evolution

   C. Natural selection works on traits that cannot be inherited

   D. Natural selection works on existing genetic variation within a population

   E. Organisms evolve due to natural selection

26. All of the following are members of the phylum Chordata EXCEPT:

   A. birds   **B.** ants   **C.** tunicates   **D.** apes   **E.** snakes

27. Genetic drift results from:

   A. environmental change   **C.** probability

   B. genetic diversity   **D.** sexual selection   **E.** genetic mutation

28. All of these statements regarding the phylum Echinodermata are true EXCEPT:

   A. echinoderms are heterotrophs   **C.** echinoderms reproduce sexually

   B. echinoderms are invertebrates   **D.** phylum includes starfish and sea urchins

                                    E. phylum includes crayfish

29. The likely result of polygenic inheritance is:

   A. human height   **C.** freckles

   B. blood type   **D.** polydactyly (extra digits)   **E.** color of the iris

30. Which of the following is the earliest form of life to evolve on Earth?

A. plants   **C.** prokaryotes

B. eukaryotes   **D.** protists   **E.** fish

31. If two animals produce viable fertile offspring under natural conditions, it can be concluded that:

A. for any given allele, they both have the same gene

B. their blood types are compatible

C. they both have haploid somatic cells

D. they both are from the same species

E. none of the above

32. The population's total collection of alleles at any one time makes up its:

A. phenotype   **C.** gene pool

B. biodiversity   **D.** genotype   **E.** genetic mutations

33. A key point in Darwin's explanation of evolution is that:

A. biological structures most likely inherited are those that have become better suited to the environment through constant use

B. mutations that occur are those that help future generations fit into their environments

C. mutations develop based on use/disuse of physical traits

D. genes change in order to help organisms cope with problems encountered within their environments

E. any trait that confers even a small increase in the probability that its possessor will survive and reproduce will be strongly favored and will spread through the population

34. If one allele's frequency in a population is 0.7, what is the frequency of the alternate allele?

A. 0.14   **B.** 0.21   **C.** 0.30   **D.** 0.4   **E.** 0.42

35. Two species of the same phylum must be members of the same:

A. order   **B.** kingdom   **C.** genus   **D.** class   **E.** family

36. Which statement about evolution is CORRECT?

A. Darwin's theory of natural selection relies solely on environmental conditions

B. Darwin's theory explains the evolution of man from present day apes

C. Darwin's theory of natural selection relies solely on genetic mutation

D. Lamarck's theory of use and disuse adequately describes why giraffes have long necks

E. Natural selection is when random mutations are selected for survival by the environment

37. A significant genetic drift may be prevented by:

A. random mutations

B. large population size

C. small population size

D. genetic variation

E. lack of migration

38. *Homo sapiens* belong to the phylum:

A. Vertebrata   **B.** Homo   **C.** Mammalia   **D.** Chordata   **E.** sapiens

39. Which of these organisms is a chordate, but not a vertebrate?

A. lizard   **B.** lancelet   **C.** shark   **D.** lamprey eel   **E.** None of the above

40. Which of the following statements would apply to a population that survived a bottleneck and recovered to its original size?

A. It is subject to genetic drift

B. It is less likely to get extinct than before the bottleneck

C. It has more genetic variation than before the bottleneck

D. It has less genetic variation than before the bottleneck

E. None of the above applies

41. Which of the following represents a correct ordering of the levels of complexity at which life is studied, from most simple to most complex?

A. cell, tissue, organ, organism, population, community

B. community, population, organism, organ, tissue, cell

C. cell, organ, tissue, organism, population, community

D. cell, tissue, organ, population, organism, community

E. tissue, organ, cell, population, organism, community

42. The difference between the founder effect and a population bottleneck is that the founder effect:

A. only occurs in island populations

B. requires the isolation of a small group from a larger population

C. requires a large gene pool

D. involves sexual selection

E. always follows genetic drift

43. Which compound was likely unnecessary for the origin of life on Earth?

A. carbon   **B.** O_2 **C.** hydrogen   **D.** H_2O **E.** nitrogen

44. Which of the following statements is TRUE for the notochord?

    A. It is present in chordates during embryological development
    B. It is always a vestigial organ in chordates
    C. It is present in all adult chordates
    D. It is present in all echinoderms
    E. It is part of the nervous system of all vertebrates

45. In a particular population of humans, there is a higher rate of polydactyly (extra fingers or toes) than in the human population as a whole. The most likely explanation is:

    A. bottleneck effect   **C.** sexual selection
    B. founder effect   **D.** natural selection   **E.** missense mutation

46. All of the following is present in all members of the phylum Chordata at some point of their life cycle, EXCEPT:

    A. pharyngeal slits   **C.** notochord
    B. tail   **D.** dorsal neural tube   **E.** backbone

47. Which of the following is an example of mutualism?

    A. nematodes   **B.** bread mold   **C.** tapeworms   **D.** lichens   **E.** epiphytes

48. Gene flow occurs through:

    A. random mutations   **C.** founder effect
    B. bottleneck effect   **D.** directional selection   **E.** migration

49. The first living organisms on Earth derived their energy from:

    A. eating dead organisms   **C.** the sun
    B. eating organic molecules   **D.** eating each other   **E.** all of the above

50. Which of the following statements is TRUE for a climax community?

    A. It is relatively stable within a given climate
    B. It is independent of the environment
    C. It consists of only one species of life
    D. It is populated mainly by so-called pioneer organisms
    E. It consists of only dead and decaying organic matter

51. Evolutionary fitness measures:

    A. population size   **C.** gene pool size
    B. reproductive success   **D.** genetic load   **E.** migration rate

52. Which of the following are most likely two organisms of the same species?

   A. Two Central American iguanas which mate in different seasons

   B. Two migratory birds nesting on different Hawaiian Islands

   C. Turnips in South Carolina and Texas that mate and produce fertile progeny only in seasons of extreme climate conditions

   D. Two fruit flies on the island of Bali with distinct courtship behaviors

   E. All of the above

53. Which of these scientists would explain the hawk's lost flying ability with a theory that "Since the hawk stopped using its wings, the wings became smaller, and this acquired trait was passed on to the offspring"?

   A. Lamarck   **B.** de Vries   **C.** Mendel   **D.** Darwin   **E.** Morgan

54. Which example is a likely result of directional selection?

   A. Increased number of cat breeds

   B. Python snakes with different color patterns exhibit different behavior when threatened

   C. Female *Drosophila* flies selecting normal yellowish-grey colored male mates over less common yellow colored males

   D. Non-poisonous butterflies evolving color changes that make them look like poisonous butterflies

   E. None of the above

55. A phenomenon where newly hatched ducklings separated from their mother, follow a human swimming in a pond is an example of:

   A. imprinting   **C.** discrimination

   B. instrumental conditioning   **D.** response to pheromones

                                      E. none of the above

56. Genetic variation would be most likely decreased by:

   A. genetic drift   **C.** directional selection

   B. balancing selection   **D.** sexual selection   **E.** stabilizing selection

57. Asexually reproducing species have a selective disadvantage to sexually reproducing species because sexual reproduction:

   A. always decreases an offspring's survival ability

   B. decreases the likelihood of mutations

   C. creates novel genetic recombination

   D. is more energy efficient

   E. always increases an offspring's survival ability

58. Which example is a likely result of stabilizing selection?

A. Female *Drosophila* flies selecting normal yellowish-grey colored male mates over less common yellow colored males

B. Increased number of different breeds within one species

C. Non-poisonous butterflies evolving color changes that make them look like poisonous butterflies

D. Female birds that lay intermediate number of eggs have the highest reproductive success

E. None of the above

59. According to one theory about origins of life, these molecules were formed by purines, pyrimidines, sugars and phosphates combining together:

A. lipids   **B.** carbohydrates   **C.** nucleosides   **D.** proteins   **E.** nucleotides

60. Which example is a likely result of sexual selection?

A. Female deer choosing to mate with males that have the biggest antlers

B. Python snakes with different color patterns exhibit different behavior when threatened

C. Non-poisonous butterflies evolving color changes that make them look like poisonous butterflies

D. Increased number of different breeds within one species

E. Lions experience a founder effect

61. What is one reason why sponges are classified as animals?

A. Sponges form partnerships with photosynthetic organisms

B. Sponges are autotrophic

C. Their cells have cell walls

D. Sponges are heterotrophic

E. Sponges are omnivores

62. Organisms that must obtain nutrients and energy by eating other organisms are:

A. heterotrophic   **C.** herbivores

B. eukaryotic   **D.** autotrophic   **E.** multicellular

63. What is a protostome?

A. outermost germ layer

B. animal whose mouth is formed from the blastopore

C. animal whose anus is formed from the blastopore

D. embryo just after fertilization

E. innermost germ layer

64. Arthropods have:

A. spiny skin and an internal skeleton
B. flattened body with an external shell
C. segmented body and an exoskeleton
D. soft body and an internal shell
E. segmented body and an endoskeleton

65. Some of the fossils that have been found from before the Cambrian Explosion are embryos, which suggests that the organisms these fossils belong to:

A. reproduce sexually
B. exhibit cephalization
C. reproduce asexually
D. have bilateral symmetry
E. protostomes

66. The Cambrian Explosion resulted in the evolution of the first:

A. bacteria
B. land animals
C. dinosaurs and mammals
D. representatives of most animal phyla
E. plants

67. A medical student looks at a slice of tissue on an unlabeled microscope slide. The student concludes the tissue is not from an animal, because the cells in the tissue have:

A. flagella
B. cell membranes
C. nuclei
D. membrane-bound organelles
E. cell walls

68. Complex animals break down food through the process of:

A. cephalization
B. intracellular digestion
C. complete metamorphosis
D. extracellular digestion
E. all of the above

69. Which of the following is true about annelids?

A. Each body segment contains several pairs of antennae
B. Annelids rely on diffusion to transport oxygen and nutrients to their tissues
C. Annelids have segmented bodies and a true coelom
D. Annelids are acoelomates
E. Annelids have segmented bodies and a pseudocoelom

70. Ancient chordates are thought to be most closely related to:

A. octopi
B. sea anemones
C. spiders
D. earthworms
E. none of the above

71. Which of the following is true about echinoderms?

    A. Echinoderms have one pair of antennae and unbranched appendages

    B. Adult echinoderms have an exoskeleton

    C. Most adult echinoderms exhibit radial symmetry

    D. Echinoderm body is divided into a head, a thorax, and an abdomen

    E. Most adult echinoderms lack radial symmetry

72. The simplest animals to have body symmetry are:

    A. cnidarians   **C.** sponges

    B. echinoderms   **D.** algae   **E.** bacteria

73. In chordates whose pharyngeal pouches develop slits that lead outside of the body, for breathing the adult most likely uses:

    A. pharynx   **B.** nose   **C.** gills   **D.** lungs   **E.** larynx

74. In a chordate embryo, nerves branch in intervals from:

    A. spinal column   **C.** pharyngeal pouches

    B. notochord   **D.** tail   **E.** hollow nerve cord

75. Each of the following is one of the essential functions animals carry out, EXCEPT:

    A. circulation   **C.** excretion

    B. cephalization   **D.** respiration   **E.** all are required

76. Which of the following animals are deuterostomes?

    A. mollusks and arthropods   **C.** annelids and arthropods

    B. arthropods and chordates   **D.** cnidarians and mollusks

                                  E. echinoderms and chordates

77. In fishes, pharyngeal pouches may develop into:

    A. gills   **B.** tails   **C.** lungs   **D.** fins   **E.** larynx

78. Which of the following is true about corals?

    A. The polyp form of corals is restricted to a small larval stage

    B. Corals only reproduce by asexual means

    C. Corals have only the medusa stage in their life cycles

    D. Coral polyps secrete an underlying skeleton of calcium carbonate

    E. One coral polyp forms a balloon-like float that keeps the entire colony afloat

79. Amphibians evolved from:

A. jawless fishes   **C.** ray-finned fishes
B. lobe-finned fishes   **D.** cartilaginous fishes   **E.** none of the above

80. The notochord is responsible for which function in an embryo?

A. respiration   **C.** processing wastes
B. processing nerve signals **D.** cognitive abilities   **E.** structural support

81. A flexible, supporting structure found only in chordates is the:

A. pharyngeal slits   **C.** nerve net
B. dorsal fin   **D.** notochord   **E.** vertebrates

82. Which of the following is NOT true about earthworms?

A. Earthworms are hermaphrodites that reproduce sexually
B. Earthworms have a digestive tract that includes a mouth and an anus
C. Earthworms have a pseudocoelom
D. Their tunnels allow for the growth of oxygen-requiring soil bacteria
E. Earthworms feed on live and dead organic matter

83. Skeletons of early vertebrates were composed of cartilage instead of bone. Which characteristic does cartilage share with notochords?

A. Oxygen can diffuse through it   **C.** It is soft and flexible
B. It contracts   **D.** It is hard and rigid
                                         E. It contains hydroxylapatite

84. Which of the following do NOT exhibit bilateral symmetry?

A. arthropods   **C.** mollusks
B. cnidarians   **D.** annelids   **E.** primates

85. Which chordate characteristic is visible on the outside of an adult cat?

A. a tail that extends beyond the anus   **C.** hollow nerve cord
B. pharyngeal pouches   **D.** notochord
                                               E. nerve plexus

86. Fewer than 5 percent of all animal species have:

A. cell membranes   **C.** a protostome development pattern
B. vertebral columns   **D.** eukaryotic cells
                            E. membrane bound organelles

87. The eggs of ovoviviparous fish are:

    A. released from the female before they are fertilized

    B. nourished by a structure similar to a placenta

    C. released from the female immediately after being fertilized

    D. retained and nourished by the female

    E. held inside the female's body as they develop

88. Which of the following statements about chordates is true?

    A. All chordates are vertebrates

    B. All chordates have paired appendages

    C. All chordates have a notochord

    D. All chordates have backbones

    E. Chordates include protostomes and deuterostomes

89. Which of the following pairs of modern chordate groups are most closely related?

    A. birds and crocodilians

    B. sharks and the coelacanth

    C. hagfishes and lungfishes

    D. lampreys and ray-finned fishes

    E. crocodilians and sharks

90. The hominoid group of primates includes:

    A. anthropoids only

    B. apes and hominines

    C. New World monkeys and Old World monkeys

    D. lemurs, lorises and anthropoids

    E. apes and anthropoids

91. What does fossil evidence indicate about the order in which these three vertebrates evolved: a bony fish with a jaw, a jawless fish, and a fish with leglike fins?

    A. The jawless fish was the last to evolve

    B. The fish with leglike fins evolved before the jawless fish

    C. The bony fish evolved before the jawless fish

    D. The fish with leglike fins was the last to evolve

    E. The bony fish was the last to evolve

92. Jaws and limbs are characteristic of:

    A. reptiles only

    B. invertebrates only

    C. all chordates

    D. worms

    E. vertebrates only

93. Which structure in the amniotic egg provides nutrients for the embryo as it grows?

    A. yolk   **B.** allantois   **C.** chorion   **D.** amnion   **E.** shell

94. A primate's ability to hold objects in its hands or feet is an evolutionary development that was necessary for it to:

A. create elaborate social systems   **C.** use simple tools

B. consume food   **D.** walk upright   **E.** escape predation

95. Which of the following trends in reproduction is evident as you follow the vertebrate groups?

A. progressively fewer openings in the digestive tract

B. closed circulatory system becomes open

C. internal fertilization evolves into external fertilization

D. endothermy evolves into ectothermy

E. external fertilization evolves into internal fertilization

96. A tail that is adapted for grasping and holding objects is:

A. prehensile   **C.** bipedal

B. radial   **D.** binocular   **E.** symmetrical

97. Suppose a paleontologist discovers a fossil skull that he believes might be distantly related to primates. Unlike true primates, the face is not quite flat and the eyes do not face completely forward. The paleontologist would most likely conclude that the animal lacked the ability to:

A. manipulate tools   **C.** form extended family groups

B. judge the location of tree branches   **D.** grip branches precisely

     E. harvest crops

98. Which of the following is NOT a characteristic of mammals?

A. hair   **C.** ability to nourish their young with milk

B. endothermy   **D.** lack of pharyngeal pouches during development

     E. three middle ear bones

99. The group that includes gibbons and humans but does not include tarsiers is the:

A. mollusks   **C.** primates

B. hominines   **D.** anthropoids   **E.** hominoids

100. When an animal's environment changes, sexual reproduction improves a species' ability to:

A. increase its numbers rapidly   **C.** react to new stimuli

B. produce genetically identical offspring   **D.** adapt to new living conditions

     E. reduce genetic diversity

101. All of the following are examples of cartilaginous fishes, EXCEPT:

    A. rays   **B.** sharks   **C.** hagfishes   **D.** skates   **E.** sturgeons

102. Old World monkeys can be distinguished from New World monkeys by observing:

    A. how the monkeys use their tails   **C.** what the monkeys eat

    B. when the monkeys are most active   **D.** how the monkeys interact with their troop

                                        E. how the monkeys gather food

103. Compared with that of a reptile, a bird's body temperature is:

    A. lower and more constant   **C.** higher and more constant

    B. lower and more variable   **D.** higher and more variable

                                        E. the same

104. Which of the following internal characteristics would you expect an earthworm to have?

    A. prostomium   **C.** pseudocoelom

    B. notochord   **D.** paired organs   **E.** blastopore

105. Mammals that lay eggs are called:

    A. reptiles   **C.** marsupials

    B. placentals   **D.** amphibians   **E.** monotremes

106. The presence of legs or other limbs indicates that the animal is:

    A. protostome   **C.** acoelomate

    B. segmented   **D.** radially symmetrical   **E.** deuterostome

107. A primatologist finds a new species of primate in a Madagascar forest. The primate has a long snout and is active at night. The primatologist would likely classify this primate as a:

    A. bush baby   **C.** anthropoid

    B. hominoid   **D.** lemur   **E.** none of the above

108. The vertebrate group with the most complex respiratory system is:

    A. fish   **B.** amphibians   **C.** mammals   **D.** reptiles   **E.** birds

109. Having a thumb that can move against the other fingers makes it possible for a primate to:

    A. judge the locations of tree branches   **C.** merge visual images

    B. display elaborate social behaviors   **D.** hold objects firmly

                                          E. develop elaborate social communications

110. Which of the following is NOT a characteristic of all chordates?

A. pharyngeal pouches

B. vertebrae

C. tail that extends past the anus

D. dorsal, hollow nerve cord

E. none of the above

111. Researchers have concluded that the Dikika Baby was most likely a better climber than modern humans. This evidence may suggest that the Dikika Baby's relatives may have spent part of their time:

A. hunting in groups

B. waging war on other troops

C. using tools

D. in trees

E. in rocky mountains

112. An animal that has body parts that extend outward from its center shows:

A. bilateral symmetry

B. radial symmetry

C. segmentation

D. several planes of symmetry

E. protostome development

113. Fossil evidence indicates that *Australopithecus afarensis:*

A. was bipedal

B. appeared later than *Homo ergaster*

C. was primarily a meat-eater

D. had a large brain

E. was a hunter and gatherer

114. Which of the following is a node that you would expect to find on the cladogram for animals?

A. leg length

B. feather arrangement

C. deuterostome development

D. fur texture

E. shape of forelimbs

115. One way in which reptiles are adapted to a fully terrestrial life is:

A. endothermy

B. viviparous development

C. external fertilization

D. placenta

E. amniotic egg

116. Which of the following variations would you expect to see in land vertebrates?

A. Varying shapes of forelimbs

B. Varying numbers of germ layers

C. Varying numbers of limbs

D. Different types of symmetry

E. Varying anterioposterior axis

117. The earliest hominine that belonged to the same genus as modern humans was probably:

A. *Australopithecus afarensis*

B. *Homo ergaster*

C. *Homo sapiens*

D. *Homo neanderthalensis*

E. *Homo habilis*

118. Researchers concluded from the leg bones of the fossil known as Lucy that it was bipedal. Which of the following would also indicate that this hominine was bipedal?

A. bowl-shaped pelvis

B. opposable thumbs

C. skull with flat face

D. broad rib cage

E. flexible spinal column

119. Which statement is true of *Homo sapiens?*

A. They replaced *Homo habilis* in the Middle East

B. They became extinct about 1 million years ago

C. They have been Earth's only hominine for the last 24,000 years

D. They evolved after the Cro-Magnons

E. They replaced *Homo habilis* in Europe

120. Which of the following was a unique characteristic of the Neanderthals?

A. Producing tools from bones and antlers

B. Burying their dead with simple rituals

C. Making sophisticated stone blades

D. Producing cave paintings

E. Members joined into groups

121. The single most important characteristic that separates birds from other chordates is:

A. amniotic egg

B. feathers

C. wings

D. four-chambered heart

E. none of the above

122. A double-loop circulatory system is present in all animals with:

A. two-chambered heart

B. lungs

C. jaws

D. vertebrae

E. bilateral symmetry

123. Which of the following statements about the cladogram of animals is true?

A. Segmentation evolved more than once in different branches of the cladogram

B. All deuterostomes have cephalization

C. Backbones evolved before segmentation

D. Radial symmetry appears only once in the cladogram

E. All deuterostomes have bilateral symmetry

Answer Key

1: D	21: D	41: A	61: D	81: D	101: C	121: B
2: D	22: B	42: B	62: A	82: C	102: A	122: B
3: A	23: A	43: B	63: B	83: C	103: C	123: A
4: E	24: E	44: A	64: C	84: B	104: D	
5: A	25: D	45: B	65: A	85: A	105: E	
6: D	26: B	46: E	66: D	86: B	106: B	
7: C	27: C	47: D	67: E	87: E	107: D	
8: A	28: E	48: E	68: D	88: C	108: E	
9: E	29: A	49: C	69: C	89: A	109: D	
10: D	30: C	50: A	70: B	90: B	110: B	
11: A	31: D	51: B	71: C	91: D	111: D	
12: B	32: C	52: B	72: A	92: E	112: B	
13: E	33: E	53: A	73: C	93: A	113: A	
14: A	34: C	54: D	74: E	94: C	114: C	
15: C	35: B	55: A	75: B	95: E	115: E	
16: C	36: E	56: E	76: E	96: A	116: A	
17: B	37: B	57: C	77: A	97: B	117: E	
18: D	38: D	58: D	78: D	98: D	118: A	
19: D	39: B	59: E	79: B	99: E	119: C	
20: E	40: D	60: A	80: E	100: D	120: B	

Explanations

Chapter 1.1: Eukaryotic Cell: Structure and Function

1. D is correct.

2. A is correct.

3. E is correct.

I: water relies on aquaporins to readily diffuse across a plasma membrane. Without aquaporins, only a small fraction of water molecules can diffuse through the cell membrane per unit time because of the polarity of water molecules.

II: small hydrophobic molecules readily diffuse through the hydrophobic tails of the plasma membrane.

III: small ions rely on ion channel transport proteins to diffuse across the membrane.

IV: neutral gas molecules (e.g., O_2, CO_2) readily diffuse through the hydrophobic tails of the plasma membrane.

4. C is correct.

5. E is correct. The lysosome is the digestive region of the cell and is a membrane bound organelle with a low pH (around 5) that stores hydrolytic enzymes.

A: vacuoles and vesicles are membrane bound sacs involved in the transport and storage of materials that are ingested, secreted, processed or digested by cells. Vacuoles are larger than vesicles and are more likely to be found in plant cells (e.g. central vacuole).

C: chloroplasts are the site of photosynthesis and are found only in algae and plant cells. Chloroplasts contain their own DNA and ribosomes and may have evolved via endosymbiosis in a similar manner to the mitochondria.

D: phagosomes are vesicles involved in the transport and storage of materials that are ingested by the cell through phagocytosis. The vesicles are formed by the fusion of the cell membrane around the particle. A phagosome is a cellular compartment in which pathogenic microorganisms can be digested. Phagosomes fuse with lysosomes in their maturation process to form phagolysosomes.

6. C is correct. The DNA damage checkpoint is a signal transduction pathway that blocks cell cycle progression in G1, G2 and metaphase and slows down the rate of S phase progression when DNA is damaged. It leads to a pause in the cell cycle, allowing the cell time to repair the damage before continuing to divide.

Cell cycle progression:

G1 → S → G2 → prophase → metaphase → anaphase → telophase

Interphase   Mitosis (PMAT)

7. A is correct. G-C base pairs are linked in the double helix by three hydrogen bonds. A-T base pairs are joined by only two hydrogen bonds. Therefore, it takes more energy to separate G-C base pairs. The less G-C rich a piece of double-stranded DNA is, the less energy is required to separate (i.e. denature) the two strands of the double helix. Because of complimentary base pairing, double-stranded DNA has equal quantities of G and C (and of A and T). This is known as Chargaff's rule.

8. A is correct. Phospholipids are a class of lipids that are a major component of all cell membranes, as they can form lipid bilayers. Phospholipids contain a glycerol backbone, a phosphate group and a simple organic molecule (e.g. choline). The 'head' is hydrophilic (attracted to water), while the 'tails' are hydrophobic (repelled by water) and the tails are forced to aggregate (via hydrophobic forces). The hydrophilic head contains the negatively charged phosphate group and glycerol.

B and C: the hydrophobic tail usually consists of 2 long fatty acid (saturated or unsaturated) hydrocarbon chains.

D: cholesterol is imbedded within the lipid bilayer in animals, but is absent in plant cell membranes.

9. B is correct.

10. B is correct. Osmolarity is determined by the total concentration of dissolved particles in solution. Compounds that dissociate into ions increase the concentration of particles and produce a higher osmolarity.

To determine which molecule (after dissociation into ions) generates the highest osmolarity, determine the number of individual ions each molecule dissociates into when dissolved in H_2O. $CaCl_2$ dissociates into 1 Ca^{2+} and 2 Cl^-. As a result, $CaCl_2$ generates the greatest osmolarity which equals 250 mOsmoles for Ca^{2+} + 500 mOsmoles for Cl^- = 750 mOsmoles when $CaCl_2$ dissociates into ions within the solution.

A: NaCl dissociates when placed into water and has an osmolarity of 600 mOsmoles because it dissociates into 1 Na^+ and 1 Cl^-. The Na^+ cations and the Cl^- anions are added together: 300 mOsmoles for Na^+ + 300 mOsmoles for Cl^- = 600 mOsmoles.

C: glucose does not dissociate when placed into water and has the same osmolarity as the starting molecule – 500 mOsmoles.

D: urea does not dissociate when placed into water and has the same osmolarity as the starting molecule – 600 mOsmoles.

11. C is correct. The two ribosomal subunits are produced in the nucleolus, a region within the nucleus. The ribosomes are the sites of protein production. Prokaryotic ribosomes (30S small + 50S large subunit = 70S complete ribosome) are smaller than eukaryotic ribosomes (40S small + 60S large subunit = 80S complete ribosome).

A: Golgi apparatus is a membrane-bound organelle that modifies (e.g. glycosylation), sorts and packages proteins synthesized by the ribosomes.

B: lysosomes have a low pH of about 5 and contain hydrolytic enzymes involved in digestion.

D: rough endoplasmic reticulum is a portion of the endomembrane system that extends from the nuclear envelope. It has ribosomes associated with its membrane and is the site of production and folding of proteins. Misfolded proteins exit the rough ER and are sent to the proteosome for degradation.

E: cell membrane is a barrier between the interior and exterior of the cell.

12. E is correct. Cyclins are phosphorylated proteins responsible for specific events during cycle division, such as microtubule formation and chromatin remodeling. Cyclins can be divided into four classes based on their behavior in the cell cycle: G1/S cyclins, S cyclins, M cyclins and G1 cyclins.

A: p53 is not a transcription factor but is a tumor suppressor gene. The p53 protein is crucial in multicellular organisms, where it regulates the cell cycle and functions as a tumor suppressor (preventing cancer). p53 is described as *the guardian of the genome* because of its role in conserving stability by preventing genome mutation.

13. B is correct. A codon is a three-nucleotide segment of an mRNA molecule that hybridizes (via complimentary base pairing) with the appropriate anticodon on the tRNA to encode for one amino acid in a polypeptide chain during protein synthesis. The tRNA molecule only interacts with the mRNA codon after the ribosomal complex binds to the mRNA.

A: tRNA molecule only interacts with the mRNA codon after (not before) the mRNA is bound to the ribosomal complex.

C: translation involves the conversion of mRNA into protein.

D: operon regulates the transcription of genes into mRNA and is not involved in translation of mRNA into proteins.

14. C is correct. Peroxisomes are organelles found in most eukaryotic cells. A major function of the peroxisome is the breakdown of very long chain fatty acids through beta-oxidation. In animal cells, the peroxisome converts the very long fatty acids to medium chain fatty acids, which are subsequently shuttled to the mitochondria, where they are eventually broken down, via oxidation, into CO_2 and H_2O.

15. E is correct. A vacuole is a membrane-bound organelle which is present in all plant and fungal cells and some protist, animal and bacterial cells. Vacuoles are enclosed compartments filled with water. They contain inorganic and organic molecules (including enzymes in solution) and may contain solids that have been engulfed. The function and significance of vacuoles varies greatly according to the type of cell in which they are present, with a much greater prominence in the cells of plants, fungi and certain protists than those in animals or bacteria.

16. C is correct. I-cell disease patients cannot properly direct newly synthesized peptides to their target organelles. The Golgi apparatus is part of the endomembrane system and serves as a cellular distribution center. The Golgi apparatus packages proteins inside the cell before they are sent to their destination and is particularly important in the processing of proteins for secretion. The Golgi apparatus is integral in modifying, sorting and packaging these macromolecules for cell secretion (i.e. exocytosis out of the cell) or use within the cell.

A: the nucleus is the largest cellular organelle that contains genetic material (i.e. DNA) and is the site of rRNA synthesis.

D: smooth ER (endoplasmic reticulum) is part of the endomembrane system, is connected to the nuclear envelope and functions in several metabolic processes. It synthesizes lipids, phospholipids and steroids. Cells that secrete these products (e.g. testes, ovaries and skin oil glands) have an extensive smooth endoplasmic reticulum. The smooth ER carries out the metabolism of carbohydrates and drug detoxification It is also responsible for attachment of receptors on cell membrane proteins and steroid metabolism.

17. D is correct.

B: the enzyme cytochrome c oxidase is a large transmembrane protein complex found in bacteria and the mitochondrion of eukaryotes. It is the last enzyme in the electron transport chain of mitochondria (or bacteria) located in the mitochondrial (or bacterial) membrane. It receives an electron from each of four cytochrome c molecules and transfers the electrons to an O_2 molecule, converting molecular oxygen to two molecules of H_2O.

18. A is correct. The cytoskeleton is integral in proper cell division because it forms the mitotic spindle and is responsible for separating sister chromatids during cell division. The cytoskeleton is composed of microtubules and microfilaments, and it gives the cell mechanical support, maintains its shape and functions in cell motility.

19. B is correct.

20. A is correct. Mitochondria divide autonomously to produce daughter mitochondria that incorporate some new nonradioactive phosphatidylcholine and also inherit radioactive phosphatidylcholine from the parent via semiconservative replication. Therefore, the daughter mitochondria have equal radioactivity.

B: mitochondria divide autonomously, and the daughter mitochondria retain the parental (original) radioactive label.

C: Mendelian inheritance from a cross between heterozygotes: Aa x Aa = AA, Aa, Aa and aa.

D: this choice requires the daughter mitochondria to be synthesized *de novo* (i.e. new) with newly-synthesized, nonradioactive phosphatidylcholine. If the mitochondria divide autonomously, the daughter mitochondria retain the radioactive label evenly via semiconservative replication.

21. C is correct.

22. B is correct.

23. E is correct. Ribosomes are composed of specific rRNA molecules and associated proteins. The ribosome is identified by the sedimentation coefficients (i.e. S units for Svedberg units). Prokaryotes have a 30S small subunit and a 50S large subunit (i.e. complete ribosome = 70S; based on density). Eukaryotes have a 40S small subunit and 60S large subunit (i.e. complete ribosome = 80S).

A: peroxisomes are organelles involved in hydrogen peroxide (H_2O_2) synthesis and degradation. They function in detoxification of the cell and contain the enzyme catalase that decomposes H_2O_2 into H_2O and O_2.

C: mitochondria are organelles that are the site of cellular respiration (i.e. oxidation of glucose to yield ATP) and are plentiful in cells that have high demands for ATP (e.g. muscle cells). The number of mitochondria within a cell varies widely by organism and tissue type. Many cells have only a single mitochondrion, whereas others can contain several thousand mitochondria.

24. A is correct. *Osmosis* is a special type of diffusion involving water and is a form of passive transport.

Hypertonic means high solute and low solvent concentrations.

Hypotonic means high solvent and low solute concentrations.

Solvents always flow spontaneously from an area of high solvent to an area of low solvent concentration. During osmosis, water flows from a hypotonic to a hypertonic environment.

25. B is correct. The complex of Cdk and cyclin B is called a maturation promoting factor or mitosis promoting factor (MPF). Cyclin B is necessary for the progression of the cells into and out of M phase of the cell cycle.

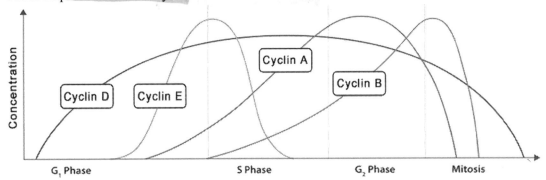

26. E is correct. Mitochondria have their own DNA genetic material and machinery to manufacture their own RNAs and proteins. In the example above, the trait must be recessive (not observed in the limited number of offspring) and is therefore encoded for by a nuclear gene.

A: mice, like all organisms that reproduce sexually, inherit the mitochondrial organelle from their mother and therefore display maternal inheritance of mitochondrial genes.

B: genes are either autosomal (i.e. non sex linked) or sex-linked (e.g. X or Y chromosome).

C: if X-linked, then selectively male (not female) progeny would display the trait.

D: mitochondrial genes cannot be recessive because all mitochondria are inherited from the mother.

27. C is correct.

28. D is correct. Plastids (e.g. chloroplast and chromoplast) are a major organelle found in plants and algae. They are the site of manufacturing and storage of important chemical compounds used by the cell. They often contain pigments used in photosynthesis, and the types of pigments present can change or determine the cell's color. Plastids, like prokaryotes, contain a circular double-stranded DNA molecule.

29. A is correct. The thyroid gland synthesizes calcitonin in response to high blood calcium levels. It acts to reduce blood calcium (Ca^{2+}), opposing the effects of parathyroid hormone (PTH). Calcitonin lowers blood Ca^{2+} levels in three ways: 1) inhibition of Ca^{2+} absorption by the intestines, 2) inhibition of osteoclast activity in bones and 3) inhibition of renal tubular cell reabsorption of Ca^{2+}, allowing it to be excreted in the urine.

B: kidney serves several essential regulatory roles. They are essential in the urinary system and also serve homeostatic functions such as the regulation of electrolytes, maintenance of acid–base balance and regulation of blood pressure (via maintaining salt and water balance). The kidney secretes renin (involved in blood pressure regulation) that induces the release of aldosterone from the adrenal cortex (it increases blood pressure via sodium reabsorption).

C: the parathyroid gland synthesizes the parathyroid hormone (PTH) which increases blood calcium. PTH acts to increase the concentration of calcium in the blood by acting upon the parathyroid hormone 1 receptor (high levels in bone and kidney) and the parathyroid hormone 2 receptor (high levels in the central nervous system, pancreas, testes and placenta).

D: the liver is the largest organ in the body. It has a wide range of functions, including detoxification, protein synthesis and the production of biomolecules necessary for digestion. The liver synthesizes bile, which is necessary for the emulsification (in the small intestine) of dietary lipids.

30. B is correct.

31. B is correct. Active transport involves a carrier protein and uses energy to move a substance across a membrane against (i.e. up) a concentration gradient: from low solute to a region of high solute concentration.

C: Donnan equilibrium refers to the fact that some ionic species can pass through the barrier, while others cannot. The presence of a different charged substance that is unable to pass through the membrane creates an uneven electrical charge. The electric potential arising between the solutions is the Donnan potential.

32. D is correct. The synthesis of cyclin D is initiated during G1 and drives the G1/S phase transition.

C: apoptosis is the process of programmed cell death (PCD), which may occur in multicellular organisms. Biochemical events lead to characteristic cell changes (morphology) and death. These changes include blebbing (i.e. irregular bulge in the plasma membrane), cell shrinkage, chromatin condensation, nuclear fragmentation and chromosomal DNA fragmentation.

In contrast to necrosis, which is a form of traumatic cell death that results from acute cellular injury, apoptosis generally confers advantages during an organism's life cycle. For example, the differentiation of fingers and toes in a developing human embryo occurs because cells between the fingers undergo apoptosis and the digits are separated. Unlike necrosis, apoptosis produces cell fragments called apoptotic bodies that phagocytic cells are able to engulf and quickly remove before the contents of the cell can spill out onto surrounding cells and cause damage.

33. E is correct. During meiosis I, homologous chromosomes separate. During meiosis II, sister chromatids (identical copies, except for recombination) separate. In Klinefelter syndrome (XXY karyotype), for a sperm to cause the defect, it must contain both an X and a Y chromosome. X and Y would be "homologous chromosomes" and would normally separate during meiosis I. Failure to do so could create a sperm containing both an X and a Y, which causes Klinefelter syndrome.

Anaphase is when the centromere splits and the homologous chromosomes / sister chromatids are drawn away (via spindle fibers) from each other toward opposite sides of the two cells. The separation of homologous chromosomes occurs during anaphase I, while the separation of sister chromatids occurs during anaphase II.

Turner's syndrome is observed in females due to the single X karyotype (single X chromosome and lacking a Y).

34. B is correct. Most proteins that are secretory, membrane-bound, or targeted to an organelle use the N-terminal signal sequence (i.e. 5 to 30 amino acids) of the protein to target the protein. The signal sequence of the polypeptide is recognized by a signal recognition particle (SRP) while the protein is still being synthesized on the ribosome. The synthesis pauses, while the ribosome-protein complex is transferred to an SRP receptor on the ER (in eukaryotes) or to the plasma membrane (in prokaryotes) before translation of the polypeptide resumes.

35. C is correct.

36. D is correct. All anterior pituitary hormones (including GH) are peptide hormones. Peptide hormones are hydrophilic and cannot pass across the hydrophobic phospholipid bilayer, and therefore peptide hormones bind to receptors on the plasma membrane. The destruction of the plasma membrane greatly reduces the concentration of GH receptors.

37. E is correct. Urea, a byproduct of amino acid metabolism, is a small uncharged molecule and therefore crosses the cell membrane by simple diffusion – a passive process that does not require energy.

A: export of Na^+ from a neuron is coupled with the import of K^+; the sodium-potassium-ATPase pump – an ATP-dependent process necessary for the maintenance of a voltage potential across the neuron membrane.

B: movement of Ca^{2+} into a muscle cell occurs against the concentration gradient, and Ca^{2+} enters cells by active transport, which requires ATP.

D: synaptic vesicles contain neurotransmitters, and their exocytosis at a nerve terminus is an ATP-dependent process that is triggered by an action potential propagating along the neuron. The fusion of the vesicle requires Ca^{2+} to enter the cell upon axon depolarization reaching the terminus.

38. C is correct. p53 is a tumor suppressor protein crucial for multicellular organisms, where it regulates the cell cycle and prevents cancer. p53 has been described as *the guardian of the genome* because of its role in conserving genetic stability by preventing genome mutations. The name p53 refers to its apparent molecular mass of 53 kDa. A dalton is defined as 1/12 the mass of carbon and is a unit of convention for expressing (i.e. often in kilodaltons or kDa) the molecular mass of proteins.

39. D is correct. The pathway of newly synthesized secretory protein is: the rough ER → Golgi → secretory vesicles → exterior of the cell (via exocytosis).

A and C: peroxisomes and lysosomes are both target destinations for proteins but do not involve exocytosis in the secretory path.

B: ribosomes synthesize proteins but the Golgi is the final organelle prior to exocytosis.

40. B is correct. Plant cell membranes have higher amounts of unsaturated fatty acids. The ratio of saturated and unsaturated fatty acids determines membrane fluidity. Unsaturated fatty acids have kinks in their tails (due to double bonds) that push the phospholipids apart, so the membrane retains its fluidity.

A: while most textbooks state that plant membranes have no cholesterol, a small amount is present in plant membranes, but it is negligible in comparison to animal cells. Cholesterol is normally found dispersed in varying amounts throughout animal cell membranes in the irregular spaces between the hydrophobic lipid tails of the membrane. Cholesterol functions as a bidirectional buffer: at high temperatures, it decreases the fluidity of the membrane because it confers stiffening and strengthening effects on the membrane; at low temperatures, cholesterol intercalates between the phospholipids to prevent them from clustering together and stiffening the membrane.

41. D is correct.

42. C is correct. The Golgi apparatus processes secretory proteins via post-translational modifications. Proteins targeted to the Golgi (from the rough ER) have three destinations: secreted out of the cell, transported into organelles or targeted to the plasma membrane (as a receptor, channel or pore).

43. C is correct.

44. A is correct. Albumin is the most abundant plasma protein. Because it is too large to pass from the circulatory system into the interstitial space, it is primarily responsible for the osmotic pressure of the circulatory system. Osmotic pressure is the force of H_2O to flow from an area of lower solute concentration to an area of higher solute concentration. If a membrane is impermeable to a particular solute, then H_2O flows across the membrane (i.e. osmosis) until the differences in the solute concentrations have been equilibrated.

In osmosis, H_2O always flows from a region of lower osmotic pressure to a region of higher osmotic pressure. The capillaries are impermeable to albumin (i.e. solute). Increasing albumin solute concentration in the arteries/capillaries increases the movement of H_2O from the interstitial fluid to reduce the osmotic pressure in the arteries/capillaries.

45. B is correct. The retinoblastoma protein (i.e. pRb or *RB1*) is a tumor suppressor protein that is dysfunctional in several major cancers. One function of pRb is to prevent excessive cell growth by inhibiting cell cycle progression until a cell is ready to divide. It also recruits several chromatin remodeling enzymes such as methylases and acetylases. pRb prevents the cell from replicating damaged DNA by preventing its progression along the cell cycle through G1 (first gap phase) into S (synthesis phase when DNA is replicated).

46. E is correct. Glycolysis occurs in the cytoplasm, while the oxidation of pyruvate to acetyl-CoA and the Krebs (TCA) cycle occur in the matrix of the mitochondria, and the electron transport chain (ETC) occurs in the inner membrane (i.e. cytochromes) / intermembrane space (i.e. H^+ proton gradient) of the mitochondria.

A: pyruvate is oxidized to acetyl-CoA as a preliminary step before joining oxaloacetate in the Krebs cycle, which occurs in the mitochondrion.

B: Krebs cycle is the second stage of cellular respiration and occurs in the matrix of the mitochondrion.

C: the electron transport chain is the final stage of cellular respiration occurring in the inner membrane (i.e. cytochromes) / intermembrane space (i.e. H^+ proton gradient) of the mitochondrion.

47. D is correct.

48. E is correct.

49. B is correct. Minerals (e.g., potassium, sodium, calcium and magnesium) are essential nutrients because they must be consumed in the diet and act as cofactors (i.e. nonorganic components) for enzymes. Minerals are not digested by lysosomes.

Organic molecules, such as nucleotides, proteins and lipids, are hydrolyzed (i.e. degraded) into monomers. Nucleotides consist of a phosphate, sugar and base; proteins consist of amino acids; lipids consist of glycerol and fatty acids.

50. D is correct.

51. B is correct. Albumins are globular proteins in the circulatory system.

A: carotenoids are organic pigments that are found in the chloroplasts of plants and some other photosynthetic organisms such as some bacteria and fungi. They are fatty acid-like carbon chains containing conjugated double bonds and sometimes have six-membered carbon rings at each end. As pigments, they produce red, yellow, orange, and brown colors in plants and animals.

C: waxes (esters of fatty acids and alcohols) are protective coatings on skin, fur, leaves of higher plants, and on the exoskeleton cuticle of many insects.

D: steroids (e.g. cholesterol, estrogen) have three fused cyclohexane rings and one fused cyclopentane ring.

E: lecithin is an example of a phospholipid that contains glycerol, two fatty acids, a phosphate group, and nitrogen-containing alcohol.

52. A is correct.

53. B is correct. Only human gametes, which are formed during meiosis, are the cells that have a single copy (1N) of the genome. Cells have a single unreplicated copy (i.e. devoid of a sister chromatid) after the second meiotic division.

54. D is correct.

55. C is correct. Centrioles are cylindrical structures composed mainly of tubulin found in most eukaryotic cells (except flowering plants and fungi). Centrioles are involved in the organization of the mitotic spindle and in the completion of cytokinesis. Centrioles contribute to the structure of centrosomes and are involved in organizing microtubules in the cytoplasm. The position of the centriole determines the position of the nucleus and plays a crucial role in the spatial arrangement of the cell.

56. E is correct. Centrioles are the organizational sites for microtubules (i.e. spindle fibers) that assemble during cell division (e.g. mitosis and meiosis).

The four phases of mitosis are: prophase, metaphase, anaphase and telophase, followed by cytokinesis, which physically divides the cell into two identical daughter cells. The condensed chromosomes are aligned along the equatorial plane in mitotic metaphase before the centromere (i.e. heterochromatin region on the DNA) splits and the two sister chromosomes begin their journey to the respective poles of the cell.

57. A is correct. cAMP is a second messenger triggered when a ligand (e.g. peptide hormone or neurotransmitter) binds to a membrane-bound receptor. Through a G-protein intermediate, the adenylate cyclase enzyme is activated and converts ATP into cAMP. Adenylate cyclase is attached to the inner layer of the phospholipid bilayer and is not located in the cytoplasm.

cAMP   ATP

58. C is correct. See explanation for question **12**.

59. A is correct. The solution is hypertonic when there is a higher concentration of solutes outside the cell than inside the cell. When a cell is immersed into a hypertonic solution, the tendency is for water to flow out of the cell in order to balance the concentration of the solutes. Osmotic pressure draws water out of the cell and the cell shrivels (i.e. crenation).

60. E is correct.

61. B is correct. Microtubules are hollow proteins composed of *tubulin* monomers. Microtubules are necessary for the 1) formation of the spindle apparatus that separates chromosomes during cell division, 2) synthesis of cilia and flagella and 3) formation of the cell cytoskeleton (within the cytoplasm).

II: actin and myosin are contractile fibers in muscle cells and are composed of microfilament (not microtubules).

62. D is correct. If a cell (e.g. pancreatic exocrine cell) is producing large amounts of proteins (i.e. enzymes) for export, this involves the rough endoplasmic reticulum (RER). Protein synthesis begins in the nucleus with the transcription of mRNA, which is translated into polypeptides with the RER. Exported proteins are packaged and modified in the Golgi. Also, this cell would have a large nucleolus for rRNA (i.e. ribosomal components) synthesis.

63. A is correct.

64. E is correct. An amoeboid moves using pseudopodia, which are bulges of cytoplasm powered by the elongation of flexible microfilaments (not microtubules).

Microtubules are long, hollow cylinders made up of polymerized α- and β-tubulin dimers. They are very important in a number of cellular processes, are involved in maintaining structure of the cell and, together with microfilaments and intermediate filaments, form the cytoskeleton.

Microtubules also make up the internal structure of cilia and flagella. They provide platforms for intracellular transport and are involved in a variety of cellular processes, including the movement of secretory vesicles, organelles and intracellular substances (see entries for dynein and kinesin). Microtubules are also involved in cell division (i.e. mitosis and meiosis) including the formation of mitotic spindles that pull apart eukaryotic chromosomes.

65. B is correct. Apoptosis is programmed cell death that occurs during both fetal development and during ageing. The synaptic cleft development, formation of separate digits in the hand of a fetus, and tadpole tail reabsorption are examples of apoptosis during development.

The synthesis of the uterine lining is an anabolic process that involves mitosis (i.e. cell division).

66. A is correct. A cell involved in active transport (e.g. intestinal epithelial cells) requires large amounts of ATP. Therefore, many mitochondria are needed to meet its cellular respiration needs (i.e. glucose $\rightarrow$ ATP).

B: high levels of DNA synthesis occur in cells using mitosis for rapid reproduction (i.e. skin cells).

C: high levels of adenylate cyclase (i.e. cAMP second messenger) are in target cells of peptide hormones.

D: polyribosomes are in cells that have a high level of protein synthesis.

E: many lysosomes would be found in phagocytic cells to enable digestion of foreign material via endocytosis.

67. E is correct. Coat-proteins, like clathrin, are used to build small vesicles to safely transport molecules within and between cells. The endocytosis and exocytosis of vesicles allow cells to transfer nutrients, import signaling receptors, mediate an immune response and degrade cell debris after tissue inflammation.

Endocytosis is the process by which the cell internalizes receptor-ligand complexes from the cell surface, such as cholesterol bound to its receptor. At the cell surface, the receptor-ligand complexes cluster in clathrin-coated pits and pinch off the vesicles that join acidic vesicles (i.e. endosomes).

68. D is correct. See explanation for question **57**.

Chapter 1.2: Molecular Biology of Eukaryotes

1. A is correct. See explanation for question **41** of *Chapter 3.1.*

2. E is correct.

3. C is correct.

4. A is correct. The addition of a 3' poly-A tail to mRNA (not proteins) is a *post transcription* event. The other *post transcription* events include the addition of a 5' cap and splicing of exons (and removal of introns) from the RNA molecule.

5. C is correct. Centromeres are regions on the chromosomes consisting of highly coiled DNA (i.e. heterochromatin) at the point of attachment of the sister chromatids.

6. D is correct. Genes supply the hereditary information for the organism, but the environment (to a degree) can determine the phenotypic pattern of proteins resulting from gene expression. Therefore, both genetic and environmental factors interact to produce the phenotype. These are examples of the environment affecting what genes are transcribed.

D: shivering is an example of a change in the environment causing a physiological (i.e. behavioral) change to maintain homeostasis. Homeostasis is achieved initially because shivering generates heat when muscles contract rapidly.

7. C is correct. The term *library* refers to organisms (e.g. bacteria) that carry a DNA molecule inserted into a cloning vector (e.g. bacterial plasmid). Complementary DNA (cDNA) is created from mRNA in a eukaryotic cell with the use of the enzyme reverse transcriptase. In eukaryotes, a poly-A tail (i.e. long sequence of A nucleotides) distinguishes mRNA from tRNA and rRNA and can therefore be used as a primer site for reverse transcription. The mRNA is purified by column chromatography using oligomeric dT nucleotide resins where only the mRNA having the poly-A tail bind (i.e. mRNA is isolated from the other RNA molecules).

Once mRNA is purified, oligo-dT is used as a complementary primer to bind the poly-A tail and extend the primer molecule by reverse transcriptase to create the cDNA strand. Now, the original mRNA templates are removed by using a RNAse enzyme leaving a single stranded cDNA (sscDNA). This single-strand cDNA is converted into a double stranded DNA by DNA polymerase for double-stranded cDNA. The cDNA is inserted and cloned in bacterial plasmids.

8. A is correct. Quaternary protein structure requires two or more polypeptide chain whereby each polypeptide is specified by different genes. Complex polysaccharides are composed of sugar monomers linked together. Many examples of complex polysaccharides include different sugar monomers such as lactose (e.g. glucose and galactose) and other molecules joined to the polysaccharide as it takes its final form. The presence of more than one monomer requires additional genes.

9. E is correct. A genomic library is a set of clones that together represents the entire genome of a given organism. Among the tools that has made recombinant DNA technology possible, are *restriction enzymes.* Restriction enzymes are endonucleases that cut DNA at specific sequences which are usually inverted repeat sequences (i.e. palindromes that read the same if the DNA strand is rotated 180°).

After cutting, the ends of some restriction digested double-stranded DNA fragments have several nucleotide overhangs of single-stranded DNA called "sticky ends." Since a restriction enzyme always cuts in the same manner, the sticky end from one fragment can be annealed via hydrogen bonding with the sticky end from another fragment cut by the same enzyme. After annealing, DNA fragments can be covalently bound together by the enzyme DNA ligase to close the plasmid into a circular DNA that can be reintroduced into bacteria.

10. D is correct. Attachment of glycoprotein side chains is a post-translational modification. The rough endoplasmic reticulum (rER) and the Golgi apparatus are the two organelles that modify proteins as a post-translational event.

I: lysosomes are organelles with low pH that function to digest intracellular molecules.

11. A is correct. Telomeres are maintained by *telomerase*, an enzyme that (during embryogenesis) adds repeats to chromosomal ends. A measure of telomerase activity in adults is an indication of one marker for cancer activity (i.e. uncontrolled cell growth) within the cell. *See explanation for question 22 (below) for more information on telomere.*

12. B is correct. cDNA library, unlike genomic library that reflects the entire DNA of the organism, uses mRNA, which corresponds to expressed genes after processing (i.e. removal of introns).

13. E is correct.

14. C is correct.

15. D is correct. Ribosomal RNA (rRNA) binds to the small and large ribosomal subunits (i.e. 30S & 50S in prokaryotic cells and 40S & 60S in eukaryotes) to create the functional ribosome (i.e. 70S in prokaryotes and 80S in eukaryotes). Ribosomes are complex assemblies of RNA molecules and several associated proteins. rRNAs are the only RNA molecules (others being mRNA and tRNA) synthesized in the nucleolus (within the nucleus).

16. A is correct.

17. B is correct.

18. E is correct. I: eukaryotes splice hnRNA (primary transcript) by removing introns and ligating together exons in the nucleus before exporting the mRNA transcript to the cytoplasm for translation.

II: after proteins are synthesized in the ER, they may be modified in the Golgi apparatus (e.g. glycosylation), but this is different than the required splicing of mRNA molecules (i.e. removal of introns) before the mRNA can pass through the nuclear pores and enter the cytoplasm.

III: prokaryotic ribosomes (30S small & 50S large subunit = 70S complete ribosome) are smaller than eukaryotic ribosomes (40S small & 60S large subunit = 80S complete ribosome).

19. C is correct.

20. B is correct. The *central dogma* of molecular biology refers to the direction of genetic information flow within a living system. The *central dogma* states that DNA is transcribed into RNA and then RNA is translated into protein: DNA → RNA → protein

The central dogma of molecular biology was violated with the discovery of retroviruses that have RNA genomes and use reverse transcriptase to make a copy of their RNA into DNA within the infected cell.

21. A is correct. See explanation for question **24**.

22. C is correct. The process of new strand synthesis prevents the loss of the terminal DNA at the coding region during replication and represents *telomere* function. Telomeres are located at chromosomal ends and consist of nucleotide repeats that provide stability and prevent the loss of the ends of the DNA coding region during DNA replication. *See explanation for question* **11** *(above) for telomerase enzyme.*

A: kinetochore is a protein collar around the centromere on the chromosome. It is the site of attachment of the spindle fiber to the chromosome during cell division. The kinetochore attaches to the spindle fibers (microtubulin) and itself is anchored to the chromosome by the centromere. The other end of the spindle fibers is attached at the poles (centromere region) of the cell to the centrioles.

B: centrosome (~*some* means body) is a region at the poles of the cell giving rise to the asters (microtubules projecting past the centrioles) of the centrioles.

D: centromeres are located on the chromosomes consisting of highly coiled DNA (i.e. heterochromatin) at the point of attachment of the sister chromatids.

23. B is correct. See explanation for question **45**.

24. C is correct. miRNA (also known as interference RNA) is a small non-coding RNA molecule (about 22 nucleotides) found in plants, animals and some viruses, which functions in transcriptional and post-transcriptional regulation of gene expression. The miRNA hybridizes to an mRNA molecule and inactivates its ability to be used as a template for the translation of protein.

25. A is correct. See explanation for question **67** of *Chapter 3.1.*

26. E is correct. The genomic library reflects the entire DNA of the organism and, unlike a cDNA library, does not use mRNA, which corresponds to expressed genes after processing (i.e. removal of introns).

IV: Retrotransposons are genetic elements that can amplify themselves in a genome and are ubiquitous components of the DNA of many eukaryotic organisms.

27. D is correct. tRNA has a cloverleaf structure and is the smallest single-stranded RNA molecule. tRNA contains the anticodon and functions to deliver individual amino acids to the growing peptide chain based on the codon specified by the mRNA.

A: hnRNA (heteronuclear RNA or primary transcript) is an RNA molecule that has not yet undergone post-transcriptional modification (i.e., 5' Guanine cap, removal of introns and ligation of exons and the addition of a 3' poly-A tail). Once these three modifications occur, the hnRNA becomes a mature mRNA and is then transported through the nuclear pores (in the nuclear envelope). The mRNA enters the cytoplasm for translation into proteins.

B: mRNA is a single-stranded RNA molecule from its genesis and is shorter than hnRNA (i.e. the primary transcript prior to post-transcriptional processing). The mRNA does not contain introns (exons are joined) and has a 5' G-cap and a 3' poly-A tail.

C: rRNA is synthesized in the nucleolus (organelle within the nucleus) and is the most numerous single-stranded RNA molecule. rRNA (and associated proteins) are necessary for the proper assembly of ribosomes that are used for protein synthesis (i.e. translation) .

28. B is correct. An inversion is a chromosome rearrangement in which a segment of a chromosome is reversed end to end. An inversion occurs when a single chromosome undergoes breakage and rearrangement within itself. Inversions usually do not cause any abnormalities in carriers as long as the rearrangement is balanced with no extra or missing DNA. However, in individuals heterozygous for an inversion, there is an increased production of abnormal chromatids, which results from crossing-over occurring within the span of the inversion. Heterozygous inversions lead to lowered fertility due to the production of unbalanced gametes.

29. E is correct. The nucleolus is the organelle within the nucleus that is responsible for the synthesis of ribosomal RNA (rRNA).

A: the Golgi apparatus is the organelle responsible for processing, packaging and distribution of proteins.

B: lysosomes are organelles with low pH that function to digest intracellular molecules.

C: mitochondrion is the organelle, bound by a double membrane, where the reactions of the Krebs (TCA) cycle, electron transport and oxidative phosphorylation occur.

30. D is correct.

31. C is correct. Glycocalyx is a general term referring to extracellular material, such as glycoprotein produced by some bacteria, endothelial, epithelial and other cells. Glycocalyx consists of several carbohydrate moieties of membrane glycolipids and glycoproteins. The glycocalyx, also located on the apical surface of endothelial cells, is composed of a negatively charged network of proteoglycans, glycoproteins and glycolipids. Generally, the carbohydrate portion of the glycolipids, found on the surface of plasma membranes, contributes to cell-cell recognition, communication and intracellular adhesion.

The glycocalyx plays a major role in endothelial vascular tissue, including the modulation of red blood cell volume in capillaries, as well as a principal role in the vasculature to maintain plasma and vessel wall homeostasis. Glycocalyx is located on the apical surface of vascular endothelial cells that line the lumen and includes a wide range of enzymes and proteins that regulate leukocyte and thrombocyte adherence.

32. A is correct. Chromatin (which comprises chromosomes) is the combination of DNA and proteins that make up the genetic contents within the nucleus of a cell. The primary functions of chromatin are: 1) to package DNA into a smaller volume; 2) to strengthen the DNA for mitosis; 3) to prevent DNA damage, and 4) to control gene expression (i.e. RNA polymerase binding) and DNA replication.

There are three levels of chromatin organization: 1) DNA wraps around histone proteins forming nucleosomes (i.e. *beads on a string*) in uncoiled euchromatin DNA; 2) multiple histones wrap into a 30 nm fiber consisting of nucleosome arrays in their most compact form (i.e. heterochromatin); 3) higher-level DNA packaging of the 30 nm fiber into the metaphase chromosome (during mitosis and meiosis).

Tandem repeats occur in DNA when a pattern of one or more nucleotides are repeated and the repetitions are directly adjacent to each other. An example would be: ATCCG ATCCG ATCCG, whereby the sequence ATCCG is repeated three times. Tandem repeats may arise from errors during DNA replication whereby the polymerase retraces its path (over a short distance) and repeats the synthesis of the same template region of DNA.

33. E is correct. DNA supercoiling is regulated by DNA topoisomerases. Gyrase (a subset of topoisomerase II) creates double-stranded breaks between the backbone of DNA and relaxes DNA supercoils by unwinding the nicked strand around the other strand.

B: helicase unwinds DNA and induces severe supercoiling in the double-stranded DNA when it hydrolyzes the hydrogen bonds between the complimentary base pairs and exposes the bases for replication.

C: DNA polymerase I is in prokaryotes and is involved in excision repair with 3'-5' and 5'-3' exonuclease activity and processing of Okazaki fragments generated during lagging strand synthesis

D: DNA polymerase III is in prokaryotes and is the primary enzyme involved in DNA replication.

34. B is correct. The 3' OH of cytosine attacks the phosphate group on the guanine.

35. A is correct. Introns are intervening sequences present in the primary transcript but excised when the RNA is processed into mRNA. The processed mRNA consists of exons that are ligated together.

C: lariat structures are the protein scaffolding used for the splicing of the hnRNA (i.e. primary transcript) into mRNA as the introns are removed and the exons are ligated together.

36. B is correct. See explanation for question **45**.

37. E is correct.

38. D is correct.

39. D is correct. Once synthesized, RNA molecules undergo three steps for post-transcriptional processing: 1) 5' G cap increases RNA stability and resistance to degradation in the cytoplasm; 2) 3' poly-A-tail functions as a *molecular clock*; and 3) introns (non-coding regions) are removed and the exons (coding regions) are spliced together.

40. B is correct.

41. A is correct. SP1 is a transcription factor that binds to nucleic acids. Therefore, SP1 binds to both RNA and DNA. Select an organelle that lacks RNA or DNA. The Golgi apparatus is the organelle responsible for processing, packaging and distributing proteins. It is not composed of nucleic acids nor does it process them. Therefore, since there are no nucleic acids within the Golgi apparatus the transcription factor SP1 would not bind to it.

B: mitochondrion is the organelle, bound by a double membrane, where the Krebs cycle, electron transport and oxidative phosphorylation occur. Mitochondria are the site of the production of the majority of the ATP molecules produced during aerobic respiration. The mitochondria (via the endosymbiotic theory) also contain their own DNA (similar in size and composition to bacterial DNA) and therefore the SP1 transcription factor would bind.

C: nucleolus is an organelle within the nucleus where rRNA synthesis occurs. Thus, SP1 would bind to nucleic acid (rRNA) within the nucleolus.

D: ribosomes are structures composed of rRNA and assembled proteins as the site of translation (mRNA is converted into proteins).

42. E is correct. A spliceosome is a large and complex molecular machine assembled from snRNP and protein complexes. The spliceosome catalyzes the removal of introns, and the ligation of the flanking exons. The snRNA is a component of the snRNP and provides specificity by *recognizing* the sequences of critical splicing signals at the 5' and 3' ends and branch site of introns.

Spliceosome removes introns from a transcribed pre-mRNA, a kind of primary transcript (i.e. heteronuclear RNA or hnRNA). Each spliceosome is composed of five small nuclear RNAs (snRNA) and a range of associated protein factors. The spliceosome occurs anew on each hnRNA (pre-mRNA). The hnRNA contains specific sequence elements that are recognized and utilized during spliceosome assembly; the 5' end splice, the branch point sequence, the polypyrimidine (i.e. cytosine and uracil) tract, and the 3' end splice site.

43. B is correct. A polynucleotide (e.g. DNA) is the only macromolecule (i.e. nucleic acids, proteins, lipids and carbohydrates) that is repaired rather than degraded. DNA repair is essential for maintaining cell function and is performed by a variety of biological repair systems (i.e., p53 tumor repressor protein). Uncontrolled cell growth via disruption to the regulation of the cell cycle may occur if mutations in somatic cells are not repaired.

44. D is correct.

45. E is correct. H1 histones are found in the linker regions of about 50 nucleotides between the nucleosomes. The nucleosome is an octomer that consists of the core histones: two H2A, two H2B, two H3 and two H4 histones.

46. B is correct. The process of splicing involves snRNA that consists of the sugar ribose (like other RNA molecules). RNA is labile (unstable) and can function as a nucleophile because of the 2'-OH.

47. C is correct. Chromatin is composed of DNA and histone proteins. DNA has an overall negative charge because of the presence of the negative oxygen (i.e. single bonded with seven valence electrons) attached to the phosphate group that is attached to the deoxyribose sugar. For electrostatic interactions, histones must be positively charged. Histones are composed of a high concentration of basic (positive charge) amino acids (e.g. lysine and arginine).

48. C is correct. Nucleic acids (e.g. DNA and RNA) elongate via nucleophilic attack by the 3'–OH of the nascent (i.e. growing strand). The 3'–OH attacks the phosphate group closest to the sugar in the incoming nucleotide. The 2'–OH is present in RNA (i.e. ribose) as compared to a 2'–H for DNA (i.e. deoxyribose, where *deoxy* signifies the absence of O as in OH).

49. A is correct. Heterochromatin is a tightly packed form of DNA while euchromatin refers to uncoiled regions of DNA actively engaged in gene expression because RNA polymerase binds to the relaxed region of DNA. Heterochromatin mainly consists of genetically inactive satellite sequences (i.e. tandem repeats of noncoding DNA). Centromeres and telomeres are heterochromatic, as is the Barr body of the second, inactivated X-chromosome in females.

50. D is correct. The RNA of the prokaryote does not undergo post-transcriptional modifications (i.e. 5' cap, removal of introns and ligation of exons or the addition of a 3' poly-A tail), as does the hnRNA in eukaryotes.

Because prokaryotes lack a nucleus, the synthesis of the RNA (i.e. transcription) occurs simultaneously with the synthesis of the polypeptide (i.e. translation).

Chapter 1.3: Cellular Metabolism and Enzymes

1. D is correct. The α-helix (alpha helix) is a common secondary structure of proteins and is a right-handed coiled or spiral conformation (helix), in which every backbone amino (N-H) group donates a partial positive hydrogen to form a hydrogen bond with one of the lone pairs of electrons from the backbone carbonyl (C=O) group of another amino acid four residues earlier.

2. A is correct. ATP is a nucleotide composed of adenosine, ribose and three phosphate groups.

3. C is correct. Hydrophobic side chain groups (e.g. phenylalanine, methionine, leucine and valine) interact through hydrophobic interactions, which exclude water from the region of attraction.

B: hydrogen bonds require hydrogen to be connected to an electronegative atom of fluorine, nitrogen, oxygen or chlorine (weaker hydrogen bond due to the larger valence shell size) to establish a δ+ and δ- (partial positive and negative regions that attract).

D: cysteine is the only amino acid capable of forming disulfide bonds because the side chain contains sulfur.

4. B is correct. Cellular respiration uses enzyme-catalyzed reactions and is a catabolic pathway to use the potential energy stored in glucose. While glycolysis yields only two ATP per glucose, all three stages of cellular respiration (i.e., glycolysis, Krebs/TCA cycle and electron transport chain) produce 30–32 ATP. Around 2004, research reduced the original estimate of 36-38 ATP due to inefficiencies of the ATP synthase.

Cellular respiration is an aerobic process, whereby O_2 is the final acceptor of electrons that are passed from cytochrome carriers during the final stage of the electron transport chain. Glycolysis occurs in the cytoplasm. The Krebs cycle occurs in the matrix (i.e., cytoplasm of mitochondria), while the electron transport chain occurs in the intermembrane space of the mitochondria for eukaryotes. For prokaryotes, both processes occur in the cytoplasm because prokaryotes lack mitochondria.

Some forms of anaerobic respiration use a different electron acceptor other than O2 (e.g., iron, cobalt or manganese reduction) at the end of the electron transport chain. In these cases, glucose molecules are completely oxidized because the products of glucose enter the Krebs cycle. In the Krebs cycle, the highly reduced chemical compounds (e.g., NADH or $FADH_2$) are used to establish an electrochemical gradient across a membrane. The reduced chemical compounds are oxidized by integral membrane proteins that transfer the electron to the final electron acceptor.

5. E is correct. K_m is the concentration of substrate at half the maximum reaction velocity, and this increases when the antibody binds to the substrate. The substrate, which binds to antibody, is not available to react with the enzyme, therefore an additional substrate is needed to bind the same amount of enzyme (as a lower concentration of substrate in the absence of antibody).

A and D: V_{max} does not change when only the amount of available substrate changes.

C: binding to substrate has no effect on V_{max} but affects K_m because the amount of available substrate is reduced, which alters the reaction kinetics because it binds to substrate.

6. C is correct. Metabolism means change and refers to life-sustaining chemical transformations within cells of living organisms. These reactions are enzyme-catalyzed and allow organisms to maintain structures, grow, reproduce and respond to their environments. Metabolism is subdivided into two categories: catabolism is the breakdown of organic matter and harvests energy from cellular respiration (e.g. glycolysis, Krebs cycle and the electron transport chain), while anabolism is the utilization of energy (ATP) for building of biomolecules (e.g. lipids, nucleic acids, proteins).

7. D is correct. Fermentation is an anaerobic process and occurs during the absence of oxygen. The purpose is to regenerate the high-energy nucleotide intermediate of NAD^+. Oxidative phosphorylation only occurs during the electron transport chain as the final stage of cellular respiration.

8. A is correct. Acid hydrolysis has several effects:

1) partial destruction of tryptophan prevents the proper estimate of the tryptophan concentration;

2) conversion of asparagine into aspartic acid prevents the direct measure of asparagine;

3) conversion of glutamine to glutamic acid.

Therefore, the concentration of glutamic acid is an indirect measure of glutamine.

9. B is correct. When plasma glucose levels are low, the body utilizes other energy sources for cellular metabolism. These sources are used in a preferential order: glucose → other carbohydrates → fats → proteins.

These molecules are first converted to either glucose or glucose intermediates, which are then degraded in the glycolytic pathway and the Krebs cycle (i.e. citric acid cycle).

Proteins are used last for energy because there is no protein storage in the body. Catabolism of protein results in muscle wasting and connective tissue breakdown, which is harmful in the long term.

10. E is correct. The activation energy is the minimum amount of energy needed for a reaction to take place. It is thought of as an "energy barrier" because at least this amount of energy must be put into the system. Enzymes act by decreasing the activation energy and lowering the energy barrier, which allows the reaction to take place more easily. This increases the rate at which the reaction takes place.

11. A is correct. A phosphate group from ATP is transferred to glucose, which creates a phosphoester bond.

12. D is correct. Second messengers are common in eukaryotes. Cyclic nucleotides (e.g. cAMP and cGMP) are second messengers that transmit extracellular signals from the cell membrane to intracellular proteins within the cell. Second messengers are used by protein hormones (e.g. insulin, adrenaline) that cannot pass across the plasma membrane.

Steroid hormones (e.g. testosterone, estrogen, progesterone) are lipid soluble and can pass through the plasma membrane and enter the cytoplasm directly without the use of second messengers.

13. E is correct. The greatest direct source of ATP synthesis involves the electron transport chain. Glycolysis occurs in the cytoplasm, the Krebs cycle occurs in the matrix of the mitochondria, and oxidative phosphorylation (i.e. the electron transport chain) occurs in the intermembrane space.

Glycolysis produces pyruvate, which is converted into acetyl CoA and joined to oxaloacetate in the Krebs cycle (The Citric Acid cycle) to form citrate as the first intermediate. The Krebs cycle produces only two GTP (i.e. ATP) per glucose molecule (or 1 GTP per pyruvate) with the remainder of ATP formed by oxidative phosphorylation when NADH and $FADH_2$ nucleotides are oxidized and donate their electrons to the cytochromes in the electron transport chain.

14. D is correct. V_{max} is the reaction velocity at a fixed enzyme concentration and depends on the total enzyme concentration. Adding more enzymes allows more enzyme reactions per minute.

II: V, not V_{max}, depends on [S]. V_{max} is a constant for a specified amount of enzyme.

III: Competitive inhibition is when the inhibitor binds reversibly to the active site. Adding enough substrate overcomes competitive inhibition and the original V_{max} is obtained.

15. B is correct. There are two major models of enzyme-substrate binding. In the induced fit mechanism the initial interactions between enzyme and substrate (in this case, hexokinase and glucose) are weak, but the weak interactions induce conformational changes that strengthen binding. The lock-and-key mechanism is when the molecules are already in the perfect conformation for binding and thus do not change.

16. D is correct. The Gibbs free energy (ΔG) of a reaction is not dependent on the amount of the reactants, enzyme or products formed. It measures the energy difference (Δ) between the reactants and products. ΔG is positive when the products are higher in energy than the reactants (absorbs energy). ΔG is negative when the products are lower in energy than the reactants (releases energy).

17. C is correct. Zymogens are inactive forms of the enzyme and have an ~ogen suffix (e.g. pepsinogen). From the question, glucokinase has a higher Michaelis constant (K_m). A higher value for K_m is due to the enzyme having a lower affinity for the reactant (e.g. glucose). Information about K_m does not answer the question and is merely a distraction.

A: hexokinase is a control point enzyme (i.e. irreversible steps) that is regulated by negative feedback inhibition.

B: from the question, hexokinase and glucokinase catalyze the same reaction. Therefore, by definition, theses enzymes are isozymes.

D: fructose is not a reactant in glycolysis.

18. B is correct. Secondary structure ($2°$) is the repetition of spirals (for α-helices) or folding (for β-pleated sheets) within the backbone of the polypeptide.

A: primary structure ($1°$) is the linear sequence of amino acids.

C: tertiary structure ($3°$) involves interactions between the side chains of the amino acids.

D: quaternary structure ($4°$) involves the interaction of two or more polypeptides. $4°$ requires more than one polypeptide chain in the mature protein (e.g. hemoglobin).

19. E is correct. Cofactors are organic or inorganic molecules and are classified depending on how tightly they bind to an enzyme. Loosely bound organic cofactors are termed coenzymes and are released from the enzyme's active site during the reaction (e.g. ATP, NADH). Tightly bound organic cofactors are termed prosthetic groups.

20. B is correct.

21. A is correct. The mitochondrion is the organelle, bound by a double membrane, where the reactions of the Krebs (TCA) cycle, electron transport and oxidative phosphorylation occur. In eukaryotes, the Krebs cycle (i.e. TCA) occurs in the matrix of the mitochondria. The matrix is an interior space in the mitochondria analogous to the cytoplasm of the cell.

B: the smooth ER (endoplasmic reticulum) is connected to the nuclear envelope and functions in several metabolic processes. It synthesizes lipids, phospholipids, and steroids. Cells that secrete these products (e.g. testes, ovaries, and skin oil glands) have a great deal of smooth endoplasmic reticulum. The smooth ER also carries out the metabolism of carbohydrates, drug detoxification, attachment of receptors on cell membrane proteins, and steroid metabolism.

C: the cytosol is common to both eukaryotes and prokaryotes, and the term is equivalent to the cytoplasm.

D: the nucleolus is the organelle within the nucleus that is responsible for the synthesis of ribosomal RNA (rRNA).

E: intermembrane space is the location of the proton gradient during the electron transport chain.

22. B is correct. Pyruvate is the product of glycolysis and is converted into acetyl Co-A as the starting reactant for the Krebs cycle. Pyruvate is not a waste product of cellular respiration.

Metabolic waste products are produced during enzymatic processes within cells. The accumulation of waste products can damage the cell and the entire organism, and therefore must be removed from the body. Pyruvate, made during glycolysis (as an intermediate of cellular respiration), is then converted to acetyl CoA and enters the Krebs cycle to be oxidized into CO_2 and H_2O, or is converted into the waste product of lactate under anaerobic conditions.

A: lactate is produced from pyruvate as the waste product of anaerobic respiration. Lactate is converted to pyruvate in the liver when O_2 becomes available. If it is not metabolized, it can lead to lactic acidosis (acidification of the blood) and death.

C and D: both CO_2 and H_2O are waste products of aerobic respiration. If CO_2 is not removed, it leads to acidosis (lowering of the pH) of the blood because CO_2 reacts with H_2O to form carbonic acid (H_2CO_3). CO_2 and H_2O are removed at the lungs via expiration during normal breathing.

E: ammonia is the waste product of protein metabolism, which is converted to the less toxic urea in the liver and removed by the kidneys. If ammonia is not converted and cleared it can cause the blood to become alkaline (higher in pH), which is potentially fatal.

23. A is correct. V_{max} is proportional to the number of active sites on the enzyme (i.e. [enzyme]). K_m remains constant because it is a measure of the active site affinity for the substrate. Regardless of how many enzyme molecules are present, each enzyme interacts with the substrate in the same manner.

Different models have been proposed for mechanisms of enzymatic catalysis. The Michaelis and Menten model describes enzyme-substrate interactions, whereby enzymatic catalysis occurs at a specific site on enzymes—the active site. Substrate (S) binds to the enzyme (E) and the active site to perform an enzyme-substrate (ES) complex. The ES complex can dissociate into enzyme and substrate once again, or can move forward in the reaction to form product (P) and the original enzyme. The rate constants for different steps in the reaction are k_1, k_2, k_3. The overall rate of product formation is a combination of all of the rate constants.

At constant enzyme concentrations, varying [S] changes the rate of product formation. At very low substrate concentrations, only a small fraction of enzyme molecules is occupied with substrate, and the rate of ES and [P] formation increases linearly with increasing substrate concentration. At very high [S], all active sites are occupied with substrate, and increasing the substrate concentration further does not increase the reaction rate, and the reaction rate is V_{max}. The equation describing the relationship between substrate concentration and reaction rate is:

$$V = \frac{[S]}{[S] + K_m} V_{max}$$

24. E is correct. An apoenzyme, together with its cofactor(s), is called a holoenzyme (an active form). Enzymes that require a cofactor, but do not have one bound, are called apoenzymes (or apoproteins). Most cofactors are not covalently attached to an enzyme, but are very tightly bound. However, organic prosthetic groups can be covalently bound. The term "holoenzyme" can also refer to enzymes that contain multiple protein subunits (e.g. DNA polymerases), whereby the holoenzyme is the complete complex with all the subunits needed for activity.

25. C is correct. Each ATP releases about –7.4 kcal/moles. As an approximation, 1 kcal/mol = ~ 4.2 kJ/mol

Structure of the nucleotide of ATP
(3 phosphates + ribose sugar + adenosine)

One molecule of ATP contains three phosphate groups. It is produced by a wide variety of enzymes (including ATP synthase) from adenosine diphosphate (ADP) or adenosine monophosphate (AMP) and various phosphate group donors. The three major mechanisms of ATP biosynthesis are: oxidative phosphorylation (in cellular respiration), substrate level phosphorylation and photophosphorylation (in photosynthesis).

Metabolic processes that use ATP as an energy source convert it back into its precursors, therefore ATP is continuously recycled in organisms.

26. D is correct. Cysteine is the only amino acid capable of forming disulfide (S–S covalent) bond and serves an important structural role in many proteins. The thiol side chain in cysteine often participates in enzymatic reactions, serving as a nucleophile. The thiol is susceptible to oxidization to give the disulfide derivative cystine.

27. C is correct. The electron transport chain (ETC) uses cytochromes in the inner mitochondrial membrane and is a complex carrier mechanism of electrons that produces ATP through oxidative phosphorylation.

A: the Krebs (i.e. TCA) cycle occurs in the matrix of the mitochondria. It begins when acetyl CoA (i.e. 2-carbon chain) combines with oxaloacetic acid (i.e. 4-carbon chain) to form citrate (i.e. 6-carbon chain). A series of additional enzyme-catalyzed reactions result in the oxidation of citrate and the release of two CO_2 and oxaloacetic acid is regenerated to begin the cycle again once it is joined to an incoming acetyl CoA.

B: long chain fatty acid degradation occurs in peroxisomes that hydrolyze fat into smaller molecules which are fed into the Krebs cycle and used for cellular energy.

D: glycolysis occurs in the cytoplasm as the oxidative breakdown of glucose (i.e. 6-carbon chain) into two molecules of pyruvate (i.e. 3-carbon chain).

E: ATP synthesis occurs in the cytoplasm (glycolysis), the mitochondrial matrix (Krebs cycle) and the inner mitochondrial membrane (ETC).

28. E is correct. See explanations to questions **19** and **24** above.

29. B is correct. A disulfide bond is a covalent bond derived by the coupling of two thiol (~S-H) groups as R–S–S–R.

C: peptide bonds are between adjacent amino acids and contribute to the linear sequence (primary structure) of the protein.

30. B is correct. Enzyme prosthetic groups attach via strong molecular forces (e.g. covalent bonds).

A: Van der Waals interactions, like dipole/dipole, are weak molecular forces.

C: ionic bonds are incredibly strong in non-aqueous solutions but are not used for attachment of prosthetic groups because they are weak in aqueous environments such as plasma. For example, NaCl dissociates when placed into water.

D: hydrogen bonds (like hydrophobic interactions) are an example of a weak molecular force.

E: dipole-dipole interactions are an intermolecular force intermediate in strength between hydrogen bonds and van der Waals interactions that require a dipole (separation in charge) within the bonding molecules.

31. E is correct. Cellular respiration begins via glycolysis in the cytoplasm but is completed in the mitochondria. Glucose (six-carbon chain) is cleaved into two molecules of pyruvate (three-carbon chain). Pyruvate is converted to acetyl CoA that enters the Krebs cycle in the mitochondria. NADH (from glycolysis and from the Krebs cycle) and $FADH_2$ (from the Krebs cycle) enter the electron transport chain on the inner membrane of the mitochondria.

32. C is correct. One of the primary activities of thyroid hormone is to increase the basal metabolic rate.

33. D is correct. See explanations to questions **19** and **24**.

34. B is correct. Primary structure is the linear sequence of amino acids within the polypeptide/protein.

35. D is correct. Denaturing the polypeptide only disrupts the 4°, 3° and 2° structure, while the 1° structure (i.e. linear sequence of amino acids) is unchanged. 1° structure determines subsequent 2°, 3° and 4° structure.

36. B is correct. $FADH_2$ is produced in the Krebs cycle (TCA) in cellular respiration.

A: the final product of glycolysis is pyruvate (i.e., pyruvic acid), which can be converted to lactic acid (in humans) or ethanol & CO_2 (in yeast) *via* anaerobic conditions. Yeast is used in the production of alcoholic beverages and for dough to rise.

C: in glycolysis, the first series of aerobic (or anaerobic) reactions occurs in the cytoplasm as the breakdown of the 6-carbon glucose into two 3-carbon pyruvate molecules.

D: two ATP are required and four ATP are produced = net of two ATP at the end of glycolysis.

E: two NADH are produced by glycolysis and are oxidized (i.e. lose electrons) to compound Q in the electron transport chain within the mitochondria.

37. A is correct. The are 13 essential (i.e., must be consumed in the diet) vitamins: A, B1 (thiamine), B2 (riboflavin), B3 (niacin), B5 (pantothenic acid), B6 (pyroxidine), B7 (biotin), B9 (folate) and B12 (cobalamin), C, D, E and K. The four fat soluble vitamins are A, D, E and K, which are stored in the body's adipose tissue.

38. D is correct. Hemoglobin is a quaternary protein with two α and two β chains with Fe^{2+} (or Fe^{3+}) as the prosthetic group. Prosthetic groups are tightly bound cofactors (i.e. nonorganic molecules such as metals).

39. E is correct. Glycine is the only amino acid that is achiral because the R group (i.e. side chain) is hydrogen. Optical activity is the ability of a molecule to rotate polarized light and depends on the presence of a chiral center (i.e. carbon atom bound to 4 different substituents).

D: cysteine contains sulfur in the side chain and is the only amino acid that forms disulfide bonds (i.e. S–S).

Amino acid structure: the α carbon is where the side chain (i.e. R) is attached.

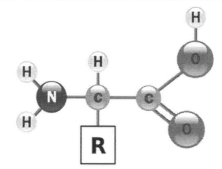

40. A is correct. During glycolysis, two net (four gross = total) molecules of ATP are produced by substrate-level phosphorylation and two molecules of NAD^+ are reduced (i.e. gain electrons) to form NADH.

B: during glycolysis, two molecules of pyruvate are formed from each starting molecule of glucose.

C: glycolysis is an anaerobic process and does not require O_2.

D: during glycolysis, glucose is partially oxidized (not reduced) into two pyruvate molecules.

E: pyruvate is produced during glycolysis (not in the Krebs cycle), which is then converted to acetyl CoA (coenzyme A) that enters the Krebs cycle.

41. B is correct.

42. C is correct. Feedback inhibition occurs when a product binds to an enzyme to prevent it from catalyzing further reactions. Since it is allosteric inhibition, this product binds to a site different than the active site, which changes the conformation of the active site and inhibits the enzyme. Although the allosteric inhibitor can be the same product of the enzyme-catalyzed reaction OR can be the final product of the metabolic pathway, it is usually the final product that shuts down the whole metabolic pathway, thus saving the energy necessary to produce the intermediates as well as the final product. When levels of the final product are low, the enzyme resumes its catalytic activity.

43. A is correct. As an obligate anaerobe, the reaction occurs in the absence of O_2.

B: oxidative phosphorylation occurs during the electron transport chain and requires a molecule of oxygen as the final acceptor of electrons that are shuttled between the cytochromes.

44. E is correct. A peptide bond is formed between two adjacent amino acids. The amino acid (written on left by convention) contributes an amino group (~NH_2) and the other amino acid (written on right by convention) contributes a carboxyl group (~COOH) that undergoes condensation (i.e. joining of two pieces to form a connected unit) *via* dehydration (i.e. removal of H_2O during bond formation).

45. A is correct. Oxidative phosphorylation (O_2 as substrate/reactant) occurs only in the electron transport chain (last step for cellular respiration) in the inner membrane (i.e. cristae) of the mitochondria. NADH and $FADH_2$ donate electrons to a series of cytochrome (i.e. protein) molecules embedded in the inner membrane and create an electron gradient, which establishes a proton (H^+) gradient within the intermembrane space of the mitochondria. This H^+ gradient drives a proton pump that is coupled to an enzyme, which produces ATP via oxidative phosphorylation.

The electrons from NADH and $FADH_2$ are transferred to ½ O_2 (i.e. oxidative phosphorylation) to generate ATP and form H_2O as a metabolic waste (along with CO_2) from cellular respiration. Both H_2O and CO_2 (as metabolic wastes) are expired from the lungs during breathing.

Each $FADH_2$ from the Krebs cycle yields two ATP, while each NADH yields three ATP during the electron transport chain. However, the NADH from glycolysis only produces two ATP because energy is expended to shuttle the NADH from glycolysis (located in the cytoplasm) through the double membrane of the mitochondria.

46. C is correct. Enzymes do not affect the free energy (ΔG) of a reaction, they only decrease the activation energy required, which increases the rate of reaction. Enzymes do not affect the position and direction of equilibrium, they only affect the speed at which equilibrium is reached.

47. C is correct.

48. B is correct. By definition, 2° structure consists of numerous local conformations, such as α-helixes and β-sheets. Secondary conformations are stabilized by many hydrogen bonds or by the covalent disulfide bonds between S–S of two cysteines to form cystine.

To be classified as 2° structure, the bonds must be formed between amino acids within about 12-15 amino acids along the polypeptide chain. Otherwise, interactions (e.g. H bonds or disulfide bonds) that connect amino acid residues more than 15 amino acids away along the polypeptide chain qualify as 3° protein structure. If the bonds occur between different polypeptide strands, then the interactions qualify as 4° protein structure (e.g. two α and two β chains in hemoglobin).

49. E is correct. Glycogen is a storage form of a polysaccharide (i.e. carbohydrate) composed of glucose monomers that are highly branched. The extensive branching provides numerous ends to the molecule to facilitate the rapid hydrolysis (i.e. cleavage and release) of individual glucose monomers when needed *via* release by epinephrine (i.e. adrenaline). Glycogen is synthesized in the liver because of high plasma glucose concentrations and is also stored in muscle cells for release during exercise.

A: glycogenesis involves the synthesis of glycogen.

B: glycogenolysis involves the degradation of glycogen.

C: glycogen is a highly branched molecule.

D: plants produce starch (i.e. analogous to glycogen) as their storage carbohydrate.

50. B is correct. The fraction of occupied active sites for an enzyme is equal to V/V_{max} which is = $[S]/([S] + K_m)$.

If $[S] = 2K_m$, then the fraction of occupied active sites is 2/3.

51. C is correct. An allosteric regulator is a molecule that binds to the enzymes at the allosteric site (i.e. other than the active site) and is, by definition, noncompetitive inhibition (in its ability to bind to the enzyme). Once bound to the allosteric site, the enzyme changes its conformational shape. If the shape change causes the ligand (i.e. molecule destined to bind to the active site) to be able to bind to the active site with less efficiency, the modulator is inhibitory. If the shape change causes the ligand to be able to bind to the active site with greater efficiency (i.e. active site becomes open and more accessible), the modulator is excitatory.

52. B is correct. Biological reactions can be either exergonic (releasing energy) or endergonic (consuming energy). All biological reactions have activation energy, which is the minimum amount of energy required for the reaction to take place. These reactions use enzymes (biological catalysts) to lower the activation energy and thus increase the rate of the reaction.

53. C is correct. Oxidative phosphorylation uses ½ O_2 as the ultimate electron acceptor. The electron transport chain (ETC) is located on the inner mitochondrial membrane and uses cytochromes (i.e. proteins) in the inner membranes of the mitochondria to pass electrons released from the oxidation of NADH & FADH$_2$.

The mitochondrial matrix (i.e. cytoplasm of mitochondria) is the site for the Krebs cycle (TCA). The outer mitochondrial membrane does not directly participate in either oxidative phosphorylation or the Krebs cycle. Glycolysis and the Krebs cycle produce ATP via substrate-level phosphorylation, while the oxidative phosphorylation requires O_2 and occurs during the ETC.

54. D is correct. Enzymes are often proteins and function as biological catalysts at an optimal temperature (physiological temperature of 36°C) and pH (7.35 is the pH of blood). At higher temperatures, the proteins denature (i.e. unfold and change shape by disruption of hydrogen and hydrophobic bonds) and lose their function. Proteins often interact with inorganic minerals (i.e. cofactors) or organic molecules (i.e. coenzymes or tightly bound prosthetic groups) for optimal activity.

Mutations affect DNA sequences that encode for proteins, resulting in a change in the amino acid sequence of the polypeptide and a change in conformation within the enzyme. A change in conformation of the enzyme often results in a change in function.

55. D is correct.

56. D is correct. Allosteric enzymes have allosteric sites that bind metabolites. These allosteric sites are distinct from the active sites that bind substrate. An enzyme must have quaternary structure (multiple subunits) in order to bind more than one molecule. Allosteric enzymes show cooperative binding of substrate, meaning that the binding of a molecule at one site affects the binding of a molecule at another site. If an enzyme binds an inhibitor at an allosteric site, this will decrease the affinity of the active site for the substrate.

57. E is correct. Monoamine oxidases (MAO) are a family of enzymes that catalyze the oxidation of monoamine neurotransmitters. Monoamine neurotransmitters and neuromodulators contain one amino group that is connected to an aromatic ring by a two-carbon chain ($-CH_2-CH_2-$). All monoamines are derived from the aromatic amino acids phenylalanine, tyrosine and tryptophan as well as the thyroid hormones by aromatic amino acid decarboxylase enzymes.

The function of monoamines is to trigger crucial components such as emotion, arousal and cognition. MAO are thought to be responsible for a number of psychiatric and neurological disorders.

For example, unusually high or low levels of MAOs in the body have been associated with schizophrenia, depression, attention deficit disorder, substance abuse and migraines. Serotonin, melatonin, norepinephrine, and epinephrine are mainly broken down by MAO. Excessive levels of catecholamines (e.g. epinephrine, norepinephrine and dopamine) may lead to a hypertensive crisis (e.g. hypertension) and excessive levels of serotonin may lead to serotonin syndrome (e.g. increased heart rate, perspiration and dilated pupils).

58. C is correct.

59. B is correct. I: prosthetic group is a tightly bound chemical compound that is required for the enzyme's catalytic activity, and it is often involved at the active site.

II: zymogens are inactive enzyme precursors that require a biochemical change to become an active enzyme. This change is usually a proteolysis reaction that reveals the active site.

60. B is correct.

61. A is correct.

62. D is correct.

63. E is correct.

Chapter 1.4: Specialized Cells and Tissues

1. E is correct.

2. A is correct. Sodium (Na^+) has a positive charge, so the anion is chloride (Cl^-), which has a negative charge.

B: potassium (K^+) is a cation with a positive charge.

C: magnesium (Mg^{2+}) has a 2+ charge.

D: lithium (Li^+) is a cation with a positive charge.

E: calcium (Ca^{2+}) has a 2+ charge.

3. A is correct.

4. D is correct.

5. B is correct.

6. E is correct.

7. B is correct.

8. D is correct.

9. C is correct. After it has acted upon the postsynaptic membrane, ACh is inactivated in the synaptic cleft by the enzyme acetylcholinesterase. Because chemical X deactivates acetylcholinesterase, ACh remains in the synaptic cleft and continues to depolarize the postsynaptic membrane.

10. D is correct.

11. A is correct. The Na^+/K^+ ATPase transports 3 Na^+ ions out for every 2 K^+ ions into the cell. When the pump is inhibited, the intracellular [Na^+] *increases* and the intracellular [K^+] *decreases*.

12. C is correct.

13. E is correct. Bone and cartilage are connective tissues that function to connect and support tissues and organs, and are related in their lineage and activities. Connective tissue functions to bind and support other tissues. Connective tissue consists of loose connective tissue and dense connective tissue (subdivided into dense regular and dense irregular). Special connective tissue consists of reticular connective tissue, adipose tissue, blood, bone and cartilage.

14. C is correct. The resting membrane potential (i.e. voltage results from differences in charge) across a nerve cell membrane depends on the unequal distribution of Na^+ and K^+ ions. The Na^+/K^+ pump is an active transport protein that maintains an electrochemical gradient across the membrane by pumping three Na^+ out and two K^+ into the cell.

This unequal pumping of ions causes: 1) cells to be negative inside relative to outside; 2) high $[Na^+]$ outside the cell relative to inside; 3) high $[K^+]$ inside relative to outside. The membrane is more permeable to K^+ ions than Na^+ ions, and the balance between the pump and the *leaky* membrane that determines the cell's resting potential (i.e. –70mV).

15. E is correct.

16. D is correct. Schwann cells (peripheral nervous system) and oligodendrocytes (central nervous system) synthesize myelin, which functions as insulation around the axon. Myelinated axons transmit nerve signals *much faster* than unmyelinated ones. Multiple sclerosis is a demyelinating disease, and patients are expected to have the disease manifest itself in highly myelinated, fast-conduction neurons.

17. D is correct.

18. B is correct.

19. A is correct. Muscle contraction occurs when the actin thin filaments and the myosin thick filaments slide past each other to shorten the muscle and produce a contraction. Neither of the filaments shortens nor elongates in size. The muscle itself decreases in size because the filaments slide past each other to cause a contraction of the muscle.

20. B is correct.

21. B is correct. Na^+/K^+ ATPase pumps three Na^+ ions out for every two K^+ ions into the cell. If activity is blocked, there is a net increase of positive charge within the cell and the inside becomes less negative. This change in ion concentrations decreases the resting potential (i.e. – 70mV). As the inside becomes more positive, it eventually reaches threshold (i.e. about – 50mV) and depolarization ensues.

Ouabain inhibits the Na^+/K^+ pump. If the pump was not working, the consumption of ATP would decrease, $[Na^+]$ increases within the cell (the pump is not moving Na^+ out of the cell), while $[K^+]$ increases in the extracellular environment (the pump is not moving K^+ into the cell).

22. B is correct.

23. E is correct. The myelin sheath covers the axon and increases conduction velocity. Saltatory conduction occurs by permitting membrane depolarization only at nodes of Ranvier. The myelin sheath is composed of lipids and is deposited by Schwann cells for peripheral nerve cells and by oligodendrocytes for central nerve cells.

24. D is correct. The synaptic cleft (i.e. synapse) is a small space between the axon terminus (end of one axon) and the abutting dendrite between two neurons. After stimulation, the presynaptic axon releases a neurotransmitter across this cleft, which diffuses and binds receptors on the postsynaptic dendrite of the next neuron. The neurotransmitter can only be released by the presynaptic axon and can only be received by the postsynaptic dendrite, and is unidirectional.

A: dendrite is where input is received and is (often) near the cell body.

B: the axon process of the neuron projects away from the cell body and extends towards the axon terminus (where neurotransmitters are released).

C: myelin sheath covers the axon and increases conduction velocity (i.e. permits saltatory conduction by preventing (via insulation) the passage of ions during depolarization except at the nodes of Ranvier). The myelin sheath, composed of lipids, is deposited by Schwann cells for peripheral nerve cells and by oligodendrocytes for central nerve cells.

E: the spinal nerve is a bundle of nerves where they enter and exit the spinal cord.

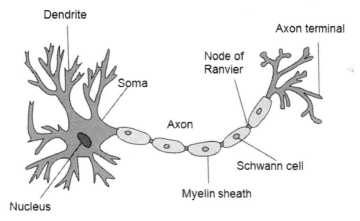

25. A is correct.

26. B is correct. Mitochondria self-replicate autonomously and the number of mitochondria in a cell varies widely by organism and tissue/type. Many cells have only a single mitochondrion, while others can contain several thousand mitochondria.

A: DNA is not an organelle.

D: ribosomes are not self-replicating. They consist of rRNA (synthesized in the nucleolus) and associated proteins. Ribosomes are assembled by the cellular machinery.

27. C is correct.

28. C is correct. In biochemistry (and pharmacology), a ligand is a molecule that forms a complex for a biological purpose. Ligands include substrates, activators, inhibitors, and neurotransmitters. In protein-ligand binding, a ligand usually is a signal triggering molecule that binds to a site on a target protein. In DNA-ligand binding studies, a ligand is usually any molecule (e.g. protein or ion) that binds to the DNA double helix.

Ligand binding to a receptor (ligand-receptor) alters a receptor protein's conformation (i.e. three dimensional shape) that determines its functional state.

29. E is correct. *Myasthenia gravis* causes the body to produce antibodies that remove acetylcholine (ACh) receptors from muscle fibers. Without these receptors, ACh cannot bind to the membrane and initiate depolarization. Without depolarization, none of the subsequent events occur, because conduction of an action potential is impaired across the sarcolemma.

For muscle contraction, a neuromuscular junction is composed of the presynaptic membrane (i.e. axon terminus) of a neuron and the postsynaptic membrane of a muscle fiber (i.e. sarcolemma). Separating the two neurons is the synapse (i.e. space between is the synaptic cleft).

From an action potential, the presynaptic membrane of a neuromuscular junction releases the neurotransmitter acetylcholine (ACh) into the synapse. ACh diffuses across the synapse and binds to acetylcholine receptors on the sarcolemma. Binding of ACh causes the postsynaptic membrane to depolarize and generate an action potential, which causes the sarcoplasmic reticulum to release calcium ions into the sarcoplasm (i.e. cytoplasm of a muscle fiber), which leads to the shortening of the sarcomeres and the resulting muscle contraction.

A: acetylcholine synthesis is not affected by *myasthenia gravis*. Acetylcholine is synthesized by the neuron itself, and this process is independent of the presence of acetylcholine receptors on the postsynaptic membrane.

Acetylcholine is one of many neurotransmitters in the autonomic nervous system (ANS). It acts on both the peripheral nervous system (PNS) and central nervous system (CNS) and is the only neurotransmitter used in the motor division of the somatic nervous system.

Acetylcholine

B: in muscle cells, calcium is stored in the sarcoplasmic reticulum, and the release of calcium is also affected by *myasthenia gravis*.

30. D is correct.

31. E is correct. Na$^+$ and K$^+$ are the major ions involved in the action potential of a neuron. Blocking Na$^+$ channels and preventing Na$^+$ from entering the cell inhibits the propagation of the action potential, because the movement of Na$^+$ creates depolarization. The –70mV (resting potential) becomes –40mV (threshold) for action potential propagation before the K$^+$ channels open to repolarize the neuron.

32. B is correct. The glial cell is a non-neuronal cell that maintains homeostasis, forms myelin, and provides support and protection for neurons in the brain and peripheral nervous system

The oligodendrocyte is a type of glial cell that myelinates the axon of neurons in the CNS.

A: astrocytes are star-shaped glial cells in the brain and spinal cord and are the most abundant cell of the human brain. They perform many functions, including biochemical support of endothelial cells that form the blood–brain barrier, facilitate nutrients to the nervous tissue, and maintain extracellular ion balance. Also, following traumatic injuries, astrocytes enlarge and proliferate to form a scar and produce inhibitory molecules that inhibit regrowth of a damaged or severed axon.

C: Schwann cells myelinate axons of neuron in the PNS.

D: choroid plexus is a plexus (i.e. branching network of vessels or nerves) in the ventricles of the brain where cerebrospinal fluid (CSF) is produced. There is a choroid plexus in each of the four ventricles within the brain.

E: chondrocytes are the only cells found in healthy cartilage and produce and maintain the cartilaginous matrix, which consists mainly of collagen and proteoglycans.

33. D is correct. The myelin sheath is not produced by the neuron but is synthesized by Schwann cells for peripheral nerve cells and by oligodendrocytes for central nerve cells. Schwann cells and oligodendrocytes wrap around the axon to create layers of insulating myelin (composed of lipids), which prevents the passage of ions along the axon process and increases the rate of depolarization along the axon.

A: axon hillock is a region in the dendrite where membrane potentials propagated from synaptic inputs (temporal and spatial summation) are aggregated before being transmitted to the axon. In its resting state, a neuron is polarized, with its inside at about –70 mV relative to the outside. It is the role of the axon hillock to "sum" the individual depolarization events to determine if a sufficient magnitude of depolarization has been achieved. Once the threshold is reached (between –50 and –40 mV), the *all or none* action potential is propagated along the axon.

C: nodes of Ranvier are openings along the axons that permit depolarization of the membrane and give rise to saltatory conduction to increase the rate of transmission along the nerve fiber.

E: oligodendrocytes are a type of glial cell that deposits myelin around axons in the central nervous system, as do Schwann cells in the peripheral nervous system.

34. E is correct. A resting muscle is not completely relaxed but experiences a slight contraction known as tonus.

A: in tetanus, there is no complete recovery before the next sustained contraction, which causes muscle contractions until fatigue results from a lack of ATP energy or waste products (i.e. lactic acid) buildup.

B: tonus is the partial sustained contraction in relaxed muscles.

C: isometric contractions involve a constant length and an increase in muscle tension.

D: isotonic contractions involve the shortening of the muscle while the tension remains constant.

35. A is correct.

36. C is correct. The Na^+/K^+-ATPase pump moves K^+ inside the cell. K^+ is positively charged, which causes the inside of the membrane to be more positive. The resting potential (i.e. –70mV) of the neuron is measured with respect to the inside.

37. B is correct.

38. D is correct. Insulin binds to the receptor on the extracellular surface of the cell. The insulin receptor (IR) is a tyrosine kinase transmembrane receptor activated by insulin, IGF-I, IGF-II. The binding of ligand (i.e. insulin) to the α-chains of the IR domain induces a conformational change within the receptor, leading to autophosphorylation of various tyrosine residues within the intracellular TK domain of the β-chain. Tyrosine kinase (e.g. insulin) receptors mediate their activity by the addition of a phosphate group to particular tyrosines on certain cellular proteins.

39. A is correct. During an action potential, Na^+ flows in and K^+ flows out. During an action potential, Na^+ flows into the cell (down its electrochemical gradient) to depolarize (make the voltage less negative) the cell membrane. Then, K^+ flows out (down its electrochemical gradient) to begin to repolarize the cell and restore the resting potential of –70mV. When this outflow is excessive, due to the slow closing K+ channels, the membrane becomes hyperpolarized (e.g. temporary value of –90mV), known as overshoot.

40. C is correct. Folic acid (also known as vitamin B_9) is a water-soluble vitamin. Folate is a naturally occurring form of the vitamin found in food, while folic acid is synthetically produced. Folic acid is not biologically active but its derivatives have a high biological importance.

Folic acid (or folate) is essential for numerous bodily functions. Humans cannot synthesize folate *de novo*, so it must be supplied through the diet. Folate is needed to synthesize, repair and methylate DNA, as well as to act as a cofactor in certain biological reactions. It is especially important in rapid cell division and growth (e.g. in infancy and pregnancy). Children and adults require folic acid to produce healthy red blood cells and prevent anemia.

A: vitamin B_6 is a water-soluble vitamin. Pyridoxal phosphate (PLP) is the active form and a cofactor in many reactions of amino acid metabolism (e.g. transamination, deamination and decarboxylation). PLP is necessary for the enzymatic reaction to release glucose from glycogen.

B: calcium is an important component of a healthy diet and a mineral necessary for life. Approximately 99 % of the body's Ca^{2+} is stored in the bones and teeth. It has other important uses in the body, such as exocytosis, neurotransmitter release and muscle contraction. In heart muscles, Ca^{2+} replaces Na^+ as the mineral that depolarizes the cell to proliferate action potentials.

D: vitamin B_{12} (also called cobalamin) is a water-soluble vitamin with a key role in the normal functioning of the brain and nervous system, and in the formation of blood. It is involved in the metabolism of every cell of the human body, especially affecting DNA synthesis and regulation, as well as fatty acid synthesis and energy production. Fungi, plants and animals are not capable of producing vitamin B_{12}. It is only produced by bacteria and archaea that have the enzymes required for its synthesis. Many foods are a natural source of B_{12} because of bacterial symbiosis. This vitamin is the largest and most structurally complicated vitamin and can be produced industrially only through bacterial fermentation-synthesis.

E: vitamin B_7 (biotin), is a water-soluble vitamin and is a coenzyme for carboxylase enzymes. It is involved in gluconeogenesis, synthesis of fatty acids and the amino acids isoleucine and valine.

41. D is correct. Local anesthetics dissolve into the hydrophobic membrane and inhibit membrane-bound proteins. The Na^+ voltage gated channel is essential for nerve conduction, but is inhibited by the local anesthetics in a nonspecific process that blocks nerve conduction.

42. E is correct.

43. D is correct. Cell surface receptors (i.e. transmembrane receptors) are specialized integral membrane proteins that facilitate communication between the cell and the outside world. Extracellular signaling molecules (e.g. hormones, neurotransmitters, cytokines or growth factors) attach to the extracellular receptor, which triggers changes in function of the cell. Signal transduction occurs when the binding of the ligand outside the cell initiates a chemical change on the intracellular side of the membrane.

44. A is correct. Piloerection is when hair stands on its end and is a sympathetic nervous system *fight-or flight* response (i.e. parasympathetic is associated with *rest and digest*). Sensory neurons bring information *toward* the central nervous system from sensory receptors, while motor neurons carry signals *from* the central nervous system to effector cells. For piloerection, a signal from the CNS is relayed to effector cells and this involves a motor neuron.

45. A is correct.

46. B is correct. Ca^{2+} binds troponin C located on the actin (thin) filament. Once Ca^{2+} binds, tropomyosin changes its conformation and exposes the myosin (thick filament) binding sites on the actin filament. When myosin binds actin, the myosin head undergoes a *power stroke* and the actin-myosin filaments slide past each other as the muscle contracts.

Plasma Ca^{2+} levels are tightly regulated by calcitonin and the parathyroid hormone (PTH). PTH raises calcium levels by stimulating osteoclasts, while calcitonin lowers plasma calcium levels by stimulating osteoblasts.

47. C is correct. See explanations for questions **24** and **49**.

48. A is correct. The axon hillock is a specialized part of the cell body (or soma) of a neuron that connects to the axon and where membrane potentials propagated from synaptic inputs are summated before being transmitted to the axon. The axon hillock is like the *accounting center*, where graded potential is summed (either spatial or temporal summation) to be sufficient to reach threshold. Both inhibitory postsynaptic potentials (IPSPs) and excitatory postsynaptic potentials (EPSPs) are summed in the axon hillock, and once a triggering threshold is exceeded, an action potential propagates – all or none – through the rest of the axon.

D: glial cell is a non-neuronal cell in the central nervous system and is involved in depositing myelin around axons.

49. E is correct. Myelin acts as an insulator, preventing ions from passing through the axon membrane, and allows axons to conduct impulses faster. Ions can only permeate through nodes of Ranvier that are channels located as small gaps in the myelin sheath. The action potential *skips* (i.e. saltatory conduction) from node to node and is much faster than conduction through a non-myelinated neuron, because the area requiring depolarization to permit Na^+ in / K^+ out is smaller.

A: action potentials are initiated by graded stimuli that cause depolarization of the axon hillock.

B: pumping of Na^+ out of the cell is achieved by the Na^+/K^+ pump.

C: resting potential is achieved by the Na^+/K^+ pump.

D: voltage regulated ion channels determine the threshold of a neuron.

50. D is correct.

51. C is correct. Nodes of Ranvier are unmyelinated regions along the axon process between Schwann cells which deposit myelin. The myelin sheath is a lipid that functions to increase the speed of propagation for the action potential to travel from the cell body to the axon terminal. Electrical impulses move by depolarization of the plasma membrane at the unmyelinated (nodes of Ranvier) and "jump" from one node to the next (i.e. *saltatory conduction*).

A: acetylcholine receptors are located on the dendrite of a postsynaptic neuron.

52. A is correct.

53. C is correct. The Purkinje fibers in cardiac tissue have large diameters and short axons. Conduction velocity is directly related to diameter and inversely related to length. A large diameter axon has more volume for the action potential to depolarize the membrane and allows more ions to migrate. An increase in the diameter and/or decrease in axon length increases the conduction velocity.

Cardiac Purkinje cell conduction must be fast and Na^+ channel dependent, because conduction is always channel dependent.

54. B is correct. The four primary tissue types are connective, muscle, nervous and epithelial.

55. E is correct. The release of a neurotransmitter requires the influx of Ca^{2+} and occurs at the axon terminus.

A: Na^+ influx causes neuron depolarization.

B: K^+ efflux (not influx) determines the neuron's resting potential and may cause hyperpolarization.

D: Cl^- ions are located inside the neuron and contribute to the negative value (i.e. $-70mV$) of the resting potential.

56. D is correct.

57. B is correct. Almost every cell in the body uses Na^+ channels to maintain osmotic gradients and prevent lysis. Skeletal, cardiac, smooth muscle cells and nerve cells are Na^+ dependent excitable cells because they are *depolarized* when Na^+ enters. Epithelial cells line all luminal structures (e.g., gastrointestinal tract, blood vessels and excretory system) and are non-excitable because they do not undergo depolarization.

58. E is correct.

Chapter 1.5: Microbiology

1. E is correct.

2. B is correct. Penicillin prevents the formation of the bacterial peptidoglycan cell wall by covalently binding to a serine residue at the transpeptidase active site.

Binding to the enzyme's active site is characteristic of an *irreversible competitive* inhibitor. It is *irreversible* because it forms covalent bond that creates a strong and permanent attachment. It is *competitive* because it competes with the substrate for the active site.

Penicillin (*β*-lactam antibiotic) inhibits the formation of peptidoglycan crosslinks in the bacterial cell wall. The enzymes that hydrolyze the peptidoglycan crosslinks continue to function, even while those that form such crosslinks do not. This weakens the cell wall of the bacterium, and osmotic pressure continues to rise—eventually causing cell death (cytolysis).

A: reversible competitive inhibitors utilize weak molecular attachment (i.e. van der Waals or hydrogen bonds) when attaching to the enzyme active site.

C: penicillin does not digest the cell wall but prevents the crosslinking of the peptidoglycan of the bacterial cell wall. The effect is to inhibit the bacteria's ability to remodel their cell wall or prevent progeny from forming a necessary cell wall to counter osmotic pressure.

E: noncompetitive inhibitors bind to an allosteric site, which is different from the active site. Binding of a molecule to the allosteric site of the enzyme causes a conformational change (i.e. three-dimensional shape change), which affects the shape (i.e. function) of the active site.

3. D is correct. Viruses are simple, non-living organisms, which take on living characteristics when they infect a host cell. Viruses contain either DNA or RNA and a protein coat (i.e. capsule). Their genetic material can be either RNA or DNA; it is not arranged into a chromosome, but is associated as a complex of nucleic acids and histone proteins. The cellular machinery and biomolecules within a host cell (either prokaryotic or eukaryotic) are required for the virus to replicate.

4. D is correct.

5. B is correct. An operon is a functional unit of genomic DNA containing a cluster of genes under the control of a single regulatory signal or promoter. The genes are transcribed together into an mRNA strand. The genes contained in the operon are either expressed together or not at all. Several genes must be both *co-transcribed* and *co-regulated* to be defined as an operon.

6. C is correct. *Neurospora* is a fungus and, for most of its life cycle, is a haploid organism. For fungus, there is only a brief diploid stage after fertilization that transitions via meiosis to produce haploid cells which repeatedly divide via mitosis before entering another sexual cycle.

A: most fungi undergo meiosis and a sexual cycle.

B: the separation of fertilization and meiosis is characteristic of the life cycle of plants.

D: fertilization immediately following meiosis is characteristic of animals that spend most of their lives as diploid organisms.

7. D is correct.

8. A is correct. The endomembrane system is an extension of the nuclear envelope and includes the endoplasmic reticulum and the Golgi apparatus. The Golgi received proteins within vesicles from the rER and then modifies, sorts and packages proteins destined for the secretory pathway (i.e. plasma membrane, exocytosis from the cell or destined for organelles within the cell).

The Golgi primarily modifies proteins delivered from the rough endoplasmic reticulum. It is also involved in lipid transport, the synthesis of lysosomes, and is the site of carbohydrate synthesis. The Golgi plays an important role in the synthesis of proteoglycans (i.e. component of connective tissue) present in the extracellular matrix of animal cells.

B: lysosomes are organelles with low pH that function to digest intracellular molecules.

C: peroxisomes are organelles found in most eukaryotic cells. A major function of the peroxisomes is the breakdown of very long chain fatty acids through beta-oxidation. In animal cells, the very long fatty acids are converted to medium chain fatty acids, which are subsequently shuttled to mitochondria, where they are eventually broken down, via oxidation, into carbon dioxide and water.

D: smooth endoplasmic reticulum connects to the nuclear envelope and functions in metabolic processes. It synthesizes lipids, phospholipids and steroids. Cells that secrete these products (e.g. testes, ovaries and skin oil glands) have a robust smooth endoplasmic reticulum. It also carries out the metabolism of carbohydrates, drug detoxification, attachment of receptors on cell membrane proteins, and steroid metabolism. The smooth endoplasmic reticulum also contains the enzyme glucose-6-phosphatase, which converts glucose-6-phosphate to glucose during gluconeogenesis.

9. C is correct. A Hfr (high-frequency recombination) cell is a bacterium with a conjugative plasmid (F factor) integrated into its genomic DNA instead of being located in an autonomous circular DNA element in the cytoplasm (i.e. a plasmid). F^+ denotes cells that contain the F plasmid, while F^- denotes cells that do not. Unlike a normal F^+ cell, Hfr strains, upon conjugation with a F^- cell, attempt to transfer their *entire* DNA through the mating bridge (pili).

The F factor has a tendency to transfer itself during conjugation, and often, the entire bacterial genome is dragged along, but the transfer is often aborted before the complete plasmid is transferred. Hfr cells are very useful for studying gene linkage and recombination. Because the genome's rate of transfer through the mating bridge is constant, investigators can use Hfr strain of bacteria to study genetic linkage and map the chromosome.

A: during conjugation, the transfer of Hfr DNA is interrupted by the spontaneous breakage of the DNA molecule at random points. F^+ denotes cells that contain the F plasmid, while F^- denotes cells that do not. Typically, the chromosome is broken before the F factor is transferred to the F^- cell. Therefore, the conjugation of an Hfr cell with an F^- cell does not usually result in an F^+ cell and the F factor usually remains in the Hfr cell.

B: Hfr cells produce the sex pili (i.e. mating bridge in this example), and the F^- cell is the recipient of the transfer.

D: the F factor is integrated into and is a part of the bacterium's chromosome, and the F factor is replicated along with cells' genome prior to conjugation.

10. E is correct. The Ames test is a method that uses bacteria to test whether a given chemical can cause cancer. It is a biological assay to assess the mutagenic potential of chemical compounds. A positive test indicates that the chemical is mutagenic and therefore may act as a carcinogen because cancer is often linked to mutation.

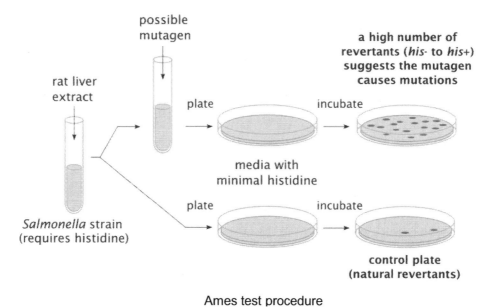

Ames test procedure

11. A is correct. Plants are generally autotrophs, animals are heterotrophs and bacteria can be either. Auxotrophs are organisms unable to synthesize a particular organic compound required for its growth. In genetics, a strain is auxotrophic if it has a mutation that renders it unable to synthesize an essential compound (i.e. *essential* is a term that means needed for growth but can't be synthesized). For example, a yeast mutant with an inactivated uracil synthesis pathway gene is a uracil auxotroph. Such a strain is unable to synthesize uracil and will only be able to grow if uracil can be taken up from the environment.

B: chemotrophs are organisms that obtain energy by the oxidation of electron donors in their environments. These molecules can be organic (e.g. chemoorganotrophs) or inorganic (e.g. chemolithotrophs). The chemotroph designation is in contrast to phototrophs, which utilize solar energy. Chemotrophs can be either autotrophic or heterotrophic.

C: heterotrophs use energy derived from another organism's metabolism (*hetero* means other).

D: prototrophs are characterized by the ability to synthesize all the compounds needed for growth.

12. B is correct. An endospore is a dormant, tough and non-reproductive structure produced by certain bacteria. Endospore formation is usually triggered by a lack of nutrients. The endospore consists of the bacterium's DNA and part of its cytoplasm, surrounded by a very tough outer coating. Endospores can survive without nutrients.

13. E is correct. Fungi are a member of a large group of eukaryotic organisms that includes other microorganisms such as mold, yeasts and mushrooms. They are classified as a kingdom (Fungi), which is separate from the other kingdoms of plants, animals, protists and bacteria. One major difference is that fungal cells have cell walls that contain chitin. The cell walls of plants and some protists contain cellulose, while the cell wall of bacteria contains peptidoglycan.

Fungi reproduce by both sexual and asexual means, and like basal plant groups (e.g. ferns and mosses) produce spores. Similar to algae and mosses, fungi typically have haploid nuclei with only a small percentage of their life cycle in a diploid phase.

14. C is correct. Cyanide is a poison that interferes with the electron transport chain (ETC) of the inner plasma membrane (for prokaryotes) or the inner mitochondrial membrane (for eukaryotes) by binding to one of the cytochromes (mainly cytochrome c oxidase) of the electron transfer complexes. Cyanide completely inhibits the flow of electrons down their reduction potential and effectively inhibits the transport chain. Consequently, the proton pump stops and ATP cannot be generated aerobically. In addition, electron carriers such as NADH and $FADH_2$ cannot be oxidized and yield their high energy electrons to the electron transport chain because it is blocked. NAD^+ and FAD are not regenerated and aerobic respiration ceases.

Energy is required for a bacteriophage (virus that infects bacteria) to replicate. DNA replication, RNA transcription and protein synthesis all require energy (either ATP or GTP) and are processes necessary for host cell functions as well as viral replication. If aerobic ATP formation is inhibited, there will be a deficiency of ATP for cell functions and for viral replication.

15. E is correct. Transformation is the genetic alteration of a cell resulting from the direct uptake, incorporation and expression of exogenous genetic material (exogenous DNA) from its surroundings and taken up through the cell membrane(s). In molecular biology, transformation may also be used to describe the insertion of new genetic material into nonbacterial cells, including animal and plant cells.

B: see explanation for question **21**.

C: transduction is the process by which DNA is transferred from one bacterium to another by a virus. In molecular biology, it also refers to the process whereby foreign DNA is introduced into another cell via a viral vector.

D: recombination is the process by which two DNA molecules exchange genetic information, resulting in the production of a new combination of alleles. In eukaryotes, genetic recombination between homologous chromosomes during prophase I of meiosis leads to a novel set of genetic information that can be passed on to progeny. Most recombination occurs spontaneously to increase genetic variation.

16. E is correct. When the genome is comprised of RNA (either single-stranded or double-stranded) the virus is referred to as a retrovirus. After infecting the host cell, the retrovirus uses enzyme *reverse transcriptase* to convert its RNA into DNA.

17. B is correct. Cells of the same strain transfer their genomes in the same order and at the same rate. The transfer is interrupted at various times, and by matching the genes transferred to the length of time necessary for the transfer, the linear order of the genes can be mapped.

A: polycistronic refers to the expression of bacterial genes on a single mRNA in an operon.

C: the rate of chromosome (i.e. F factor) transfer must be constant because differences in the transfer rate would obscure the results.

D: F factors and the bacterial chromosome are replicated by the same method during conjugation.

18. C is correct. Prokaryotic ribosomes (30S small & 50S large subunit = 70S complete ribosome) are smaller than eukaryotic ribosomes (40S small & 60S large subunit = 80S complete ribosome).

19. A is correct. Retroviruses are a family of enveloped viruses that use reverse transcription to replicate within a host cell. A retrovirus is a single-stranded RNA virus with its nucleic acid in the form of a single-stranded mRNA genome (including the 5' cap and 3' poly-A tail) and is an obligate parasite within the host cell.

Once inside the host cell cytoplasm, the virus uses its own reverse transcriptase enzyme to produce DNA from its RNA genome (*retro* to describe this backwards flow of genetic information). This new DNA is then incorporated into the host cell genome by an integrase enzyme. The integrated retroviral DNA is referred to as a *provirus*. The host cell then treats the viral DNA as part of its own genome, translating and transcribing the viral genes along with the cell's own genes, producing the proteins required to assemble new copies of the virus.

20. C is correct. A virus is a simple non-living organism that takes on living characteristics when it enters host cells. Viruses contain genetic material as either DNA (single-stranded or double-stranded) or RNA (single-stranded or double-stranded) and have a protein coat. Since viruses are not free-living organisms, they must replicate within a host cell.

A bacteriophage (virus that infects bacteria) injects its DNA into the bacterium while leaving the protein coat on the cell surface. However, in eukaryotes, the entire virus (including the protein capsid) may enter the host cell and then the protein coat is removed (i.e. virus is unencapsulated) after it enters the cytoplasm.

21. A is correct. Bacterial conjugation is the transfer of genetic material (plasmid) between bacterial cells by direct cell-to-cell contact or by a bridge-like connection (i.e. pili) between two cells. Conjugation is a mechanism of horizontal gene transfer, as are transformation and transduction, although these other two mechanisms do not involve cell-to-cell contact (only for conjugation is a pili required).

B, C and D: see explanation for question **15**.

22. D is correct. *E. coli* on the soft agar produce a solid growth (i.e. lawn) covering the agar. Phages lyse the cells which results in the release of more phages (i.e. viruses). The clear spots on the agar plate correspond to where the bacteria were lysed and are known as plaques (i.e. clear spots).

A: viral replication produces clear spots in the lawn of *E. coli* growth.

B: colonies on the agar surface indicate *E. coli* growth, not locations where the virus lysed the bacteria cells.

C: growth of a smooth layer of bacteria across the agar plate would indicate the absence of virus.

23. E is correct. See explanations for questions **40** and **46** for more on Gram staining method.

Gram-negative bacteria are more resistant to antibiotics than Gram-positive bacteria, despite their thinner peptidoglycan layer. The pathogenic capability of Gram-negative bacteria is often associated with certain components of their membrane, in particular the lipopolysaccharide layer (LPS).

All penicillins are β-lactam antibiotics and are used in the treatment of bacterial infections caused by susceptible, usually Gram-positive, organisms.

Penicillin core structure, where "R" is the variable group.

24. B is correct. The viral genome can be single stranded DNA, double stranded DNA, single stranded RNA or double stranded RNA. By definition, retroviruses contain a single-stranded RNA (i.e. mRNA as its genome). Retroviruses contain the enzymes reverse transcriptase and integrase to facilitate their infective ability within the host cell.

25. E is correct. Operons are only found in prokaryotes and encode for a cluster of related (i.e. similar function) genes. The *lac* operon of *E. coli* allows the digestion of lactose and consists of a set of control and structural genes. There are three structural genes controlled by an operator found on another part of the genome. The *lac* genes encoding enzymes are *lacZ*, *lacY* and *lacA*. The fourth *lac* gene is *lacI*, encoding the lactose repressor—"I" stands for *inducibility*.

In the absence of lactose a repressor protein (*lacI*,) is bound to the operator, which prevents RNA polymerase from binding to the DNA. The binding of the repressor protein prevents translation of the structural genes (*lacZ*, *lacY* and *lacA*). However, when lactose is present (i.e. glucose is absent), *lacI* binds to the repressor that dissociates from the operator region, and RNA polymerase (i.e. *inducible system*) attaches to the promoter, and translation occurs.

26. D is correct. A prophage is a phage (i.e. viral) genome inserted and integrated into the circular bacterial DNA chromosome or existing as an extrachromosomal plasmid. This is a lysogenic (i.e. latent form) of a bacteriophage, in which the viral genes are present in the bacterium without causing disruption of the bacterial cell.

The lytic cycle is one of the two cycles of viral reproduction, the other being the lysogenic cycle. The lytic cycle results in the destruction of the infected cell and its membrane.

A key difference between the lytic and lysogenic phage cycles is that in the lytic phage, the viral DNA exists as a separate molecule within the bacterial cell, and replicates separately from the host bacterial DNA.

The location of viral DNA in the lysogenic phage cycle is within the host DNA, therefore in both cases the virus/phage replicates using the host DNA machinery. In the lytic phage cycle, the phage is a free floating separate molecule to the host DNA.

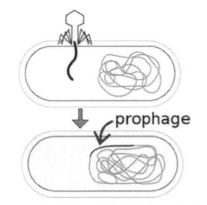

Formation of a prophase: virus is shown at the top attached to the bacterial cell before injection of viral genome

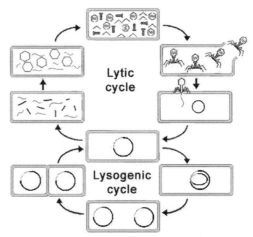

Comparison of viral lifecycle: lytic vs. lysogenic

27. C is correct. By definition, a retrovirus has an RNA genome that is reverse transcribed (by the virus encoded reverse transcriptase enzyme) into single- and then double-stranded DNA that becomes incorporated into the genome of the host cell.

B: reverse transcriptase is used to convert the RNA of the virus into DNA when infected into the host cell.

28. B is correct. *E. coli* are normal flora of the human gut and are adapted to live in the human intestine. Therefore, the optimal temperature for growth is 37°C whereby enzyme activity is optimal.

A: the temperature of the human intestine is 37°C. Bacteria growing optimally at this temperature exhibit the normal phenotype.

C: conjugation by bacteria is not necessary for growth. Conjugation is mediated by a small genetic element (i.e. fertility or F factor) as either independent or integrated into the bacterial chromosome. The F factor encodes for the F pili, which forms a conjugation bridge and allows for the transfer of genetic material between mating cells. Cells carrying the F factor are F$^+$ and transfer it to an F$^-$ cell. During conjugation, part of the bacterial chromosome can be transferred, but the point of origin and the gene order are always the same.

D: oxygen utilization is irrelevant to the growth temperature of bacteria.

29. A is correct.

30. C is correct. Viruses attach to the host by their host specific tail fibers. Once attached to the host cell, the virus enters via endocytosis and the protein capsid remains outside the host cell.

31. C is correct. Bacteria are prokaryotes and they have no membrane-bound organelles such as peroxisomes or nucleolus.

32. E is correct. A lysogen is a bacterial cell in which a phage (i.e. virus) exists as DNA in its dormant state (prophage). A prophage is either integrated into the host bacteria's chromosome (i.e. lysogenic cycle) or (rarely) exists as a stable plasmid within the host cell. The prophage expresses gene(s) that repress the phage's lytic action, until this repression is disrupted and the virus enters the lytic life cycle that ultimately ruptures the cell and releases more virons.

B: temperate refers the ability of some bacteriophages to enter the lysogenic lifecycle.

33. D is correct. For translation, hydrogen bonds form between the codon on the mRNA and the anticodon on the tRNA, not between the amino acid and mRNA. For prokaryotes, the cellular machinery of translation is the 70S ribosome (consisting of the 30S small and 50S large subunits) and protein synthesis (like for eukaryotes) occurs in three phases: initiation, elongation and termination.

First, an initiation complex forms with the 30S subunit, mRNA, initiation factors and a special initiator formyl methionine F-met tRNA. F-met is used only in prokaryotic translation initiation. Then, the 50S subunit binds the initiation complex to form the 70S ribosome. The complete large subunit of the ribosome complex has two binding sites: the P (peptidyl transferase) site and the A (amino acyl) site. The initiator formyl methionine F-met tRNA is located in the P site.

Elongation begins with the binding of a second tRNA (charged with its corresponding amino acid) to the vacant A site of the large ribosomal subunit. The appropriate amino acid is determined by the complimentary hydrogen bonding between the anticodon of the tRNA and the next codon of mRNA. The orientation with respect to the mRNA (from 5' to 3') is E-P-A (ribosomes move towards the 3' end of mRNA). Peptide-bond formation is catalyzed peptidyl transferase (at the P site) which participates in a nucleophile attack from the lone pair of electrons on the N of the amino terminus on the carbonyl of the C terminus (positioned in the A site).

The tRNA in the P site, now uncharged because it lacks an amino acid, moves to the E site and dissociates from the complex. The tRNA that was in the A site (after peptide bond formation and the ribosome translocates) is now in the P site, the A site is vacant and this is where the incoming tRNA (carrying the next amino acid to join the polypeptide) binds. This cycle of binding to the A site, peptide bond formation, and translocation continues to create a growing polypeptide chain.

Termination occurs when one of the three tRNA stop codons (i.e. not charged with an amino acid) hybridizes to the mRNA in the A site. *Termination factors* catalyze the hydrolysis of the polypeptide chain from the tRNA and the ribosomal complex dissociates.

N-Formylmethionine (fMet) is a derivative the methionine amino acid whereby a formyl group has been added to the amino group.

A: prokaryote mRNA does not undergo splicing to remove introns.

B: translation in prokaryotes uses N-terminal formyl methionine (F-met) as the initiator amino acid.

C: prokaryotes have no nucleus and therefore the mRNA has no nuclear membrane to cross and translation begins even before its synthesis is complete. Eukaryotic mRNA must be processed (5' G-cap, 3' poly-A tail and splicing together of exons with the removal of introns) and transported across the nuclear membrane before translation begins in the cytoplasm.

34. B is correct. Surface-to-volume ratio limits the size of biological cells due to the fact that when the volume increases, so does the surface area, but the surface area increases at a slower rate. Eukaryotes are able to grow in size beyond the apparent limitation of the surface-to-volume ratio because of organelles (that allow for a specialized environment) and the extensive cytoplasmic matrix which utilizes microfilaments and motor proteins for cargo transport.

35. D is correct. Prokaryotes have a peptidoglycan cell wall, contain 30S and 50S ribosomes and have a plasma membrane that lacks cholesterol.

36. E is correct. Viruses consist of genetic material in the form of single-stranded DNA, double-stranded DNA, single-stranded RNA or double-stranded RNA packaged within a protein coat.

A: viruses do not have membrane-bound organelles but consist entirely of nucleic acid (RNA or DNA) and, for retroviruses, the reverse transcriptase enzyme.

C: bacteria have a peptidoglycan (i.e. N-linked glucose polymers) cell wall.

D: viruses do not have a phospholipid bilayer membrane.

37. A is correct.

38. B is correct. Reverse transcriptase is used to convert the RNA of the virus into ds-DNA when infected into the host cell. This new DNA is then incorporated into the host cell genome by an integrase enzyme, at which point the retroviral DNA is referred to as a provirus. The host cell treats the viral DNA as part of its own genome, translating and transcribing the viral genes along with the cell's own genes, producing the proteins required to assemble new copies of the virus.

Retrovirus flow of genetic information: RNA (virus) → DNA (from virus template) → RNA (host cell machinery) → polypeptide

39. A is correct. T4 has a DNA genome, which is transcribed and translated by the host's machinery. Since the bacterial cell does not contain a nucleus, the location of transcription and translation are both in the cytoplasm. Therefore, both processes occur in the same locations and are concurrent in time.

B: lysozyme enzymes are encoded by a late gene because the host is not lysed until the viruses are assembled and packaged into their protein capsid to be released for infection of other host cells.

C: assembly must be complete before lysis.

D: bacterial cells have a peptidoglycan cell wall and therefore budding cannot occur. The cell wall is lysed by a lysozyme and the host cell ruptures, which causes the release of the virus.

40. C is correct. Gram-positive bacteria take up the violet stain used in the Gram staining method. This distinguishes them from the other large group of bacteria, the Gram-negative bacteria that cannot retain the crystal violet stain. Gram-negative bacteria take up the counterstain (e.g. safranin or fuchsine) and appear red or pink. Gram-positive bacteria are able to retain the crystal violet stain due to their thick peptidoglycan layer. This peptidoglycan layer is superficial to the cell membrane, whereas in Gram-negative bacteria, this peptidoglycan layer is much thinner and is located between two cell membranes. The cell membrane, peptidoglycan layer and cell wall are three distinct structures. Cell walls provide structural support, protection and rigidity to the cell.

D: teichoic acids are found within the cell wall of Gram-positive bacteria, such as species in *Staphylococcus*, *Streptococcus*, *Bacillus*, *Clostridium* and *Listeria*, and appear to extend to the surface of the peptidoglycan layer. Teichoic acids are not found in Gram-negative bacteria.

41. C is correct. A bacterium with an outer lipopolysaccharide layer is Gram negative and protects against certain antibiotics such as penicillin.

A: fimbriae, a proteinaceous appendage in many Gram-negative bacteria, is thinner and shorter than a flagellum, and allows a bacterium to attach to solid objects.

B: cell membrane of the bacteria is composed of a phospholipid bilayer as in eukaryotes, except it lacks cholesterol.

D: Gram-negative bacteria have a thinner peptidoglycan cell wall that does not retain Gram stain.

42. E is correct. DNase is an enzyme that degrades DNA by hydrolysis of the DNA molecule. Bacterial cells treated with DNase eventually die because all bacteria have DNA genomes while RNA viruses are unaffected by DNase and continue to synthesize proteins following treatment with DNase.

A: having multiple copies of a gene is not enough to prevent DNase from degrading the DNA.

C: viral genomes typically contain multiple reading frames to efficiently use their limited nucleic acid, but multiple reading frames do not prevent DNA hydrolysis.

D: a viral protein coat is not able to denature DNase. Denaturation involves the breaking of weak (e.g. hydrogen, dipole and hydrophobic) bonds from heat or chemical treatment.

43. C is correct.

44. A is correct. Translation in prokaryotes uses the initiation amino acid of N-terminal formyl methionine (F-met). Certain cells in the human immune system can identify f-MET and release local toxins to inhibit bacterial (i.e. prokaryote) infections.

45. C is correct. As a large polysaccharide, dextran does not pass through the cell membrane. Therefore, the osmotic pressure of the solution increases and H_2O moves out of the cell to reduce the osmotic pressure difference and the cell undergoes crenation (i.e. shrinking).

46. A is correct. Gram staining is a method of differentiating bacterial species into two large groups: Gram-positive and Gram-negative. Gram staining differentiates bacteria by the chemical and physical properties of their peptidoglycan cell walls. Gram-positive bacteria have a thicker peptidoglycan cell wall which gives Gram-positive bacteria a purple-blue color from staining while a Gram-negative results in a pink-red color. The Gram stain is usually the first assay in the identification of a bacterial organism.

47. A is correct. DNA polymerase is the enzyme for replicating strands of DNA by synthesizing a new DNA strand using a template strand. Thus, DNA polymerase must be responsible for the replication of the DNA F factor in F$^+$ cells before conjugation.

B: reverse transcriptase is a retroviral enzyme that synthesizes DNA from an RNA (virus) template.

C: DNA ligase catalyzes formation of phosphodiester bonds that link adjacent DNA bases.

D: integrase is a retroviral enzyme that integrates provirus DNA into host genomes.

48. B is correct. *Conjugation* occurs between bacterial cells of different mating types. "Maleness" in bacteria is determined by the presence of a small extra piece of DNA that can replicate independently of the larger chromosome. Male bacteria having this *sex factor* (the *F factor*) are denoted F$^+$ if the sex factor exists extrachromosomally. F$^+$ bacteria can conjugate only with F$^-$ bacteria, the "female" that do not possess the F factor.

Genes on the F factor determine the formation of *sex pili* (hair-like projections) on the surface of the F⁺ bacterium which form cytoplasmic bridges for the transfer of genetic material. The pili also aid the F⁺ cell in adhering to the F⁻ cell during conjugation. During conjugation of an F⁺ cell with an F⁻ cell, and prior to transfer, the F factor replicates and the F factor itself is the DNA most likely to be transferred to the female. The F⁻ thus becomes an F⁺ by receiving one copy of the F factor, while the original F⁺ retains a copy.

If this were the only type of genetic exchange in conjugation, all bacteria would become F⁺ and conjugation would eventually cease. However, in F⁺ bacterial cultures, a few bacteria can be isolated that have the F factor incorporated into their chromosome and are referred to as *Hfr* bacteria, which may also conjugate with F⁻ cells. They do not transfer their F factor during conjugation but often transfer linear portions of their chromosomes. The transfer is interrupted by the spontaneous breakage of the DNA molecule at random sites, usually before the F factor crosses to the F⁻ cell. This process is unidirectional and no genetic material from the F⁻ cell is transferred to the *Hfr* cell.

49. E is correct. Retrotransposons (also known as transposons via RNA intermediates) are a subclass of transposons. They are genetic elements that can amplify themselves in a genome and are ubiquitous components of the DNA of many eukaryotic organisms. Around 42% of the human genome is made up of retrotransposons.

Retrotransposons copy themselves to RNA and then back to DNA that may integrate back to the genome. The second step of forming DNA may be carried out by a reverse transcriptase, which is encoded by the retrotransposon.

See the image on the following page.

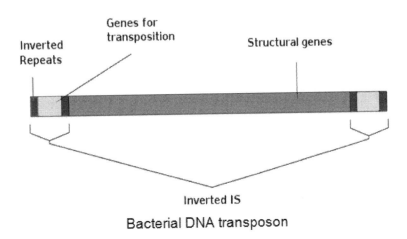

Bacterial DNA transposon

50. B is correct. Plasmids are in bacteria as small circular double-stranded DNA that carry extrachromosomal genetic information. Plasmids (with an origin of replication) are replicated by bacterial proteins and are inherited to progeny. The proteins encoded by genes on plasmids often provide resistance to antibiotics by degrading the antibiotic.

Through recombinant DNA technology, plasmids are engineered to carry other genes not normally found in bacteria. Plasmids can be introduced into cells by transformation techniques that allow DNA to cross through the cell wall and plasma membrane without killing bacterial cells. After transformation, exposure to the specific antibiotic allows selection of bacteria that received plasmid. Bacteriophages such as bacteriophage λ are viruses that infect bacterial cells. Bacteriophage λ can also be engineered to carry novel genes not normally present in either the bacteria or bacteriophages.

I: plasmids are extra-chromosomal circular DNA molecules and are not organelles because organelles are membrane-bound cellular components present only in eukaryotes.

II: like the bacterial genome, plasmids are found in the cytoplasm. Without a nucleus, prokaryotic ribosomes translate plasmid mRNA into proteins while it is being transcribed into mRNA from DNA.

III: plasmids rely on bacterial machinery for metabolic processes (e.g. replication, transcription and translation).

51. B is correct. Fungi are members of a large group of eukaryotic organisms that includes microorganisms such as yeasts, molds and mushrooms. These organisms are classified as a Fungi kingdom (which is separate from plants, animals, protists and bacteria). One major difference is that fungal cells have cell walls that contain chitin, unlike the cell walls of plants and some protists (contain cellulose), and unlike bacteria (contain peptidoglycan).

A: protoplasts are plant, bacterial or fungal cells that have lost their cell wall by either mechanical or enzymatic means.

C: L forms are strains of bacteria that lack a cell wall.

D: viruses do not have cell walls. A complete virus particle (virion) consists of nucleic acid surrounded by a protective coat of protein known as capsid.

E: mycoplasma refers to a genus of bacteria that lack a cell wall (they have call membranes only). Without a cell wall, they are unaffected by many common antibiotics such as penicillin or other beta-lactam antibiotics that target peptidoglycan cell wall synthesis.

52. E is correct. Prokaryotes do not have membrane-bound organelles (e.g. nucleus, mitochondria, lysosomes etc).

C: Prokaryotes have peptidoglycan cell walls while fungi have cell walls composed of chitin.

53. C is correct. Bacteriophages are viruses that infect bacteria and typically consist of a protein coat (i.e. head) and a core containing nucleic acid. Like all viruses, bacteriophages contain host-specific protein tail fibers specialized for attaching to bacteria. Upon infection, bacteriophage can enter one of two life cycles.

In the lytic cycle, the viral nucleic acid enters the bacterial cell and begins using host machinery to produce new virions (i.e. virus particles), which eventually lyse the host cell and infect other cells. In the lysogenic cycle, the viral DNA integrates into the bacterial chromosome, replicating with it and being passed to daughter cells (integrated into their genome) in this inactive form. Once an integrated virus (i.e. prophage) becomes activated, it exits the lysogenic cycle and enters the lytic cycle. Activation triggers for the virus to enter the lytic cycle include cellular stresses such as UV light, depleted nutrients or other cellular abnormalities.

Retroviruses are RNA-containing viruses that replicate by reverse transcriptase through a DNA intermediate by a viral-coded RNA-dependent DNA polymerase. Human immunodeficiency virus (HIV), the causative agent of AIDS, is an example of a retrovirus. Retroviral life-cycle consists of four main events: 1) the virus binds to its host and injects its RNA and a few viral enzymes; 2) the RNA is then converted to DNA by reverse transcriptase; 3) the DNA then integrates into the host cell's genome; 4) the viral genes are expressed and virions are assembled and released from the cell by budding.

Both bacteriophage and retrovirus can integrate their genetic material into the host cell genome (i.e. lysogenic cycle) or lyse the cell (i.e. lytic cycle).

A: bacteriophages have host-specific protein tail fibers and only infect bacteria.

B: bacteria (target of bacteriophage) have no immune system. Bacteria do have a defense against infection from nucleotide-specific restriction enzymes used in the laboratory for molecular biology and biotechnology.

D: retroviruses contain an RNA genome (either ss-RNA or ds-RNA) and contain the enzyme reverse transcriptase, which is not present in bacteriophage.

54. E is correct.

55. A is correct. The sequences recognized by most restriction enzymes are inverted repeats (i.e. palindromes) that read the same if inverted (i.e. rotated by180°). Therefore, the sticky ends at each end of a DNA fragment are the same in either orientation, so ligation occurs from either orientation.

B: DNA strands do serve as primers for DNA polymerase

C: DNA ligase creates a phosphodiester bond that covalently links sticky ends of DNA.

D: plasmid DNA, like bacterial DNA, is double stranded.

56. D is correct. The lytic lifecycle is used by virulent viruses. Virulent describes either disease severity or a pathogen's infectivity. The virulence factors of bacteria are typically proteins or other molecules that are synthesized by enzymes. These proteins are coded for by genes in chromosomal DNA, bacteriophage DNA or plasmids.

57. C is correct. The promoter is a region on DNA that is recognized and bound by RNA polymerase as the initiation site for transcription.

B: the inducer is the molecule which inactivates the repressor. When the inducer binds to the repressor, the bound repressor dissociates from the DNA and allows the polymerase to bind and turns on the operon.

D: the repressor binds to the operator (i.e. segment of DNA within the operon).

Chapter 1.6: Photosynthesis

1. E is correct. C_4 plants are those plants that use C_4 carbon fixation. There are the three carbon fixation mechanisms: C_3, C_4 and CAM. C_4 is named for the 4-carbon molecule present in the first product of carbon fixation compared to a more common 3-carbon molecule products in C_3 plants. C_4 plants have a competitive advantage over plants that use C_3 carbon fixation pathway under conditions of drought, high temperatures, and nitrogen or CO_2 limitation. About 7,600 plant species use C_4 carbon fixation, which represents about 3% of all terrestrial species of plants.

C_4 fixation is believed to have evolved more recently. C_4 and CAM overcome the tendency of the enzyme RuBisCO to wastefully fix oxygen rather than carbon dioxide in what is called photorespiration. This is achieved by using a more efficient enzyme to fix CO_2 in mesophyll cells and shuttling this fixed carbon to bundle-sheath cells. In these bundle-sheath cells, RuBisCO is isolated from atmospheric oxygen and saturated with the CO_2 released by decarboxylation of the malate or oxaloacetate. Because these additional steps require more energy in the form of ATP, C_4 plants are able to more efficiently fix carbon in only certain conditions, with the C_3 pathway being more efficient in other conditions.

2. D is correct. CAM (crassulacean acid metabolism) photosynthesis is a carbon fixation pathway that evolved in some plants as an adaptation to arid conditions. In a plant using full CAM, the stomata in the leaves remain shut during the day to reduce evapotranspiration, but open at night to collect carbon dioxide (CO_2). The CO_2 is stored as the four-carbon acid malate, and then used during photosynthesis during the day. The pre-collected CO_2 is concentrated around the enzyme RuBisCO, increasing photosynthetic efficiency.

3. A is correct.

5. C is correct.

4. B is correct.

6. D is correct.

Chloroplast

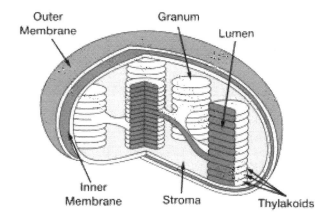

7. A is correct. Stroma is the colourless fluid surrounding the thylakoids within the chloroplast. Within the stroma are grana (i.e. stacks of thylakoids) – the sub-organelles, where photosynthesis is commenced before the chemical changes are completed in the stroma.

8. B is correct.

9. A is correct.

10. E is correct.

11. C is correct.

12. C is correct.

13. E is correct. Thylakoid is a membrane-bound compartment inside chloroplasts and cyanobacteria. They are the site of photosynthesis. Thylakoids consist of a thylakoid membrane surrounding a thylakoid lumen. Chloroplast thylakoids frequently form stacks of disks (i.e. grana).

14. B is correct.

15. A is correct. See explanation for question **13**.

16. A is correct.

17. B is correct.

18. E is correct.

19. D is correct.

20. A is correct.

21. A is correct.

22. B is correct.

See illustration →

23. E is correct.

24. C is correct.

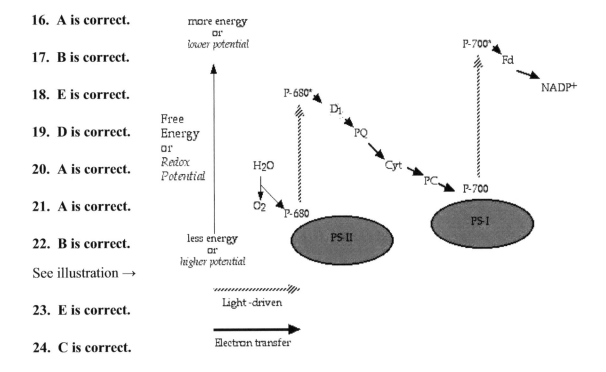

25. B is correct. The Calvin cycle (i.e. C_3 cycle) is a series of biochemical redox reactions that take place in the stroma of chloroplasts in photosynthetic organisms. It is one of the light-independent reactions used for carbon fixation. The key enzyme of the cycle is RuBisCO.

Photosynthesis occurs in two stages in a cell. In the first stage, *light-dependent reactions* capture the energy of light and use it to make the energy storage and transport molecules ATP and NADPH. The *light-independent* Calvin cycle uses the energy from short-lived electronically excited carriers to convert carbon dioxide and water into organic compounds that can be used by the organism (and by animals that feed on it). This set of reactions is known as carbon fixation.

26. E is correct.

27. D is correct.

28. D is correct.

29. C is correct.

30. A is correct.

31. C is correct.

32. A is correct.

33. E is correct.

34. C is correct. An autotroph (i.e. *producer*) is an organism that produces complex organic compounds (i.e. carbohydrates, fats and proteins) from simple substances present in its environment, generally using energy from light (photosynthesis) or inorganic chemical reactions (chemosynthesis). Autotrophs are the producers in a food chain (plants on land and algae in water), in contrast to heterotrophs that consume autotrophs. They are able to make their own food, and do not need a living energy or organic carbon source. Autotrophs can reduce carbon dioxide to make organic compounds. Most autotrophs use water as the reducing agent, but some can use other hydrogen compounds. *Phototrophs* (i.e. green plants and algae), a type of autotroph, convert physical energy from sunlight into chemical energy in the form of reduced carbon.

35. A is correct.

36. D is correct.

37. E is correct.

38. B is correct.

39. A is correct.

40. B is correct.

41. A is correct.

42. B is correct.

43. B is correct.

44. C is correct.

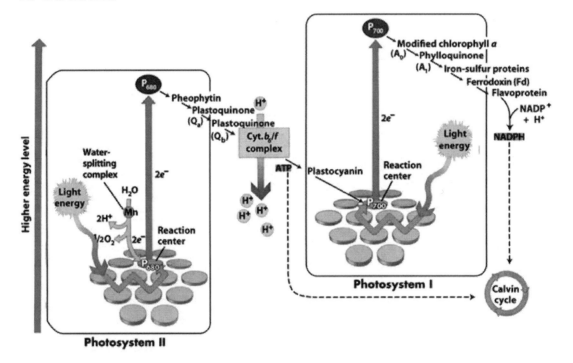

45. E is correct.

46. A is correct.

47. D is correct.

48. A is correct.

49. E is correct. Chlorophyll *a* is a specific form of chlorophyll used in oxygenic photosynthesis. It absorbs most energy from wavelengths of violet-blue and orange-red light. It also reflects green/yellow light, and therefore contributes to the observed green color of most plants. This photosynthetic pigment is essential for photosynthesis in eukaryotes, cyanobacteria and prochlorophytes because of its role as primary electron donor in the electron transport chain.

D: Chlorophyll *b* is another form of chlorophyll that helps in photosynthesis by absorbing light energy. It is more soluble than chlorophyll *a* in polar solvents because of its carbonyl group. Its color is yellow and it primarily absorbs blue light.

50. C is correct.

51. A is correct.

52. B is correct. See explanation for question **25**.

53. D is correct.

54. C is correct. See explanation for question **25**.

55. B is correct.

56. E is correct.

57. C is correct.

58. A is correct.

UNIT 2. ECOLOGY

Chapter 2.1: Energy Flow, Nutrient Cycles, Ecosystems, Biomes

1. A is correct. An ecosystem is a community of living organisms (e.g. plants, animals and microbes) in conjunction with the nonliving components of their environment (e.g. air, water and minerals). These living (i.e. biotic) and nonliving (i.e. abiotic) components are linked together through nutrient cycles and energy flows. Ecosystems are defined by the network of interactions among organisms, and between organisms and their environment.

2. B is correct.

3. A is correct. Symbiosis is close and often long-term interaction between two or more different species. The definition of symbiosis involves a controversy among biologists whereby some believe that the term should refer only to persistent mutualisms (where all organisms derive benefits from the relationship), while others believe it should apply to all types of persistent biological interactions whether mutualistic (+/+), commensalistic (+/0) or parasitic (+/−). After decades of debate, modern biology and ecology now use the latter definition.

4. C is correct.

5. E is correct. By their percentage contribution to the greenhouse effect on Earth, the four major gases are: water vapor (36–70%), carbon dioxide (9–26%), methane (4–9%) and ozone (3–7%).

6. D is correct.

7. E is correct.

8. C is correct. Biological aspects of an organism's niche involve the biotic factors required for an organism's survival. Soil, sunlight, water and minerals are abiotic factors. Niche is a term that describes a species' way of life with each species having a separate, unique niche. The ecological niche describes how an organism (or population) responds to the distribution of resources and competitors (e.g. growing with abundant resources while predators are scarce) and how it alters those same factors (e.g., limiting access by other organisms, being a food source for predators and a consumer of prey).

9. C is correct.

10. B is correct. The competitive exclusion principle (also known as Gause's law of competitive exclusion or Gause's law) is a concept that states that two species competing for the same resources cannot coexist if other ecological factors are constant. When one species has even the slightest advantage over another, the one with the advantage will dominate in the long term. One of the two competitors will always overcome the other, leading to either the extinction of this competitor or an evolutionary or behavioral shift toward a different ecological niche.

11. D is correct.

12. E is correct. Primary producers (i.e. autotrophs) are organisms in an ecosystem that produce energy from inorganic compounds. Often, they use photosynthesis (e.g. plants and cyanobacteria). Archea (i.e. unicellular organisms) may produce biomass from the oxidation of inorganic chemical compounds (i.e. chemoautotrophs) such as in hydrothermal vents in the deep ocean.

Fungi and other organisms that produce biomass from oxidizing organic materials are decomposers and are not primary producers.

13. A is correct.

14. B is correct. Habitat is an environmental (or ecological) area that is inhabited by a species. It is the natural environment in which an organism lives or the physical environment that encompasses a population.

15. A is correct.

16. E is correct. See explanation for question **20**.

17. A is correct.

18. B is correct. Biomass is the mass of living biological organisms in a given area or ecosystem at a given time. The term can refer to *species biomass* (i.e. the mass of one or more species), or to *community biomass* (i.e. the mass of all species in the community). It can include microorganisms, plants or animals. The mass can be expressed as the average mass per unit area, or as the total mass in the community. How biomass is measured depends on why it is being measured.

19. C is correct.

20. D is correct. Parasitism is a non-mutual symbiotic relationship between species, where one species (i.e. the parasite) benefits at the expense of the other (i.e. the host).

A: synnecrosis is a rare type of symbiosis in which the interaction between species is detrimental to both organisms. Therefore, it is a short-lived condition because the interaction eventually causes death. Evolution selects against synnecrosis and it is uncommon in nature.

B: see explanation to question **40**.

C: mutualism is the way two organisms of different species exist in a relationship in which each individual benefits.

E: commensalism is a class of relationship between two organisms where one organism benefits without affecting the other.

21. C is correct.

22. D is correct.

23. E is correct.

24. C is correct.   Primary succession is one of two types of biological and ecological succession of plant life, occurring in an environment devoid of vegetation and usually lacking soil (e.g. lava flow). It is the gradual growth of an ecosystem over a longer period.

In contrast, secondary succession occurs on substrate that previously supported vegetation before an ecological disturbance from less cataclysmic events (e.g. floods, hurricanes or tornadoes) which destroyed the plant life.

25. B is correct.

26. E is correct. See explanation for question **24**.

27. B is correct. Snail is a typical detritivore. Detritivores are heterotrophs that obtain nutrition from decomposing plants, dead parts and feces (i.e. detritus) and contribute to decomposition and the nutrient cycles. Detritivores are not to be confused with other decomposers such as bacteria, fungi and protists, which are unable to ingest minute lumps of matter. Decomposers absorb and metabolize nutrients on a molecular scale (i.e. saprotrophic nutrition is the processing of dead or decaying material).

28. A is correct.

29. D is correct. In ecology, climax community describes a biological community of plants, animals and fungi which, through the process of ecological succession (i.e. vegetation in an area over time) has reached a steady state. This equilibrium occurs because the climax community is composed of species best adapted to average conditions in that area.

30. A is correct.

31. B is correct. See explanation for question **33**.

32. C is correct.

33. B is correct. Tundra is a biome where tree growth is hindered by low temperatures and short growing seasons. The vegetation is composed of dwarf shrubs, grasses, mosses and lichens.

Dessert is a barren area of land where little precipitation occurs and consequently living conditions are hostile for plant and animal life.

34. A is correct.

35. D is correct.

36. E is correct. Nitrogen fixation is when atmospheric nitrogen (N_2) is converted into ammonium (NH_4). Nitrogen fixation is necessary for life because nitrogen is required for nucleotides of DNA and RNA, and amino acids of proteins.

Microorganisms that can fix nitrogen are prokaryotes (i.e. both bacteria and archaea) called diazotrophs. Some higher plants and some animals (e.g. termites) have symbiotic relationships with diazotrophs.

37. C is correct.

38. B is correct. Biomes are climatically and geographically defined as contiguous areas with similar climatic conditions such as communities of animals, plants and soil organisms, and are often called ecosystems. Biomes are defined by factors such as plant structures (e.g. shrubs, trees, and grasses), leaf types (e.g. broadleaf and needleleaf), plant spacing (e.g. forest, woodland, savanna) and climate.

Biomes, unlike ecozones, are not defined by genetic, taxonomic, or historical similarities. Biomes are often identified with particular patterns of ecological succession and climax vegetation.

39. A is correct.

40. E is correct. In ecology, predation is a biological interaction where a predator (a hunting organism) feeds on its prey. Predators may or may not kill their prey prior to feeding on them, but the act of predation most often results in the prey's death and the eventual absorption of the prey's tissue through consumption. Often the remains of the prey become food source for scavengers – organisms that feed on dead organisms that were not killed by the scavenger. However, organisms that are most often considered to be predators, may frequently engage in scavenging feeding behavior when given a chance (lions, tigers, wolves etc.).

41. B is correct.

42. C is correct. Transpiration is the process of water movement through a plant and its evaporation from aerial parts, such as from leaves but also from stems and flowers.

43. D is correct.

44. A is correct. Phytoplankton (microalgae) contain chlorophyll and require sunlight to live and grow, like terrestrial plants. Most phytoplankton are buoyant and float in the upper layers of the ocean, where sunlight penetrates the water. Phytoplankton is the base of many aquatic food webs whereby shrimp, jellyfish and whales feed on them.

Zooplankton are heterotrophic (sometimes detritivorous) plankton – organisms drifting in oceans, seas and bodies of fresh water. Individual zooplankton are usually microscopic, but some (e.g. jellyfish) are larger and visible with the naked eye.

45. C is correct.

46. B is correct. Boreal forest (aka taiga) is a biome characterized by coniferous (i.e. softwood) forests consisting mostly of pines, spruces and larches and is the world's largest terrestrial biome.

47. E is correct.

48. A is correct.

49. D is correct. See explanation for question **52**.

50. A is correct.

51. C is correct.

52. D is correct. Estuaries are partly enclosed coastal bodies of brackish water (i.e. saline levels of about 0.05 to 3% which is between fresh and salt water). An estuary forms when one or more rivers/streams flow in and has a free connection to the open sea.

Estuaries form a transition zone between river and maritime environments and have both marine influences (e.g. tides, waves and saline water) and river influences (e.g. fresh water and sediment). The inflows of both sea and fresh water provide high levels of nutrients in both the water column and sediment, making estuaries among the most productive natural habitats.

53. E is correct.

54. A is correct.

55. B is correct. A: paleontology is the scientific study of prehistoric life and includes the study of fossils to determine organisms' evolution.

C: microbiology is the scientific study of microscopic organisms, either unicellular (i.e. single cell), multicellular (i.e. cell colony), or acellular (i.e. lacking cells). Microbiology encompasses sub-disciplines such as mycology (i.e. study of fungi), virology, parasitology and bacteriology.

D: entomology is the study of insects (branch of arthropodology) which is a branch of zoology.

E: zoology is the branch of biology that studies the animal kingdom, including the structure, embryology, evolution, habits, classification and distribution of both living and extinct animals.

56. B is correct.

57. D is correct. The photic zone (sunlight zone) is the depth of the water in a lake or ocean that is exposed to sufficient sunlight for photosynthesis to occur. It extends from the surface down to a depth where light intensity falls to one percent. The thickness of the photic zone depends on the extent of intensity loss in the water column and can vary from only a few inches in highly turbid lakes (i.e. seasonal affects) to around 600 feet in the ocean.

Chapter 2.2: Populations, Communities, Conservation Biology

1. D is correct.

2. A is correct. In ecology, the term community refers to populations of two or more different species occupying the same geographical area and at a particular time.

3. B is correct.

4. D is correct. Sustainable development is the practice of meeting the needs of people in the present without compromising the ability of future generations to meet their own needs. Preventing long-term harm to the environment ensures that people in the future can benefit from its use.

5. C is correct. The range (or distribution) of population is the geographical area within which that population can be found. Within that range, dispersion is variation in local density.

6. E is correct.

7. A is correct. The population growth rate refers to the rate at which the number of people increases in a given period of time. The flow of people in and out of the country, as well as the number of births and deaths, affect the number of individuals in a population.

8. B is correct.

9. C is correct. A biotic factor is any living component that affects another organism, including animals that consume the organism in question, and the living food that the organism consumes. Each biotic factor needs energy to do work and food for proper growth. Biotic factors include human influence.

10. E is correct. Coal, natural gas, oil and other commodities derived from fossil fuels, as well as minerals like copper and others, are non-renewable resources - resources that do not renew themselves at a sufficient rate for sustainable economic extraction in meaningful human time-frames (require tens of thousands to millions of years and certain conditions to form).

A renewable resource is a natural resource which can replenish with the passage of time, either through biological reproduction or other naturally recurring processes. Renewable resources are a part of Earth's natural environment and the largest components of its ecosphere. Renewable resources may be the source of power for renewable energy. However, if the rate at which the renewable resource is consumed exceeds its renewal rate, renewal and sustainability will not be ensured.

11. A is correct.

12. B is correct.

13. C is correct.

14. D is correct.

15. E is correct. A slowed growth rate means that there are fewer new individuals in a population. Increased birthrate and decreased death rate lead to a rise in growth rate, whereas decreased birthrate slows down the population growth rate.

16. E is correct. Desertification is a type of land degradation in which a relatively dry land region becomes increasingly arid, typically losing its bodies of water as well as vegetation and wildlife. It is caused by a variety of factors, such as climate change and human activities. The immediate cause is the removal of most vegetation. This is driven by a number of factors, alone or in combination, such as drought, climatic shifts, tillage for agriculture, overgrazing and deforestation for fuel or construction materials. Desertification is a significant global ecological and environmental problem.

B: Monoculture is the agricultural practice of growing a single crop over a wide area and for a large number of consecutive years. It is widely used in modern industrial agriculture and its implementation has allowed for large harvests from minimal labor. However, monocultures can lead to the quick spread of pests and diseases whereby a uniform crop is susceptible to a pathogen.

17. A is correct.

18. C is correct. See explanation for question **16**.

19. B is correct.

20. D is correct. Soil erosion is the gradual wearing away of a field's top layer of soil by natural elements such as water and wind, and by farming activities. Irrigation involves controlling the amount of water released by supplying it at regular intervals for farming, which is an effective way of maintaining the quality of the soil.

21. D is correct. See explanation for question **66**.

22. A is correct.

23. E is correct.

24. A is correct.

25. B is correct.

26. C is correct. DDT was a pesticide used in the United States to control the amount of insects on food crops and in buildings for pest control. It was discontinued because of the harmful effects on wildlife.

27. D is correct. See explanation for question **10**.

28. C is correct. Density-dependent limiting factors affect the size or growth of a population due to variation in population density. Dense populations are more strongly affected than less crowded ones. Examples include availability of food, disease, living space, predation and migration.

29. E is correct. Carrying capacity is the maximum size of a population that the environment can sustain indefinitely based on the resources available. If a population grows larger than the carrying capacity, species may die due to lack of environmental resources.

30. B is correct.

31. B is correct.

32. E is correct.

33. A is correct. Acid rain is caused by a chemical reaction in which sulfur dioxide and nitrogen oxides are released into the air. These compounds rise into the atmosphere, where they react with water and oxygen, and eventually fall to the ground as precipitation.

34. C is correct. A logistic curve (or logistic function) is a common sigmoid function (S-shape) in relation to population growth. The initial stage of growth is approximately exponential; then, as saturation begins, the growth slows, and at maturity, growth stops.

35. C is correct.

36. A is correct. See explanation to question **33**.

37. D is correct.

38. D is correct. See explanation for question **28**.

39. E is correct.

40. B is correct. See explanation for question **29**.

41. A is correct. Species diversity is the effective number of different species that are represented in a collection of individuals.

B: genetic diversity refers to the total number of genetic characteristics in the genetic makeup of a species. It is distinguished from genetic variability, which describes the tendency of genetic characteristics to vary. It serves as a way for populations to adapt to changing environments.

C: ecosystem diversity refers to the diversity of a place at the level of ecosystems (i.e. biotic and abiotic factors).

D: biodiversity refers to variation in species rather than ecosystems.

42. B is correct.

43. D is correct. See explanation for question **41**.

44. E is correct.

45. A is correct.

46. B is correct.

47. D is correct. Higher birthrate and lower death rate both increase the number of individuals in a population, thus increasing competition. Fewer resources and higher population density also increase competition. A decrease in population size can reduce competition.

48. C is correct. Habitat fragmentation is the process of division of large habitats into smaller, isolated patches due to loss of habitat through human activity and natural causes. This negatively affects biodiversity because it reduces the amount of suitable habitat available for certain species.

49. C is correct.

50. E is correct.

51. D is correct. See explanation for question **28**.

52. A is correct.

53. B is correct.

54. A is correct.

55. C is correct.

56. E is correct. See explanation for question **66**.

57. B is correct.

58. D is correct.

59. E is correct.

60. C is correct.

61. C is correct. See explanation for question **66**.

62. D is correct.

63. B is correct. Monoculture farming is the practice of growing the same single crop or plant in a field year after year. This requires standardized planting, maintenance and harvesting, which results in greater yields and lower costs. However, over time, monocultures can lead to quicker buildup and spread of pests and diseases.

64. E is correct.

65. C is correct.

66. D is correct. Demographic transition refers to the shift from high birth and high death rates to low birth and low death rates as a country develops from a pre-industrial to an industrialized economic system.

67. A is correct.

68. E is correct. Global warming refers to the rise in the average temperature of the Earth's climate system. The graph shows that the land-surface air temperature over the last century has been gradually increasing at a fast rate.

69. B is correct.

70. A is correct.

UNIT 3. GENETICS

Chapter 3.1: DNA and Protein Synthesis

1. D is correct.

2. D is correct. In 1928, Frederick Griffith reported one of the first experiments suggesting that bacteria are able to transfer genetic information through a process known as transformation. Griffith observed a "transforming principle", where heat killed S bacterial strain was destroyed, and what we now know to be its DNA had survived the process and was taken up by the R strain bacteria. Equipped with the S genetic fragments, the R strain bacteria were able to protect themselves from the host's immune system and could kill the host.

3. E is correct. Depurination is the hydrolysis of the glycosidic bond of DNA and RNA purine nucleotides guanine or adenine. Nucleotides are composed of a sugar, phosphate and base (A, C, G, T/U). For DNA, the sugar is deoxyribose, while RNA contains ribose.

A glycosidic bond is formed between the hemiacetal group of a saccharide (or an amino acid) and the hydroxyl group of some organic compound, such as an alcohol.

4. A is correct. Unequal crossing over is a type of gene duplication (or deletion) event that deletes a sequence in one strand of the chromatid and replaces it with a duplication from its sister chromatid in mitosis, or when homologous chromosomes recombine during prophase I in meiosis.

5. D is correct. Sulfur is found in the amino acid cysteine, but is absent in nucleic acids. The Hershey-Chase (i.e. blender) experiment used radiolabeled molecules of phosphorus (^{32}P for nucleic acids) and sulfur (^{35}S for proteins) to determine whether nucleic acids (phosphorus) or protein (sulfur) carried the genetic information.

Nucleic acids contain the elements C, H, O, N and P. Nucleic acids are polymers of nucleotide subunits and encode all the information needed by the cell of the organism to synthesize proteins and replicate via cell division.

6. D is correct.

7. A is correct.

8. C is correct. Translation is initiated on ribosomes located within the cytoplasm.

A: transcription (not translation) occurs in the nucleus.

B: Golgi receives proteins from the rough endoplasmic reticulum for processing and sorting. The Golgi modifies (e.g. adds carbohydrate groups) and sorts proteins in the secretory pathway.

D: proteins secreted into the lumen of the rough endoplasmic reticulum have at their amino terminus a special sequence of amino acids referred to as a *leader sequence* (about 6 to 10 amino acids long). This sequence is recognized by the signal recognition protein (SRP) that binds to a receptor on the rough ER, attaching the ribosome and the nascent polypeptide to the endoplasmic reticulum membrane.

9. E is correct. See explanation for question **4** above.

10. C is correct. To show that it was DNA, rather than some small amount of RNA, protein or some other cell component, that was responsible for transformation, Avery, McLeod and McCarty used a number of biochemical tests. They found that trypsin, chymotrypsin and ribonuclease (enzymes that destroy proteins or RNA) did not affect the transforming agent that caused disease, but a DNAse treatment that could degrade DNA destroyed the extract's ability to cause disease.

Streptococcus pneumoniae (i.e. pneumococcus) is a Gram-positive pathogenic bacterium. *S. pneumoniae* was recognized as a major cause of pneumonia in the late 19th century and has been the subject of many humoral immunity (i.e. antibody mediated) studies.

11. B is correct. Western blotting separates proteins by electrophoresis and is commonly used to identify the presence of HIV antibodies (proteins).

All blotting techniques rely on the use of gel electrophoresis to separate DNA, RNA or proteins based on size. After resolution by electrophoresis, the gel (containing the resolved products) is placed next to blotting paper (i.e. nitrocellulose). After transfer of the macromolecules (DNA, RNA or proteins) onto the gel by capillary action, the blotting paper is probed by specific markers that hybridize to regions with complimentary sequences on the blotting paper.

A: Eastern blotting does not exist.

C: Northern blotting uses RNA in gel electrophoresis.

D: Southern blotting uses DNA in gel electrophoresis.

12. C is correct. Position C represents the 3' hydroxyl (~OH) of the ribose sugar. As in DNA, the 3' hydroxyl is the site of attachment. RNA and DNA require a free 3' OH for the nucleic acid to increase in length.

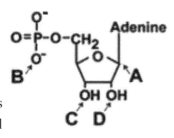

Position D contains a 2' hydroxyl that distinguishes this sugar as ribose, as opposed to DNA which would lack the 2' hydroxyl (deoxyribose = without oxygen) at the 2' position of the sugar.

13. E is correct. Adenine and guanine are purines. Cytosine and thymine/uracil are pyrimidines; note the presence of y for the pyrimidines.

A: the strands of DNA are antiparallel: one strand has a 5'→ 3' polarity and its complementary strand has a 3' → 5' polarity.

B: DNA consists of nucleotides, which are composed of a phosphate group, a deoxyribose sugar, and a base (A, C, G, T).

C: the sugar molecule in DNA is deoxyribose (lacking a 3' ~OH), while RNA uses ribose.

D: cytosine binds to guanine with three hydrogen bonds. Adenine binds to either thymine (in DNA) or uracil (in RNA) with two hydrogen bonds.

14. C is correct. A tumor suppressor gene is a gene that protects a cell from one step on the path to cancer. When this gene mutates to cause a loss (or reduction) in its function, the cell can progress to cancer, usually in combination with other genetic changes. The loss of these genes may be even more important than proto-oncogene/activation for the formation of many kinds of human cancer cells.

Apoptosis is the process of programmed cell death (PCD) that may occur in multicellular organisms. Biochemical events lead to characteristic cell changes (morphology) and death, which include blebbing, cell shrinkage, nuclear fragmentation, chromatin condensation and chromosomal DNA fragmentation. In contrast to necrosis (i.e. traumatic cell death that results from acute cellular injury), apoptosis confers advantages during an organism's lifecycle. For example, the separation of fingers and toes in a developing human embryo occurs because cells between the digits undergo apoptosis. Unlike necrosis, apoptosis produces cell fragments called apoptotic bodies that phagocytic cells are able to engulf and quickly remove before the contents of the cell can spill out onto surrounding cells and cause damage.

A: telomerase is an enzyme that adds DNA sequence repeats (i.e. TTAGGG) to the 3' end of DNA strands in the telomere regions that are found at the ends of eukaryotic chromosomes. This region of repeated nucleotide called telomeres contains noncoding DNA and hinders the loss of important DNA from chromosome ends. Consequently, when the chromosome is copied, only 100–200 nucleotides are lost, which causes no damage to the coding region of the DNA.

Telomerase is a reverse transcriptase that carries its own RNA molecule that is used as a template when it elongates telomeres that have been shortened after each replication cycle. Embryonic stem cells express telomerase, which allows them to divide repeatedly. In adults, telomerase is highly expressed in cells that divide regularly (e.g. male germ cells, lymphocytes and certain adult stem cells), whereas telomerase is not expressed in most adult somatic cells.

15. B is correct.

16. C is correct. The fact that DNA was the transforming principle was verified by other types of experiments done by Avery, McLeod and McCarty (1944) and by Hershey and Chase (1952).

17. D is correct. There are 4 different nucleotides (adenine, cytosine, guanine, and thymine/uracil). Each codon is composed of 3 nucleotides, and therefore there must be 64 (4^3) possible variations of codons to encode for the 20 amino acids. The genetic code is degenerate (i.e. redundant) because several different codons encode for the same amino acid. There are 61 codons that encode for amino acids and 3 stop codons that terminate translation.

18. E is correct.

19. C is correct. Adenine pairs with thymine *via* 2 hydrogen bonds, while guanine pairs with cytosine *via* 3 hydrogen bonds. Treatment of DNA with 2-aminopurine causes the adenine-thymine (A-T) base pair to be replaced with a guanine-thymine (G-T) base pair (before replication). Therefore, after replication the thymine is replaced by cytosine: G-C base pair.

This single point mutation is then incorporated into future generations. If the mutation had been corrected before replication (via proofreading mechanisms during replication), there would be no change in the DNA base sequence.

20. D is correct. Codons consisting of only two bases would be insufficient because the four bases in a two-base codon would only form $4^2 = 16$ pairs, which is less than the 20 combinations needed to specify the amino acids. A triplet would be sufficient because four bases in a three-base codon can form $4^3 = 64$ pairs, which are enough to encode for the 20 amino acids.

21. B is correct. After one replication, the DNA was found to have intermediate density between ^{14}N and ^{15}N. After two replications, there were two densities – one band in the centrifuge tube was an intermediate between the ^{14}N and ^{15}N, while the other consisted of only ^{14}N.

The semiconservative model mechanism of replication was one of three models originally proposed for DNA replication and tested by the Meselson-Stahl experiment.

- *Semiconservative replication* would produce two copies that each contained one of the original strands and one new strand.

- *Conservative replication* would leave the two original template DNA strands together in a double helix and would produce a copy composed of two new strands containing all of the new DNA base pairs.

- *Dispersive replication* would produce two copies of the DNA, both containing distinct regions of DNA composed of either both original strands or both new strands.

In the Meselson-Stahl experiment, *E. coli* were grown for several generations in a medium with ^{15}N. When DNA was extracted and separated by centrifugation, the DNA separated according to density. The DNA of the cells grown in ^{15}N medium had a higher density than cells grown in normal ^{14}N medium. Then, *E. coli* cells with only ^{15}N in their DNA were transferred to a ^{14}N medium and were allowed to divide.

Since conservative replication would result in equal amounts of DNA of the higher and lower densities (but no DNA of an intermediate density), conservative replication was excluded. However, this result was consistent with both semiconservative and dispersive replication.

Semiconservative replication would result in double-stranded DNA with one strand of ^{15}N DNA and one strand of ^{14}N DNA, while dispersive replication would result in double-stranded DNA with both strands having mixtures of ^{15}N and ^{14}N DNA, either of which would have appeared as DNA of an intermediate density.

22. C is correct. A DNA microarray (known as DNA chip) is a collection of microscopic DNA spots attached to a solid surface. DNA microarrays are used to measure the expression levels of large numbers of genes simultaneously or to genotype multiple regions of a genome. Each DNA spot contains picomoles (10^{-12} moles) of a specific DNA sequence, known as *probes* (or *reporters or oligos*). These short sections of a gene (or other DNA element) are used to hybridize a cDNA (or anti-sense RNA) probe. Since an array can simultaneously use tens of thousands of probes, the microarray evaluates many genetic tests in parallel.

23. B is correct.

24. A is correct. Codon-anticodon hybridization (i.e. bonding interaction) occurs between mRNA (codon) and tRNA (anticodon) during translation for protein synthesis.

25. B is correct. A Northern blot is a molecular biology technique used to study gene expression via the detection of RNA (or isolated mRNA) in a sample. Northern blotting uses electrophoresis to separate RNA samples by size. It uses a hybridization probe via complementary hydrogen bonding to target expressed RNA fragments (i.e. expressed genes). It involves the capillary transfer of RNA from the electrophoresis gel to the blotting membrane. Eukaryotic mRNA can be isolated through the use of oligo (dT) cellulose chromatography to isolate only mRNA with a poly-A tail, which are separated by gel electrophoresis. Since the electrophoresis gels are fragile and the probes are unable to enter the gel matrix, the separated (by size during electrophoresis) RNA sample is transferred to a positively charged nylon (or nitrocellulose) membrane through capillary blotting, whereby the negative charged mRNA adheres to the positive charge on the nylon.

In situ hybridization is a type of hybridization that uses labeled probes of complementary DNA (cDNA) or RNA strands to localize a specific DNA or RNA sequence in a section of tissue (*in situ*). The probe hybridizes to the target sequence at an elevated temperature, and then the excess probe is washed away.

26. E is correct. Chromosomes replicate during the synthesis (S) phase of interphase.

27. E is correct. DNA contains T, which is replaced with U in RNA.

28. B is correct.

29. D is correct. Hershey and Chase showed that when bacteriophages (i.e. viruses), composed of DNA and protein, infect bacteria, their DNA enters the host bacterial cell while their protein does not.

Hershey and Chase grew separate populations of viruses and incorporated either radioactive sulfur (^{35}S to label protein) or phosphorus (^{32}P to label DNA) into the bacteriophages. Two groups of viral progeny contained either ^{32}P or ^{35}S radioactive isotopes.

Separate aliquots of the labeled progeny were allowed to infect unlabeled bacteria. The viral ^{35}S protein coats remained outside the bacteria, while the ^{32}P DNA entered the bacteria.

Centrifugation separated the phage protein coats from the bacteria. The bacteria were then lysed to release the infective phages. The experiment demonstrated that the DNA, not protein, was the *transforming* molecule that entered the bacteria from viral infection.

30. C is correct. The replication fork opens by the disruption of hydrogen bonds between complementary nucleotide base pairs (e.g. A bonded to T and C bonded to G). Gyrase cuts one of the strands of the DNA backbone and relaxes the positive supercoil that accumulates as helicase separates the two strands of DNA. Ligase seals the backbone of DNA (i.e. joins the Okazaki fragments) by forming phosphodiester bonds between deoxynucleotides in DNA.

31. A is correct. Three amino acids with the peptide bond are highlighted with the box. Note that the oxygen on the carbonyl is oriented 180° from the H (antiperiplanar) on the nitrogen because the lone pair on the nitrogen participates in resonance and the peptide bond is rigid (i.e. double-bond like character)

32. D is correct. DNA and ribozymes of RNA are capable of self-replication. Some functions of proteins include: 1) peptide hormones as chemical messengers transported within the blood, 2) enzymes that catalyze chemical reactions by lowering the energy of activation, 3) structural proteins for physical support within the cells, tissues and organs, 4) transport proteins as carriers of important materials, and 5) antibodies of the immune system that bind foreign particles (antigens).

33. B is correct.

34. E is correct. The percent of adenine cannot be determined because RNA is a single-stranded molecule and the base pairing rules of DNA (i.e. Chargaff's rule) that make the determination possible for double-stranded DNA are not applicable to single-stranded RNA.

35. A is correct. I: not all AUG sequences are the original start codon and can be contained within the mRNA downstream for the initiating start codon AUG. Assuming that this sequence is not the start codon for this polypeptide, this change could result in a stop codon (UAA).

II: the genetic code is read from 5′ to 3′ and AUG (encoding for methionine) is a start codon. A change in the start sequence in the mRNA (AUG to AAG) causes a failure in initiating translation.

III: depending on the reading frame, the change occurs in either the first, second, or third amino acid and changing the U to A changes the encoding of the resultant amino acid. There are certain nucleotide changes, which do not change the specified amino acid. The third position of the codon is considered the *wobble* position, because the specified amino acid often does not change.

For example, AUU and AUA codons both encode for isoleucine, and the genetic code is considered *degenerate*, whereby changing a nucleotide (often in the 3rd position) does not change the amino acid which is encoded for by the triplet codon. However, each codon (three nucleotides) encodes for only one amino acid and therefore there is no ambiguity in the genetic code.

36. B is correct.

37. D is correct. Polymerase I (Pol I) adds nucleotides at the RNA primer-template junction (i.e. origin of replication) and is involved in excision repair with 3'-5' and 5'-3' exonuclease activity and processing of Okazaki fragments generated during lagging strand synthesis.

B: primase adds the first two RNA primers at the start of DNA replication because the DNA polymerase must bond to double-stranded molecules. The RNA primer is removed by DNA polymerase I after the newly synthesized DNA strand has been replicated via DNA polymerase III.

DNA replication

a: template strand of parental DNA

b: leading strand of newly synthesized DNA

c: lagging strand (Okazaki fragments) of newly synthesized DNA

d: replication fork with helicase opening and unwinding the double-stranded DNA

e: RNA primer synthesized by primase

f: direction of DNA strand synthesis

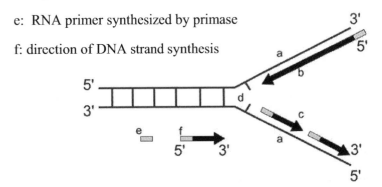

38. E is correct. Okazaki fragments are associated with the lagging DNA strand and (like all DNA) are synthesized in a 5'→ 3' direction by DNA polymerase III.

A: DNA polymerase I removes the short sequence RNA primers deposited by primase that are needed for the anchoring of the polymerase III to DNA for synthesis of DNA in a 5'→ 3' direction.

B: Okazaki fragments are used for replication of the lagging strand and are covalently linked by *DNA ligase* (not DNA polymerase I), forming a continuous DNA strand.

D: Okazaki fragments are not synthesized to fill in gaps after the removal of the RNA primer by DNA polymerase I.

39. C is correct. The ribosomal RNAs form a large subunit and a small subunit. During translation, mRNA is located between the small and large subunits, and the ribosome catalyzes the formation of a peptide bond between the two amino acids that are held by the rRNA. A single mRNA can be translated simultaneously by multiple ribosomes.

The ribosome catalyzes the formation of a peptide bond between the two amino acids that are contained in the rRNA. A ribosome also has three binding sites: A, P, and E. The A site binds to an aminoacyl-tRNA (a tRNA bound to an amino acid). The amino (NH_2) group of the aminoacyl-tRNA, which contains the new amino acid, attacks the ester linkage of peptidyl-tRNA (contained within the P site), which contains the last amino acid of the growing chain, forming a new peptide bond. This reaction is catalyzed by peptidyl transferase. The tRNA that was holding on the last amino acid is moved to the E site, and the aminoacyl-tRNA is now the peptidyl-tRNA.

40. A is correct. Protein synthesis does require energy.

B: rRNA is part of the ribosome and is necessary for proper binding of the ribosome to mRNA.

C: tRNA brings an amino acid to the ribosome, where it interacts with the mRNA of the proper sequence.

D: tRNA does have the amino acid bound to its 3' end.

E: mRNA (like all nucleic acids) is synthesized from $5' \rightarrow 3'$.

41. B is correct. PCR requires knowledge of the sequence at the ends of the fragment trying to be amplified. It is from the ends that complimentary base pairing is used to design the primers that anneal to the target fragment and permit amplification. No knowledge of the region between the ends is required, because the parent strands will be the template used by the DNA polymerase.

42. D is correct.

43. E is correct.

44. A is correct. Magnesium is a divalent mineral that DNA and RNA polymerases use as a cofactor (i.e., catalyst not consumed in the reaction) to help stabilize the interaction between the polymerase and the negative charge on the nucleic acid backbone.

45. C is correct.

46. A is correct. There are three main types of RNA: mRNA, tRNA and rRNA, which are all encoded for by DNA. rRNA is synthesized in the nucleolus within the nucleus. tRNA functions as a carrier of amino acid molecules. Unlike mRNA, tRNA is a comparatively short ribonucleotide polymer of RNA subunits.

Although tRNA is mostly single-stranded, there are some double-stranded segments where the nucleotide chain loops back upon itself (i.e. hairpin turns) that results from hydrogen bonds between complementary base pairs (e.g. 2 hydrogen bonds between A and U; 3 hydrogen bonds between C and G). mRNA is the template for protein synthesis and has a poly-A tail, which functions as a *molecular clock* for the degradation of mRNA.

47. D is correct.

48. E is correct. Puromycin is an analog because it has a similar shape to tRNA. Puromycin joins the ribosome, forms one peptide bond and becomes covalently attached to the nascent protein. But, since it lacks a carboxyl group, it cannot be linked to the next amino acid and protein synthesis terminates prematurely.

A: initiation requires the binding of only a single aminoacyl-tRNA (the initiator) to the ribosome and is not affected.

Puromycin

B: aminoacyl-tRNA enters the large subunit of ribosome at the A site during elongation.

D: puromycin lacks a carboxyl group and therefore can only form one bond, so peptide synthesis stops prematurely.

49. C is correct.

50. D is correct. DNA polymerase I proofreading increases the fidelity of replication by monitoring for mismatched pairs originating from the high processivity of polymerase III that rapidly replicates DNA. Bacteria have a much lower DNA replication rate of about 1 in 1,000, and this increases the mutation rate of bacteria.

51. C is correct. DNA is a nucleotide polymer of the deoxyribose sugar, a phosphate group and a nitrogenous base (e.g. A, C, G, T). The nucleotides in DNA's backbone are joined by phosphodiester bonds.

52. A is correct.

53. E is correct. All of the molecules, except cysteine, are nitrogenous bases – the component molecules of DNA and RNA (e.g. mRNA, rRNA & tRNA). The nitrogenous bases guanine (G) and adenine (A) are purines, while cytosine (C) and thymine (T is in DNA only) or uracil (U is in RNA only) are pyrimidines. Cysteine is an amino acid (not a nitrogenous base), and amino acids are the monomers for proteins.

54. A is correct. A restriction enzyme (or restriction endonuclease) is an enzyme that cuts DNA at specific nucleotide sequences (i.e. restriction sites). These enzymes, found in bacteria and archaea, provide a defense mechanism against invading viruses. Inside a prokaryote, the restriction enzymes selectively cut up *foreign* DNA, while host DNA is protected by a modification enzyme (i.e. methylase) that modifies the prokaryotic DNA and prevents cleavage by the endogenous restriction enzyme.

55. C is correct. RNA polymerase is an enzyme that produces primary transcript RNA in the process of transcription. Molecule C has a phosphodiester bond from the 3' of the base to the 5' of the downstream base and a triphosphate at the 5' end of the molecule.

A: represents DNA because of the absence of a hydroxyl at the 2' position of the sugar.

B: contains a monophosphate at the 5' position.

D: shows a phosphodiester bond at the 2' position (not 3' position).

E: shows a phosphodiester bond between two 5' ends.

56. B is correct. In the first step, the two strands of the DNA double helix are physically separated at a high temperature in a process of DNA melting. In the second step, the temperature is lowered and the two DNA strands become templates for DNA polymerase to selectively amplify the target DNA. The selectivity of PCR results from the use of primers that are complementary to the DNA region targeted for amplification under specific thermal cycling conditions. The process continues for 30 to 40 cycles, doubling the amount of DNA each cycle.

57. A is correct. DNA is double stranded with A, C, G and T, while RNA is a single stranded molecule with U replacing the T of DNA. DNA uses the sugar of deoxyribose (i.e. absence of ~OH group at the 2' in RNA), while RNA uses ribose (i.e. presence of ~OH group at the 2').

58. B is correct. Ligase is an enzyme used by the cell during DNA replication (and also in other biochemical processes) that catalyzes the joining of two large molecules (e.g. DNA nucleotides) by forming a new chemical bond. The newly formed bond is *via* a condensation reaction (joining) and usually involves dehydration with the loss of H_2O when the molecules (e.g. DNA, amino acids) are linked together.

59. E is correct. In DNA, thymine (T) base-pairs with adenine (A) with two hydrogen bonds, while cytosine (C) base-pairs with guanine (G) with three hydrogen bonds. More energy is required to break three hydrogen bonds than to break two hydrogen bonds. A DNA sequence with more C-G pairs has a higher melting point and therefore requires more energy to denature (i.e. separate) the DNA strand.

When bonded to its complementary strand, the DNA strand with GCCAGTCG has two T-A pairs and six C-G pairs (2 pairs x 2 H bonds = 4) + (6 pairs x 3 bonds = 18) = 22 H bonds. Thus, this DNA strand has the most hydrogen bonds and therefore the highest melting point.

A: has five T-A pairs and three C-G pairs:

(5 pairs x 2 H bonds = 10) + (3 pairs x 3 bonds = 9) = 19 H bonds.

C: has four A-T pairs and four C-G pairs:

(4 pairs x 2 H bonds = 8) + (4 pairs x 3 bonds = 12) = 20 H bonds.

D: has four A-T pairs and four C-G pairs:

(4 pairs x 2 H bonds = 8) + (4 pairs x 3 bonds = 12) = 20 H bonds.

60. C is correct. Reverse genetics is an approach used to discover the function of a gene by analyzing the phenotypic effects of specific gene sequences obtained by DNA sequencing. This process proceeds in the opposite direction of classical genetics that investigates the genetic basis of a phenotype. Reverse genetics seeks to find which phenotypes arise as a result of particular genetic sequences.

61. D is correct. Protein synthesis requires biochemical energy in the form of ATP or GTP. Two high-energy phosphate bonds (from ATP) provide the energy required for the formation of one aminoacyl-tRNA involving the attachment of each amino acid to its tRNA.

Formation of the initiation complex requires the energy from one GTP.

Elongation with the delivery of each new tRNA to the A site requires one GTP.

Translocation of the peptidyl-tRNA requires one GTP.

Termination does not require the hydrolysis of a high energy phosphate bond (ATP nor GTP).

For initiation, one high-energy phosphate bond (ATP) is required.

For each amino acid added to the polypeptide chain, two high-energy phosphate bonds are required for "charging" the tRNA with the correct amino acid (2 GTP x 50 amino acids = 100).

For chain formation, two high-energy phosphate bonds are required – one to carry the amino acid to the ribosome and another to translocate the ribosome (2 GTP x 49 peptide bonds = 98).

Total: 1 + 100 + 98 = 199.

62. E is correct.

63. B is correct. DNA repair is a collection of processes by which a cell identifies and corrects damage to the DNA molecules that encodes its genome. Before replication of DNA, the cell methylates the parental strand for reference during any potential errors introduced during replication. Additionally, both normal metabolic activities and environmental factors such as UV light and radiation can cause DNA damage, resulting in as many as one million individual lesions per cell per day. Many of these lesions cause structural damage to the DNA molecule and can alter the cell's ability to transcribe genes for the survival of its daughter cells after mitosis.

The DNA repair process is constantly active as it responds to damage in the DNA structure, and methylation provides a reference of the original strand when mismatches are detected during replication.

64. A is correct. Methionine is specified by the *start codon* of mRNA and is the first amino acid in eukaryotic proteins. The mature protein may excise a portion of the original polypeptide, so methionine is not always present as the first amino acid in the mature protein after modification in the Golgi.

65. D is correct.

66. B is correct. 3'–CAGUCGUACUUU–5' anticodon of the tRNA (known from question stem)

5'–GUCAGCAUGAAA–3' codon of the mRNA

3'–CAGTCGTACTTT–5' DNA

There is a polarity when nucleic acids base pair, whereby the 3' end of the tRNA anticodon corresponds to the 5' end of the mRNA codon.

Also, C pairs with G and U pairs with A (base pairing is U to A in RNA).

Sequence of mRNA is 5'–GUCAGCAUGAAA–3'. For complementary DNA, A pairs with T (not U as in RNA) and the polarity of the strands is antiparallel. Therefore, the 3' end of the RNA hybridizes with the 5' end of the DNA.

Another approach to solve the problem: since both the 3' end of tRNA and the 3' end of DNA hybridize to the 5' of mRNA, the DNA sequence is the same orientation and similar to the tRNA (i.e. replace T in DNA with U in tRNA).

By convention, nucleic acids are always written with the 5' end on the left (top left in a double-stranded molecule) and are also read in the 5' to 3' direction.

67. A is correct. A promoter is a region of 100–1000 base pairs long on the DNA that initiates transcription of a particular gene. Promoters are located near the genes they transcribe, on the same strand and upstream on the DNA (towards the 3' region of the anti-sense strand – also called template strand and non-coding strand). Promoters contain specific DNA sequences and response elements that provide a secure initial binding site for RNA polymerase and for transcription factors proteins. Transcription factors have specific activator (or repressor) sequences of nucleotides that attach to specific promoters and regulate gene expressions.

In bacteria, the promoter is recognized by RNA polymerase and an associated sigma factor (i.e. protein needed only for initiation of RNA synthesis), which are often brought to the promoter DNA by an activator protein's binding to its own nearby DNA binding site. In eukaryotes, the process is more complicated, and at least seven different factors are necessary for the binding of an RNA polymerase II to the promoter.

68. A is correct. Only isoleucine-glycine is composed of two amino acids and therefore is a dipeptide.

69. E is correct.

70. D is correct. RNA polymerase synthesizes the new strand in the anti-parallel orientation with new nucleotides adding to the growing chain in the 5' to 3' direction.

71. B is correct. Adenosine (A) bonds with Thymine (T) with two hydrogen bonds; Guanine (G) bonds with Cytosine (C) with three hydrogen bonds.

72. B is correct. The lagging strand is composed of Okazaki fragments. All nucleotides (e.g., DNA and RNA) are extended from the 3'-OH group. Therefore, both the leading (i.e. continuous) and lagging (i.e., discontinuous) strands use the 3'-OH end as a nucleophile during the condensation (i.e., via dehydration) reaction for chain elongation.

A: 3' end of the template strand.

C: 5' end of a lagging strand.

D: 5' end of the template strand

73. C is correct. The ribosomal subunits, which are radio-labeled with *heavy* carbon and *heavy* nitrogen, were placed in a test tube during bacterial protein synthesis. During translation, the small and large subunits assemble to form a complete ribosome. After translation ceases, the complete ribosome dissociates back into individual small and large subunits. Since the sample used in centrifugation was taken after translation finished, the individual ribosomal subunits (not the assembled ribosomes) were present. Centrifugation separates cellular components based on density and since the subunits are different sizes, two different bands are expected in the centrifuge tube.

An understanding of the size of the two ribosomal subunits in bacteria is required. Bacteria have two subunits of 30S and 50S, which assemble to form a 70S complex. Eukaryotes have two subunits of 40S and 60s, which assemble to form an 80S complex.

74. E is correct. Translation is the process of protein production, whereby one amino acid at a time is added to the end of a protein. This operation is performed by a ribosome, whereby the sequence of nucleotides in the template mRNA chain determines the sequence of amino acids in the generated amino acid chain. Addition of an amino acid occurs at the C-terminus of the peptide, and thus, translation is said to be amino-to-carboxyl directed.

75. D is correct.

76. B is correct.

77. A is correct. In PCR amplification, a primer hybridizes to the end of a DNA fragment and acts as the initiation site for DNA polymerase to bind for replication of the entire strand.

Since DNA is replicated from $5' \rightarrow 3'$, the primer must be the complement of the $3'$ end of the DNA fragment, because DNA polymerase can only read the template strand $3' \rightarrow 5'$.

78. C is correct.

79. C is correct.

80. B is correct.

81. E is correct.

Chapter 3.2: Mendelian Genetics, Inheritance Patterns

1. A is correct.

2. C is correct.

3. B is correct. r/K selection theory refers to the selection of traits in a species, whereby the parental investment is related to the quantity/quality of offspring. A higher quantity of offspring with decreased parental investment, or a lower quantity of offspring with increased parental investment, promotes reproductive success in different environments.

Species that are r-selected species emphasize high growth rates and typically exploit less-crowded ecological niches. They produce many offspring, each of which may die before adulthood (i.e. high *r*, low *K*). This strategy fares better in an environment with density-independent factors (e.g. harsh environment, short seasons), and the species are able to better withstand massive predation, because they produce so many more offspring than needed for their survival.

K-selected species display traits associated with living at densities close to carrying capacity, and are typically strong competitors in crowded niches. They invest more heavily in fewer offspring, most of which will mature (i.e. low *r*, high *K*). *K*-strategists fare better with density-dependent factors (e.g. limited resources) and are able to exploit limited resources by specializing.

Species that use r-selection are referred to as *opportunistic*, while K-selected species are described as *equilibrium*.

4. C is correct. Eye color is sex-linked in *Drosophila*. Determine the phenotype of the parents. A red eyed fly with red eyed and sepia eyed parents must be heterozygous, because a sepia eyed parent only contributes the recessive sepia allele.

When the heterozygous (Rr) red eyed fly is crossed with a homozygous recessive (rr) sepia eyed fly, ½ the offspring are red eyed (Rr) because of the dominant (red) allele from the heterozygous fly.

The Punnett square is:

Red eyed parent

	R	r
r	Rr (red)	rr (sepia)
r	Rr (red)	rr (sepia)

Sepia eyed Parent

Since the question does not assign the gender to the sepia and red eyed parents, the Punnett squares for two possible combinations for sex-linked traits are:

		R	r
		Red eyed female (♀)	
Sepia eyed male (♂)	r	Rr (red)	rr (sepia)
	y	Ry (red)	ry (sepia)

		R	y
		Red eyed male (♂)	
Sepia eyed female (♀)	r	Rr (red)	ry (sepia)
	r	Rr (red)	ry (sepia)

5. E is correct. Color blindness always pertains to the cone photoreceptors in retinas, as the cones are capable of detecting the color frequencies of light. About 8 percent of males, but only 0.5 percent of females, are colorblind, whether it is one color, a color combination, or another mutation.

Males are at a greater risk of inheriting an X-linked mutation, because males have only one X chromosome (XY), while females have two (XX). If a woman inherits a normal X chromosome in addition to the one that carries the mutation, she does not display the mutation. Men lack a second X chromosome to compensate for the X chromosome that carries the gene mutation.

6. D is correct. The decimating fire was a random event unrelated to the apparent fitness of the fly in its normal environment. Genetic drift is the random change over time in the allele frequency within a population, such as the one caused by the decimating fire in the loss of allele(s) for the altered structure.

A: reproduction is not involved in the loss of the advantageous modification.

B: natural selection is not the cause, since the death of the flies was apparently unrelated to their fitness, and it is likely that a decimating fire would have killed all flies, regardless of their advantageous modification.

C: Hardy-Weinberg describes ideal circumstances which do not apply to this situation.

7. C is correct.

8. E is correct. AAbbCc produces 2 gametes: AbC and Abc = 1/2

AaBbCc produces 8 gametes: ABC, ABc, AbC, Abc, aBC, aBc, abC, abc = 2/8 possible = 1/4

From probability: 1/2 x 1/4 = 1/8

	ABC	ABc	AbC	Abc	aBC	aBc	abC	abc
AbC	X	X	X	X	X	**Yes**	X	X
Abc	X	X	X	X	**Yes**	X	X	X

9. B is correct.

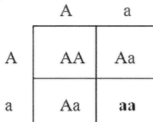

Afflicted children are aa = 1 of 4 possibilities

= ¼ or 25%.

10. B is correct. If all four of these criteria are met, the gene frequencies remain constant. If one or more of these criteria are not met, the gene frequencies (and allele frequencies) change and evolution occurs.

Hardy-Weinberg law states that the gene ratios (P + Q =1) and the allelic frequencies (p^2 + 2pq + q^2 = 1) remain constant between generations in a population that is not evolving.

The Hardy-Weinberg law requires four criteria: 1) a large population, 2) random mating, 3) no migration into or out of the population, and 4) no mutations. All four of these conditions are required for the gene frequency to remain constant. If one (or more) of these four conditions are not satisfied, the gene frequencies changes over the generations and the population evolves.

11. C is correct. A tumor suppressor gene (or antioncogene) is a gene that protects cells from one step on the path to cancer. When this gene's function is lost or reduced due to mutation, the cell can progress to cancer (usually in combination with other genetic mutations). The loss of the tumor suppressor gene may be even more important than proto-oncogene/oncogene activation for the formation of many kinds of human cancer cells. Both alleles of the tumor suppressor gene that code for a particular protein must be affected before an effect is manifested, because if only one allele is damaged, the second can still produce the correct protein.

12. D is correct. Color blindness is a sex-linked trait, because the gene is on the X chromosome. The mother is a carrier (not afflicted with the condition) and is heterozygous for the recessive allele (color blindness), while the father has the allele on his X chromosome (Y chromosome lacks the gene). The genotype and phenotype of a son depends entirely on the mother (afflicted vs. carrier), since the afflicted father transmits the gene on his X. Since the mother is heterozygous, a son would have a 50% chance of receiving the color blindness allele from his mother.

13. B is correct.

14. C is correct. AaBbCcDdEe x AaBbCcDdEe   From probability: ½ x ½ x ½ x ½ x ½ = 1/32

15. A is correct. Autosomal recessive inheritance is the product of mating two carriers (i.e. heterozygous parents). In the mating of two heterozygotes for an autosomal recessive gene, there is a:

1) 25% (1/4) probability of a homozygous unaffected child

2) 25% (1/4) probability of a homozygous affected child

3) 50% (1/2) probability of a heterozygous (carrier) child

Overall, 75% of children are phenotypically normal (25% AA and 50% Aa). Out of all children, 50% are phenotypically normal, but carry the mutant gene (Aa).

16. D is correct. If the frequency of the dominant allele is three times that of the recessive allele, then $p = 3q$. According to the Hardy-Weinberg equilibrium, $p + q = 1$ so $3q + q = 1$.

Solving for q, $4q = 1$; $q = 0.25$ and $p = 0.75$ Allele frequency: $p^2 + 2pq + q^2 = 1$.

Heterozygote allele $= 2pq$ Substituting for p and q, $2(0.75)(0.25) = 0.375$ or 37.5%.

17. E is correct. Loss of heterozygosity is a chromosomal event that results in loss of the entire gene and the surrounding chromosomal region. Most diploid cells (e.g. human somatic cells) contain two copies of the genome, one from each parent. Each copy contains approximately 3 billion bases, and for the majority of positions in the genome, the base present is consistent between individuals. However, a small percentage may contain different bases. These positions are called single nucleotide polymorphisms (or SNP). When the genomic copies from each parent have different bases for these regions, the region is heterozygous. Most of the chromosomes within somatic cells of individuals are paired, allowing for SNP locations to be potentially heterozygous. However, one parental copy of a region can sometimes be lost, which results in the region having just one copy. If the copy lost contained dominant allele of the gene, the remaining recessive allele will appear in a phenotype.

18. A is correct. Maternal inheritance involves all the progeny exhibiting the phenotype of the female parent.

B: not maternal inheritance, because the progeny exhibit the phenotype of only the male parent.

C and D: normal Mendelian 1:1 segregation and not maternal inheritance. Maternal inheritance is a type of uniparental inheritance by which all progeny have the genotype and phenotype of the female parent.

19. E is correct.

20. B is correct.

21. A is correct. Mutations affect proteins, but not lipids or carbohydrates. In proteins, the possible effects on the protein are no change (i.e. silent mutation), abnormal protein production, loss of protein (enzyme) function, or gain of protein (enzyme) function.

Loss of function of a gene product may result from mutations encoding a regulatory element or from the loss of critical amino acid sequences.

Gain-of-function mutations are due to changes in the amino acid sequence resulting in enhancing of the protein function. There may be either an increase in the level of protein expression (affecting the operator region of the gene) or an increase in each protein molecule's ability (change in the shape of the protein) to perform its function.

22. A is correct. The desired phenotype is green smooth peas and both green and smooth are dominant phenotypes. Therefore, the genotypes selected for the cross must avoid the two recessive alleles (g and s). For GgSs x GGSS, one parent (GGSS) is a double dominant, and therefore all offspring have the dominant phenotype (G and S) regardless of the genotype of the other parent.

B: Gg x gg yields 1/2 yellow (g) phenotype offspring.

C: ss x Ss yields 1/2 wrinkled (s) phenotype offspring.

D: Gg x Gg yields 1/4 yellow (g) phenotype offspring.

23. C is correct. Retinoblastoma (Rb) is a rapidly developing cancer that develops from the immature cells of a retina, the light-detecting tissue of the eye. It is the most common malignant tumor of the eye in children. Only a single allele needs to be inherited (i.e. dominant) for the phenotype to be shown.

24. E is correct. The probability for the second child is not influenced by the phenotype of the first child. For example, the chance of getting a tail on the first toss of a coin has no influence on the chance of getting a head on the second toss.

A Punnett square determines all possible gametes and their combinations. If ½ of the woman's gametes carry the trait and ½ of the father's gametes carry the trait, the probability of a child receiving the allele from both parents is ½ × ½ = ¼.

25. C is correct.

26. B is correct.

27. E is correct. The probability (p) of rolling any one number (e.g. 4) on a 6-sided die is 1/6, since there are 6 sides and only one side carries a number 4. p (of a 4) = 1/6.

Using two dice, what is the probability of rolling a pair of 4s? Multiply the individual probabilities together, the probability of a 4 appearing on both dice is:

p (of two 4s) = 1/6 x 1/6 = 1/36.

28. D is correct. The degree of genetic linkage measures the physical distance of two genes that are on the same chromosome. The probability of a crossover and corresponding exchange that occurs between gene loci (location on the chromosome) is generally directly proportional to the distance between the loci.

Pairs of genes that are far apart from each other on a chromosome have a higher probability of being separated during crossover compared to genes that are located physically close together. Thus, the frequency of genetic recombination between two genes is related to the distance between them. Recombination frequencies can be used to construct a genetic map. One map unit (Morgan units) is defined as a 1 percent recombinant frequency.

Recombination frequencies are roughly additive, but are only a good approximation for small percentages. The largest percentage of recombinants cannot exceed 50%, which results when the two genes are at the extreme opposite ends of the same chromosome. In this situation, any crossover events would result in an exchange of genes, but only an odd number of crossover events (a 50/50 chance between an even and odd number of crossover events) results in a recombinant product.

29. A is correct. D: epigenetic inheritance is a result of the changes in gene activity, which are *not* caused by changes in the DNA sequence. It can be used to describe the study of stable, long-term alterations in the transcriptional potential of a cell that are not necessarily heritable. Unlike simple genetics based on changes to the DNA sequence (genotype), the changes in gene expression or cellular phenotype of epigenetics have other causes.

One example of an epigenetic change in eukaryotic biology is the process of cellular differentiation. During morphogenesis, totipotent (i.e. all potent) stem cells become the various pluripotent (i.e. highly potent, but limited in possible determinate potential) cell lines of the embryo, which in turn become fully differentiated cells. As a single fertilized egg cell (zygote) continues to divide, the resulting daughter cells change into all the different cell types in an organism (e.g. neurons, muscle cells, epithelium and endothelium of blood vessels) by activating some genes, while inhibiting the expression of others.

30. C is correct. Female children receive one X from their mother and one X from their father. The X from the father must carry the color blindness allele, because the father is colorblind. The X from the mother has a wild type and a color blindness allele, because she is heterozygous recessive. Thus, 50% of female children are homozygous colorblind, and 50% are heterozygous carriers of the color blindness trait.

31. D is correct.

32. C is correct.

	A	A
a	Aa	Aa
a	Aa	Aa

	A	a
A	AA	Aa
a	Aa	aa

F_1: all tall F_2: ¾ are tall and ¼ is short

33. B is correct. A frameshift mutation occurs when there is an addition or deletion of 1 or 2 base pairs. A 3 base pair addition or deletion causes an in-frame mutation because each codon is encoded for by 3 nucleotides. A frameshift mutation – addition/deletion of other than multiples of 3 nucleotides – causes the ribosome to read all downstream codons in the wrong frame. Frameshift mutations usually result in truncated (nonsense mutation) or non-functional proteins (missense mutation).

An altered base pair (point mutation) is not a frameshift mutation, because it substitutes (not adds or deletes), and therefore does not cause the ribosome to read out of frame. Point mutations can also result in nonsense (premature stop codon) or missense (improperly folded protein) mutations.

Base pair additions/deletions (other than in multiples of 3) cause a frameshift mutation.

34. D is correct. From Mendel's law of independent assortment, the probability of a cross resulting in a particular genotype is equal to the *product* of the individual probabilities. This cross involves two individuals who are heterozygous for the three genes (A, B, and C) to produce an offspring that is homozygous dominant for each trait.

Both parents are heterozygous for A, genotype = Aa. The ratio of their offspring = 1/4 AA, 1/2 Aa and 1/4 aa. This is a typical 1:2:1 ratio for all heterozygous crosses.

Both parents are also heterozygous for genes B and C, the probability of their offspring being BB is ¼, and being CC is also 1/4.

Therefore, the probability that a particular offspring will have the genotype AABBCC = the product of the individual probabilities: $1/4 \times 1/4 \times 1/4 = 1/64$.

35. E is correct. Nonsense mutation is a DNA point mutation that results in a premature stop codon in the transcribed mRNA and in a truncated (i.e. incomplete) protein that is usually nonfunctional.

A missense mutation is a point mutation where a single nucleotide is changed to cause substitution of a different amino acid. Some genetic disorders such as sickle cell anemia, thalassemia and Duchenne muscular dystrophy result from nonsense mutations.

36. B is correct. A recessive trait is expressed only when it is present in both copies or is the only copy of the gene present. A recessive X-linked allele would only be expressed in unaffected women who have two (homozygous) copies of the allele. If only one X is present, then all recessive (single copy) alleles on the X are expressed (e.g. hemophilia).

The presence of a Y chromosome (in humans) confers maleness on the organism.

37. B is correct.

38. A is correct. The two traits are unlinked either because they are inherited as if on separate chromosomes or because the genes are greater than 50 centimorgans (i.e. double crossing over occurs and the genes appear as unlinked).

39. E is correct. Transduction involves a virus and is one of three methods (along with conjugation and transformation) that bacterial cells use to introduce genetic variability into their genomes. Transduction is when a virus introduces novel genetic material while infecting its host.

Mitosis occurs in all somatic cells for growth, repair and tissue replacement, producing two identical diploid (2N) daughter cells.

E: tissue repair involves fibroblasts and may cause scar tissue formation if the basement membrane has been breached.

40. D is correct. Let T = tall and t = short; B = brown eyes and b = blue eyes. The father is homozygous tall and homozygous blue-eyed, his genotype is TTbb. The mother is heterozygous tall and heterozygous brown-eyed, her genotype is TtBb.

Determine the probability that these parents could produce a tall child with blue eyes (T_bb). The genes for height and eye color are unlinked. The father (TTbb) contributes T and b alleles so his gametes have both T and b alleles. The mother (TtBb) contributes either T or t and either B or b, so her gametes are (in equal amounts): TB, tB, Tb, or tb.

Possible genotypes of the offspring: TTBb, TTbb, TtBb, Ttbb. Half the offspring are tall and brown eyed (T_B_), and half are tall and blue-eyed (T_bb). Therefore, the probability of a tall child with blue eyes is ½.

A faster method is calculating phenotype ratios for height and eye-color separately and then combining them. The mating of TT x Tt = 100% tall. The mating of Bb x Bb = ½ blue and ½ brown. Multiplying 1 tall x ½ blue = ½ tall blue.

41. A is correct.

42. C is correct. There are two possible alleles for each of the three genes. If the genes assort independently (not linked), then there are $2^3 = 8$ possible combinations.

43. A is correct.

44. C is correct. Parents are Aa x Aa (carriers, but not afflicted with the disease). Progeny could be AA, Aa, Aa, or aa. From the Punnett square, eliminate aa because this is the disease state. The question asks what is the chance she is heterozygous (Aa), but not homozygous (AA) = 2/3.

45. E is correct. In genetics, anticipation is associated with an earlier onset of symptoms as well as an increase in disease severity with every generation. It is usually seen in autosomal dominant diseases associated with trinucleotide repeat expansion (e.g. Huntington's disease or myotonic dystrophy), because triple repeat sequences of increased length are unstable during cell division.

A: codominance is a phenomenon in which a single gene has more than one dominant allele. An individual who is heterozygous for two codominant alleles expresses the phenotypes associated with both alleles. In the ABO blood group system, the I^A and I^B alleles are codominant to each other. Individuals who are heterozygous for the I^A and I^B alleles express the AB blood group phenotype, in which both A- and B-type antigens are present on the surface of red blood cells. Another example occurs at the locus for the beta globin component of hemoglobin, where the three molecular phenotypes of Hb^A/Hb^A, Hb^A/Hb^S, and Hb^S/Hb^S are detectable by protein electrophoresis.

B: penetrance is the proportion of individuals carrying a particular variant of a gene (allele or genotype) that also expresses an associated trait (phenotype). In medical genetics, the penetrance of a disease-causing mutation is the proportion of individuals with the mutation who exhibit clinical symptoms. For example, if a mutation in the gene responsible for a particular autosomal dominant disorder has 95% penetrance, then 95% of those with the mutation develop the disease, while 5% do not develop the disease.

C: heterozygous advantage occurs when possessing both alleles is a benefit. For example, individuals that are heterozygous for the sickle cell trait are less likely to suffer from malaria.

D: gain of function mutations change the gene product such that it gains a new and abnormal function. These mutations usually have dominant phenotypes and are expressed with the presence of a single allele.

46. B is correct. The Hardy-Weinberg equation is $p^2 + 2pq + q^2 = 1$, where p equals the gene frequency of the dominant allele, and q equals the gene frequency of the recessive allele. Hence, p^2 is the frequency of homozygous dominants in the population, 2pq is the frequency of heterozygotes, and q^2 is the frequency of homozygous recessives. For a trait with only two alleles, p + q must equal 1, since the combined frequencies of the alleles = 100%.

If the frequency of the recessive allele for a particular trait is 0.6, q = 0.6. Since p + q = 1, p = 0.4.

To calculate the frequency of individuals expressing the dominant phenotype (not the dominant genotype), determine the number of individuals homozygous for the dominant trait (p^2) and add it to the number of heterozygotes (2pq) that also exhibit the dominant phenotype:

$p^2 = (0.4) \times (0.4) = 0.16$

$2pq = 2 \times (0.6) \times (0.4) = 0.48$

So, $p^2 + 2pq = 0.16 + 0.48, = 0.64$

A: 0.48 = frequency of heterozygous individuals.

C: 0.16 = frequency of homozygous dominant individuals.

D: 0.36 = frequency of homozygous recessive individuals.

47. C is correct.

48. B is correct.

49. D is correct.

50. E is correct. The 3 pyrimidines are cytosine, thymine (and uracil replaces thymine in RNA). Note that the word pyrimidine contains a "y" as do the nucleotides which are pyrimidines. The pyrimidines (longer word than purine) consist of one-ring structures, while purines (shorter word) are adenine and guanine, which are larger structures consisting of two rings.

51. D is correct. Point mutations occur when a single nucleotide base (A, C, G, T) is substituted by another. A silent mutation is a point mutation that either 1) occurs in a noncoding region, or 2) does not change the amino acid sequence due to the degeneracy of the genetic code.

A frameshift mutation is either an insertion or deletion of a number of nucleotides. These mutations have serious effects on the coded protein, since nucleotides are read as series of triplets. The addition or loss of nucleotides (except in multiples of three) changes the reading frame of the mRNA and often gives rise to premature polypeptide termination (i.e. nonsense mutation).

A missense mutation results from the insertion of a single nucleotide that changes the amino acid sequence of the specified polypeptide.

52. E is correct.

53. C is correct. EEBB × eebb produces offspring of single genotype EeBb, as determined by the Punnett square for the cross between a homozygous dominant by a homozygous recessive.

54. D is correct.

55. E is correct.

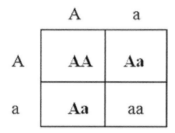

Afflicted children are either AA, Aa or Aa (not aa) =

3 of 4 possibilities = ¾ or 75%.

56. D is correct. Recombinant frequencies of linked genes map the relative locations of genes on a single chromosome. Recombinant frequencies are determined by crossing individuals that differ in alleles for all the genes in question and then determining the genotypes of their progeny. The recombinant frequencies equal frequencies of nonparent genotypes, since these genotypes arise only by genetic crossover.

Mapping is based on the *probability* of a crossover between two points. The probability of crossover *increases* as the distance between the genes *increases*. Therefore, the *farther* away two genes are, the *greater* their recombinant frequency. The chance that two genes are inherited together (i.e. or will be linked) *decreases* as the distance between them on a chromosome increases.

One map unit = 1% recombination frequency, and recombinant frequencies are (roughly) additive. However, if the genes are very far apart, then the recombination frequency reaches a maximum of 50%, at which point the genes are considered to be sorting independently.

In this problem, there are four genes (D, E, F, and G) and the recombinant frequencies between each possible pair are given. To make the map, start with the allele pair that has the highest recombinant frequency: between G and E (23%), which means that G and E are 23 map units apart and therefore are on the two ends.

Determine the intervening genes by finding genes that are closest to the two endpoints. G and D are 8 map units apart, which is closest to G. Thus, D must be next to G. By elimination, the genes on this chromosome must be G, D, F, E. The sequence EFDG is equally correct if the map started from the opposite direction, but this is not an answer choice.

As a check, D and E are 15 map units apart, because the distance from G to D, which is 8, plus the distance from D to E, which is 15, is the distance from G to E, which equals 23. Also, G and F are 15 map units apart, while F and E are 8 units apart. The numbers add up exactly, whereby the distance from G to E equals the distance from G to D + the distance from D to E. The observed numbers may be off by one or two map units (not a mistake), because map distances are only roughly additive (i.e. based on rounding for the probabilities).

Questions **57** through **63** are based on the following:

The pedigree illustrated by the schematic shows the inheritance of albinism, a homozygous recessive condition manifested in a total lack of pigment. Specify the following genotypes using *A* and *a* to indicate dominant and recessive alleles, respectively.

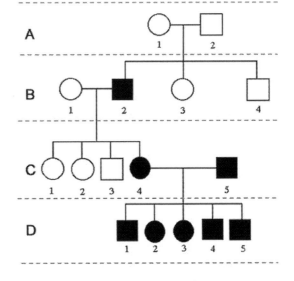

Note: solid figures are albino individuals.

57. C is correct.

58. C is correct.

59. C is correct.

60. B is correct.

61. C is correct.

62. B is correct.

63. B is correct.

64. E is correct. BBss × bbss produces gametes Bbss, which are then crossed: Bbss × Bbss and produce the four gametes: BBss, Bbss, bBss and bbss.

¾ (BBss, Bbss, bBss) are black and spotted, while ¼ (bbss) is red and spotted.

65. A is correct. An X-linked gene mutation is more common in males (i.e. a single copy of the X chromosome) than in females (i.e. two X chromosomes). The gene appears to skip generations because it is being transmitted through heterozygous females that are phenotypically normal.

Affected males transmit the gene to their daughters, who are carriers for the trait. There is no father-to-son transmission, because the sons inherit the father's Y chromosome. Males that carry the mutant gene show the trait.

66. C is correct. X-inactivation (i.e. lyonization) is a process whereby one of the two copies of the X chromosome present in female mammals is inactivated. The inactive X chromosome is silenced by transcriptional inactivity within heterochromatin (i.e. tightly condensed DNA). Females have two X chromosomes, and X-inactivation prevents them from having twice as many X chromosome gene products as males (XY) for dosage compensation of the X chromosome. The choice of which X chromosome is inactivated is random in humans, but once an X chromosome becomes inactivated, it remains inactive throughout the lifetime of the cell and its descendants within the organism.

Calico cats and tortoiseshell colorization are phenotypic examples of X-inactivation, because the alleles for black and orange fur coloration reside on the X chromosome. For any given patch of fur, the inactivation of an X chromosome that carries one gene results in the fur color of the allele for the active gene. Calico cats are almost always female because the X chromosome determines the color of the cat, and female cats (like all female mammals) have two X chromosomes. Male mammals have one X and one Y chromosome. Since the Y chromosome does not have any color genes, there is no chance a male cat could have both orange and non-orange together. One main exception to this is when, in rare cases, a male has XXY chromosomes (Klinefelter syndrome).

67. D is correct. Autosomal dominant is the most common type of inheritance. For autosomal dominant traits, a person needs a single copy of the mutant gene to exhibit the disease. Usually, equal numbers of males and females are affected, traits do not skip generations, and father-to-son transmission is observed.

68. E is correct.

69. B is correct. Catastrophic events do not cause significant genetic drift to a large homogeneous population. Emigration, selective predation and mutation all affect the Hardy-Weinberg equilibrium.

70. C is correct.

71. D is correct.

72. C is correct. See explanation for question **81**.

73. A is correct. A genetic marker is a gene or DNA sequence with a known location on a chromosome that can be used to identify individuals or organisms. It can be a variation that resulted from mutation or alteration in the genomic loci. A genetic marker may be a short DNA sequence, such as a sequence surrounding a single base-pair change (single nucleotide polymorphism – SNP), or a long one (e.g. minisatellites).

74. D is correct. See explanation for question **76**.

75. E is correct. Transformation is one of three mechanisms of "horizontal gene transfer," in which foreign genes transfer from one bacterium to another. The other two mechanisms are conjugation (i.e., transfer of genetic material between two bacteria in direct contact) and transduction (i.e., injection of exogenous DNA by a bacteriophage into the host bacterium).

In transformation, foreign DNA is taken up from the medium and incorporated directly through the cell membrane. Transformation is dependent on the recipient bacterium which must be in a state of natural or artificial competence to take up exogenous genetic material.

76. D is correct. Plasmid is a small DNA molecule that is physically separate from chromosomal DNA within a cell and can replicate independently. Most commonly found as small, circular, double-stranded DNA molecules in bacteria, plasmids are sometimes present in archaea and eukaryotic organisms. In nature, plasmids carry genes that may be beneficial to the survival of the organism (e.g. antibiotic resistance in bacteria) and can be transmitted from one bacterium to another, even of another species. Artificial plasmids are commonly used as vectors in molecular cloning to drive the replication of recombinant DNA sequences within host organisms.

77. B is correct.

78. B is correct.

79. A is correct. Recombinant DNA are DNA molecules formed in the laboratory through genetic recombination (e.g. molecular cloning) to combine together genetic material from multiple sources, thereby creating sequences that cannot be found in biological organisms. Recombinant DNA is possible because DNA molecules from of all organisms share the same chemical structure. They differ only in the nucleotide sequence within that identical overall structure.

80. C is correct.

81. B is correct. Transgenesis is the process of introducing an exogenous gene – a transgene – into a living organism, so that the organism will exhibit a new property and transmit that property to its offspring.

The term hybrid in genetics has several meanings, the most common of which is the offspring resulting from the interbreeding between two animals or plants of different species.

Polyploid cells and organisms are those containing more than two paired (i.e. homologous) sets of chromosomes. Polyploidy refers to a numerical change in a whole set of chromosomes. Most eukaryotic organisms are diploid, which means they have two sets of chromosomes – one set inherited from each parent. However, polyploidy is found in some organisms and is especially common in plants. In addition, polyploidy also occurs in some tissues of animals that are otherwise diploid (e.g. human muscle tissues).

82. A is correct.

83. D is correct.

84. E is correct.

85. C is correct.

86. A is correct.

87. B is correct. DNA profiling (also called DNA testing, DNA typing, genetic fingerprinting) is a technique employed by scientists to assist in the identification of individuals by their DNA profiles. DNA profiles are encrypted sets of bases that reflect a person's DNA makeup, which can also be used as the person's identifier. DNA profiling/fingerprinting should not be confused with full genome sequencing. It is often used in parental testing and criminal investigation. Although 99.9% of human DNA sequences are the same in every person, enough of the DNA is different to distinguish one individual from another, unless they are monozygotic (i.e. identical) twins.

UNIT 4. ORGANISMAL BIOLOGY

Chapter 4.1: Plants: Structure, Function, Reproduction

1. E is correct. The stamen (plural is stamina) is the pollen-producing reproductive organ of a flower. Stamens typically consist of a stalk called the filament and an anther which contains microsporangia. Anthers are often two-lobed and are attached to the filament either at the base or in the middle portion.

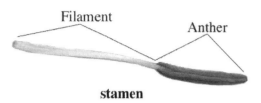

A: ovule is the structure that gives rise to and contains the female reproductive cells with the megaspore-derived female gametophyte (megagametophyte) in the center. The megagametophyte (i.e. embryo sac in flowering plants) produces an egg cell (or several egg cells) for fertilization. After fertilization, the ovule develops into a seed.

B: sepals are a part of the flower of angiosperms (flowering plants) and, morphologically, both sepals and petals are modified leaves.

D: carpels are (categorically) the appendages that contain and enclose ovules.

2. A is correct.

3. B is correct.

4. A is correct. Tracheids are elongated cells in the xylem of vascular plants that function to transport water and mineral salts. Tracheids and vessel elements are the two types of tracheary elements. Tracheids, unlike vessel elements, do not have perforations.

D: albuminous cells have a similar role to companion cells, but are associated with sieve cells only and are therefore found only in seedless vascular plants and gymnosperms.

E: companion cells carry out all of the cellular functions of a sieve-tube element in nucleated plant cells. They usually have a larger number of ribosomes and mitochondria. The cytoplasm of a companion cell is connected to the sieve-tube element by plasmodesmata.

5. E is correct.

6. C is correct.

7. D is correct. Filament is part of a stamen which is the male plant reproductive structure.

8. C is correct.

399

9. B is correct.

10. D is correct. Xylem and phloem are the two types of transport tissue in vascular plants. Xylem can be found: in vascular bundles, present in non-woody plants and non-woody parts of woody plants, in secondary xylem, laid down by a meristem called the vascular cambium in woody plants, and as part of a stellar arrangement not divided into bundles (i.e. in many ferns). The xylem transports water and soluble mineral nutrients from the roots throughout the plant. It is also used to replace water lost during transpiration and photosynthesis.

In vascular plants, phloem is the living tissue that carries organic nutrients such as sucrose to all parts of the plant where needed. In trees, the phloem is the innermost layer of the bark. The phloem is involved mainly with the transport of soluble organic material (known as translocation) made during photosynthesis. Phloem tissue consists of conducting cells (i.e. sieve elements), parenchyma cells (both specialized companion and albuminous cells, and unspecialized cells) and supportive cells (i.e. fibers and sclereids).

D: meristem is the tissue in most plants that contains undifferentiated cells found in growth zones of the plant. Meristematic cells give rise to various organs of the plant and keep the plant growing.

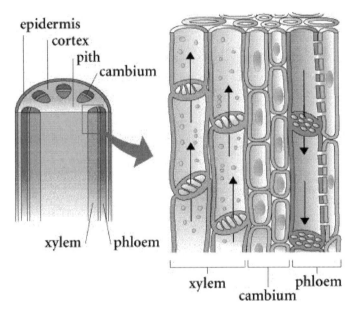

11. C is correct.

12. B is correct.

13. E is correct. Plants that rely on animals to spread their seeds usually have the seeds enclosed in fleshy fruits, which encourages animals to eat them. Fruit flesh is digested by the animal, but the seed enclosed in a thick indigestible coating passes through an animal's digestive tract without being digested. When animal excretes undigested seeds, they normally land in a place distant from their parent and can germinate when conditions are right.

14. D is correct.

15. C is correct.

16. A is correct. See explanation for question **10**.

17. D is correct.

18. E is correct.

19. A is correct. See illustration →

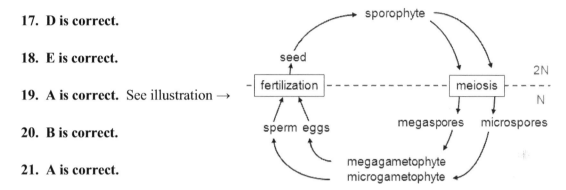

20. B is correct.

21. A is correct.

22. B is correct. Dormant seeds are ripe seeds that do not germinate because they are subject to external environmental conditions that prevent the initiation of metabolic processes and cell growth. When conditions become favorable, the seed begins to germinate and the embryonic tissues resume growth, developing into a seedling.

23. E is correct.

24. D is correct.

25. C is correct. Rhizoids are simple hair-like protuberances that extend from the lower epidermal cells of bryophytes (see explanation for question **42**). They are similar in structure and function to the root hairs of vascular land plants. Rhizoids are formed from single cells, whereas roots are multicellular organs composed of multiple tissues that collectively carry out a common function.

26. E is correct.

27. A is correct.

28. C is correct.

29. B is correct. Germination refers to the growth of an embryonic plant contained within a seed, which results in the formation of the seedling. The seed of a vascular plant is a small package produced in a fruit or cone after the union of gamete male and female cells. All fully developed seeds contain an embryo (and in most plant species, some store nutrient reserves) encased in a seed coat.

Some plants produce varying numbers of seeds that lack embryos; these are empty seeds that do not germinate. Most seeds undergo a period of dormancy when there is no active growth and the seed can be transported (e.g. animals) to a new location and/or survive adverse climate conditions until favorable circumstances develop.

30. D is correct.

31. D is correct.

32. E is correct.

33. B is correct. Mosses are a phylum of small soft plants that are typically 0.4–4 inches tall, though some species can grow much larger. They commonly grow close together in clumps or mats in damp or shady locations. They do not have flowers or seeds, and their simple leaves cover the thin wiry stems. At certain times, mosses produce spore capsules. Mosses are non-vascular plants in the land plant division Bryophyta (see explanation for question **42**). They are small non-woody plants that absorb water and nutrients mainly through their leaves and harvest sunlight to create food by photosynthesis. They differ from vascular plants because they lack water-bearing xylem tracheids or vessels.

34. A is correct.

35. D is correct.

36. C is correct. See explanation for question **10**.

37. A is correct.

38. E is correct.

39. C is correct. Auxins coordinate development in plants from the cellular level, through organs and ultimately to the whole plant. Auxin molecules present in cells may trigger responses directly through stimulation (or inhibition) of gene expression. On the cellular level, auxin is essential for cell growth, affecting both cell division and cellular expansion. Auxin concentration level (along with other local factors), result in cell differentiation and specification.

40. B is correct.

41. D is correct.

42. E is correct. The division Bryophyta includes mosses, liverworts and hornworts. Bryophyte is an embryophyte (i.e. land plant) that does not have true vascular tissue and is therefore called a "non-vascular plant." Most kinds of plants have two sets of chromosomes in their vegetative cells and are said to be diploid (i.e. each chromosome has a partner that contains the same, or similar, genetic information). By contrast, bryophytes have only a single set of chromosomes and are therefore haploid (i.e. each chromosome exists in a unique copy within the cell). There are periods in their life cycle when they do have a double set of paired chromosomes, but this happens only during the sporophyte stage whereby each sporophyte contains twice the number of paired chromosomes.

43. A is correct.

44. C is correct.

45. E is correct. Ethylene is produced from almost all parts (e.g. leaves, stems, roots, flowers, fruits, tubers, and seeds) of higher plants. Developmental and environmental factors regulate ethylene production. Ethylene production is induced during certain stages of growth such as germination, ripening of fruits, cutting of leaves and senescence of flowers. Ethylene can be induced by wounding, environmental stresses and certain chemicals, including auxin and other regulators.

46. C is correct.

47. B is correct.

48. A is correct. Stomata (singular *stoma*) is a pore, found in the epidermis of leaves, stems and other organs that is used to control gas exchange. The pore is bordered by a pair of specialized parenchyma cells (i.e. guard cells) that are responsible for regulating the size of the opening. The term stoma is also used to refer to an entire stomatal complex, both the pore itself and its accompanying guard cells.

49. D is correct.

50. B is correct. See explanation for question **10**.

51. C is correct. Mesophyll tissue (i.e. chlorenchyma or parenchyma) makes up most of the interior of the leaf between the upper and lower layers of epidermis. This assimilation (i.e. nutrient absorbing) tissue is the primary location of plant photosynthesis with the products of photosynthesis called *assimilates*.

52. A is correct.

53. E is correct. A tropism is a biological phenomenon, indicating growth or turning movement of a biological organism (usually a plant), in response to an environmental stimulus. There are many types of tropisms. Thigmotropism is a movement or growth in response to touch or contact.

A: photoperiodism is not a type of tropism, but the physiological reaction of organisms to the length of day or night. It occurs in plants and animals.

B: chemotropism is a movement or growth in response to chemicals.

C: phototropism is a movement or growth in response to lights or colors of light.

D: gravitropism is a movement or growth in response to gravity.

54. B is correct.

55. D is correct. See explanation for question **10**.

56. B is correct.

57. A is correct. Guard cells are specialized cells located in the lower leaf epidermis of plants. Pairs of guard cells surround tiny stomata airway pores and regulate the plants ability to take in CO_2 from the atmosphere for photosynthesis. They also regulate how much water plants lose to the atmosphere. Opening and closure of the stomata pore is mediated by changes in the turgor (i.e. osmotic) pressure of the two guard cells.

58. E is correct.

59. C is correct. The gymnosperms are a group of seed-producing plants that includes conifers, cycads, Ginkgo and Gnetales. Their seeds are not enclosed within an ovary, in contrast to flowering plants (angiosperms). Gymnosperm seeds develop either on the surface of scales or leaves, often modified to form cones, or at the end of short stalks as in Ginkgo. The gymnosperms and angiosperms together compose the spermatophytes (i.e. seed plants).

60. D is correct.

61. A is correct.

62. B is correct.

63. A is correct. The pressure flow hypothesis (i.e. mass flow hypothesis) is the best-supported theory to explain the movement of food through the phloem. A high concentration of organic substance inside cells of the phloem at a source (e.g. a leaf), creates a diffusion gradient (osmotic gradient) that draws water into the cells. Movement occurs by bulk flow (mass flow); phloem sap moves from sugar sources to sugar sinks via turgor pressure (i.e. hydrostatic pressure). A sugar source is any part of the plant that is producing or releasing sugar. During the plant's growth period (usually during the spring), storage organs (e.g. roots) are sugar sources, while the plant's many growing areas are sugar sinks. The movement in phloem is bidirectional, while in xylem cells, it is unidirectional (upward).

64. C is correct. See explanation for question **59**.

65. E is correct.

66. D is correct. Pollen originates from plants and is a fine to coarse powder containing the microgametophytes of seed plants that produce the male gamete (i.e. sperm cell). Pollen grains have a hard coat that protects the sperm cells during their movement from the stamens (i.e. pollen-producing reproductive organ of a flower) to the pistil (i.e. tube-like portion between the stigma and the ovary) of flowering plants or from the male cone (i.e. organ containing the reproductive structure) to the female cone of coniferous plants.

Pollination occurs when pollen lands on a compatible pistil or female cone where it germinates and produces a pollen tube that transfers the sperm to the ovule (that contains the female gametophyte).

Individual pollen grains are microscopic and require magnification to visualize. The study of pollen is highly useful in forensics, archeology, paleoecology and paleontology (i.e. scientific study of prehistoric life).

67. D is correct.

68. B is correct. Ground tissue of plants includes all tissues that are neither dermal nor vascular. It can be divided into three classes based on the type of the cell walls. *Parenchyma* cells have thin primary walls and usually remain alive after they become mature, which forms the "filler" tissue in the soft parts of plants. *Collenchyma* cells have thin primary walls with some areas of secondary thickening and provide extra structural support (i.e. regions of new growth). *Sclerenchyma* cells have thick lignified secondary walls and often die when mature, which provides the main structural support to a plant.

69. E is correct.

Chapter 4.2: Endocrine System

1. A is correct. Mevacor is a drug that competitively inhibits the enzyme HMG-CoA reductase. This enzyme catalyzes the committed step in cholesterol biosynthesis. If the committed step is blocked, cholesterol cannot be made. Cholesterol is the necessary precursor of the five major classes of steroid hormones: androgens, estrogens, progestogens, glucocorticoids and mineralocorticoids.

Insulin is a peptide hormone that causes cells in the liver, skeletal muscles, and fat tissue to absorb glucose from the blood. In the liver and skeletal muscles, glucose is stored as glycogen, and in fat cells (adipocytes) – as triglycerides. When blood glucose levels fall below a certain level, the body begins to use stored sugar as an energy source through glycogenolysis, which breaks down the glycogen stored in the liver and muscles into glucose, which is then utilized as an energy source.

B: cortisol is a steroid hormone.

C, D and E: testosterone, aldosterone and progesterone have the same (~one) suffix as steroids.

2. D is correct. See explanations to questions **20** and **55**.

3. D is correct. Hormones are either steroid or peptide molecules, released from a gland, that travel through the blood to affect a distant target.

A: paracrine signals are hormones released by a cell to induce a nearby cell's behavior or differentiation.

C: autocrine signals are hormones that are released from a cell and then affect the same cell that synthesized the molecule. Autocrine signals are common during development.

4. A is correct. Steroid hormones are able to pass through the plasma membrane of the cell and enter the nucleus. Once in the cytoplasm of the cell, the steroid hormone binds to its cytosolic receptor. Once bound, the complex migrates into the nucleus of the cell and binds to the DNA. The hormone/receptor complex functions as a transcription factor to activate or inactivate gene transcription.

B: eicosanoids are signaling molecules made by the oxidation of 20-carbon fatty acids. They exert complex control over many bodily systems, mainly in inflammation and immunity, and as messengers in the central nervous system. Examples include prostaglandins and arachidonic acid.

C: peptide (similar to amino acid) hormones are not able to enter through the cell's plasma membrane but bind (as a ligand) to receptors on the cell's surface via second messengers.

D: amino acid (similar to peptide) hormones are not able to enter through the cell's plasma membrane but bind (as a ligand) to receptors on the cell's surface via second messengers.

5. D is correct. Acromegaly is a condition that results from an oversecretion by the anterior pituitary of growth hormone (GH) that decreases the sensitivity of insulin receptors. Insulin is secreted by the pancreas in response to high blood glucose. Acromegaly patients do not bind insulin as well as do normal cells, and the effects of insulin are diminished. If insulin cannot exert its effects on the cells, then excess glucose is not converted into glycogen. Hence, the patient has high blood glucose concentrations.

A: high blood glucose concentrations lead to the excretion of glucose, along with the loss of water in the urine, meaning that patients have increased (*not decreased*) urine volume.

B: cardiac output is defined as the volume of blood pumped by the heart / unit time, which is unrelated to the sensitivity of insulin receptors.

C: low blood glucose concentration is the opposite effect.

E: glucose in the urine increases the osmolarity (number of particles) in the urine that is compensated by increased urine volume.

6. A is correct. The two hormones of the posterior pituitary are oxytocin (uterine contraction and lactation) and vasopressin – antidiuretic hormone or ADH – (stimulates water retention, raises blood pressure by contracting arterioles and induces male aggression).

7. E is correct. The primary exception to negative feedback control of hormones is the female reproduction cycle. See explanations for questions **3** and **4** for more information about hormones.

D: ducts are present in the exocrine (not endocrine) system, whereby hormones are released from a gland and distributed throughout the body via the bloodstream.

8. B is correct. The hypothalamus is responsible for certain metabolic processes and other activities of the autonomic nervous system. It synthesizes and secretes certain neurohormones (i.e. releasing hormones), and these, in turn, stimulate or inhibit the secretion of anterior/posterior pituitary hormones. The hypothalamus is involved in the regulation of a large number of basic human functions, including temperature regulation, sleep/wake cycles, water and salt balance, hunger and others. It produces hormones such as vasopressin (aka ADH) and oxytocin (stored in the posterior pituitary). The hypothalamus also produces releasing factors that control secretions of the anterior pituitary.

A: cerebrum regulates such functions as memory, conscious thought, voluntary motor activity and the interpretation of sensation. The cerebrum refers to the part of the brain comprising the cerebral cortex (two cerebral hemispheres); it includes several subcortical structures such as the basal ganglia, hippocampus and olfactory bulb. In humans, the cerebrum is the superior-most region of the central nervous system (CNS).

C: medulla oblongata (aka medulla) is the lower half of the brainstem. The medulla contains the respiratory, cardiac, vomiting and vasomotor centers and regulates autonomic (i.e. involuntary) functions, such as heart rate, breathing and blood pressure.

D: pons is part of the brainstem that links the medulla oblongata to the thalamus. The pons relays signals from the forebrain to the cerebellum. It also is involved primarily with sleep, respiration, posture, equilibrium, bladder control, hearing, swallowing, taste, eye movement, facial expressions and facial sensation.

E: pineal gland secretes the hormone melatonin, which is involved in the control of circadian rhythms and may be involved in sexual maturation.

9. A is correct. Adrenocorticotropic hormone (ACTH) is a polypeptide tropic hormone produced and secreted by the anterior pituitary gland. It is an important component of the hypothalamic-pituitary-adrenal axis and is often produced in response to biological stress (e.g. long-distance run).

Glucagon is a storage form of glucose that increases blood sugar by releasing glucose molecules in the blood. An increased heart rate and sympathetic blood shunting away from the intestines are expected during a long-distance run.

10. E is correct.

11. D is correct. Hormones (both steroid and amino acid/peptide) are substances released from glands and enter the bloodstream to affect target tissues.

12. E is correct. Parathyroid hormone controls blood Ca^{2+} levels via two mechanisms by both stimulating osteoclasts activity (i.e. bone resorption) and decreasing the Ca^{2+} loss in urine by the kidney.

See explanations for questions **32** and **54** about parathyroid hormone.

13. D is correct. See explanations for questions **18** and **58**.

14. C is correct. The posterior pituitary releases two hormones: oxytocin and antidiuretic hormone (ADH - also known as vasopressin).

A: TSH (thyroid-stimulating hormone) is released by the anterior pituitary.

B: prolactin is produced in the pituitary and also in decidua (uterine lining – endometrium), myometrium (the middle layer of the uterine wall), breasts, lymphocytes, leukocytes and the prostate.

D: progesterone is produced in the ovaries (corpus luteum), the adrenal glands and, during pregnancy, in the placenta. It is synthesized from the precursor – pregnenolone.

E: calcitonin is released by the thyroid gland.

15. C is correct. The anterior pituitary secretes three gonadotropic hormones (e.g. FSH, LH and placental chorionic gonadotropins hCG). FSH causes the ova to mature and enlarge.

A: oxytocin is synthesized by the hypothalamus and stimulates uterine contractions during birth and milk letdown during lactation. Prolactin stimulates the production of milk for lactation.

D: LH stimulates ovulation.

16. E is correct. Estrogen is a steroid hormone (versus peptide hormone). Because steroids are lipid-soluble, they diffuse from the blood through the cell membrane and into the cytoplasm of target cells. In the cytoplasm, the steroid binds to the specific receptor. Upon steroid binding, many kinds of steroid receptors dimerize to form one functional DNA-binding unit that enters the cell nucleus. Once in the nucleus, the steroid-receptor ligand complex binds to specific DNA sequences and regulates transcription (activates or inactivates) of its target genes.

17. C is correct. Adrenocorticotropic hormone (ACTH) is produced and released by the anterior pituitary gland to increase production and release of glucocorticoid steroid hormones (corticosteroids) by adrenal cortex cells.

18. B is correct. Most second messenger systems are initiated by peptide hormones (i.e. ligands) which cannot pass through the plasma membrane and instead bind to receptors on the cell surface.

Cholesterol-derived steroid hormones (e.g., estrogen) enter the cell and bind to a cytosolic receptor. Together with the receptor, the hormone-receptor complex passes through the nuclear envelope and enters the nucleus to function as a transcription factor and increase/ decrease gene expression.

All anterior pituitary (e.g., TSH) and pancreatic (e.g., insulin) hormones are peptides. Peptide hormones are hydrophilic and cannot cross the hydrophobic phospholipid bilayer. As a result, these endocrine mediators bind as ligands to cell membrane receptors and activate intracellular second-messengers (e.g. cAMP).

19. C is correct.

20. B is correct. The thyroid hormones T_3 and T_4 (triiodothyronine and thyroxine) are derived from the amino acid tyrosine and accelerate oxidative metabolism throughout the body. Thyroid-stimulating hormone (TSH) is secreted by the anterior pituitary to stimulate the thyroid gland to produce T_4 and then T_3, which stimulate the metabolism of almost every tissue in the body.

21. D is correct. *IGF*-1 is the primary protein synthesized in response to growth hormone. Growth hormone (GH) is synthesized by the anterior pituitary and stimulates tissue growth through the *IGF*-1 protein. A deletion of both alleles of the IGF-1 gene causes a deficiency in growth.

A: parathyroid hormone deficiency leads to decreased Ca^{2+} levels and brittle bones.

B: anemia is an abnormally low level of hemoglobin in the blood.

C: deficiency of pancreatic lipase (or bile salts) may result in the inability to digest lipids.

22. E is correct. See explanations for questions **4**, **16**, **25** and **58** about steroid hormones.

23. B is correct. See the explanation for question **45**.

24. B is correct. Thyroid hormone increases the basal metabolic rate by stimulating protein synthesis and increasing the activity of the Na^+/K^+ ATPase pump.

There are two types of thyroid dysfunction: gland *hyperthyroidism* and *hypothyroidism*. Under normal circumstances, the anterior pituitary gland produces TSH that binds to TSH receptors and stimulates the release of thyroid hormone. Auto-antibodies bind to TSH receptors and over-stimulate the thyroid gland, causing abnormally high thyroid hormone levels.

25. C is correct. If the hormone accumulates inside the cell without the process of endocytosis, the hormone must diffuse through the plasma membrane. Steroid hormones are hydrophobic and freely diffuse through the plasma membrane.

A: neurotransmitters are charged molecules that bind to a receptor at the cell surface.

B: second messengers transmit signals inside cells but are not hormones.

D: polypeptides cannot diffuse through membranes because they are large, hydrophilic molecules.

E: amines are derived from single amino acids (e.g. tyrosine) and include epinephrine, norepinephrine and dopamine.

26. B is correct. The pancreas is both an exocrine (via ducts) and endocrine (via bloodstream) gland. The exocrine function (via a series of ducts) is performed by the cells that secrete digestive enzymes (e.g. amylase, lipase and maltase) and bicarbonate into the small intestine.

The endocrine function of the pancreas is performed by small glandular structures called the islets of Langerhans (i.e. alpha, beta and delta cells). Alpha cells produce and secrete glucagon. Beta cells produce and secrete insulin. Delta cells produce and secrete somatostatin.

27. A is correct. The surge of luteinizing hormone (LH) causes ovulation. If the anterior pituitary secretes low levels of LH, this would inhibit the release of an oocyte and cause infertility. Oral contraceptives inhibit the release of LH by altering the ratio of estrogen/progesterone levels.

C: follicle-stimulating hormone (FSH) stimulates oocyte maturation.

28. C is correct.

29. E is correct. Cortisol (also known as hydrocortisone) is a steroid hormone. It is a glucocorticoid named from its role in the regulation of metabolism (glucose levels). Cortisol is produced by the adrenal cortex and is released in response to stress and a low level of blood glucocorticoids. Its primary functions are to increase blood sugar through gluconeogenesis, to suppress the immune system and to aid in carbohydrate, protein and fat metabolism. It also decreases bone formation. Cortisol, the most important human glucocorticoid, is essential for life and regulates/supports a variety of important metabolic, cardiovascular, immunologic and homeostatic functions.

A: adrenaline is also known as epinephrine.

B: norepinephrine (aka noradrenaline) is a catecholamine (i.e. derived from the amino acid tyrosine) with multiple roles, including being both a hormone and a neurotransmitter. It is the hormone and neurotransmitter most responsible for vigilant concentration. This is in contrast to its most chemically similar hormone, dopamine, responsible for cognitive alertness. The primary function of norepinephrine is its role as the neurotransmitter released from the sympathetic neurons to affect the heart, which increases the rate of heart contractions.

Norepinephrine (as a stress hormone) affects parts of the brain (e.g. amygdala) where attention and responses are controlled. Norepinephrine also underlies the fight-or-flight response (in conjunction with epinephrine), directly increases the heart rate, triggers the release of glucose from energy stores, increases the blood flow to skeletal muscle and increases oxygen supply to the brain.

30. D is correct. Excessive production of epinephrine results in clinical manifestations affecting the sympathetic nervous system (*fight or flight* response) with: 1) elevated blood pressure, 2) increased in heart rate, 3) dilation of the pupils and 4) inhibition of gastrointestinal tract function and motility.

31. C is correct. A releasing hormone (or releasing factor) is a hormone whose sole purpose is to control the release of another hormone. The hypothalamus is located just below the thalamus. One of the most important functions of the hypothalamus is to link the nervous system to the endocrine system via the pituitary gland (hypophysis). The hypothalamus is responsible for certain metabolic processes and other activities of the autonomic nervous system. It synthesizes and secretes releasing hormones, and these, in turn, stimulate or inhibit the secretion of pituitary hormones. The hypothalamus controls body temperature, hunger, thirst, fatigue, sleep, circadian rhythms and important aspects of parenting and attachment behaviors.

Examples of releasing factors synthesized by the hypothalamus are: thyrotropin-releasing hormone (TRH), Corticotropin-releasing hormone (CRH), Gonadotropin-releasing hormone (GnRH), Growth hormone-releasing hormone (GHRH). Two other factors are also classified as releasing hormones, although they inhibit pituitary hormone release: somatostatin and dopamine.

32. B is correct. Parathyroid hormone is secreted by the parathyroid gland and functions to increase blood calcium by the removal of calcium from bones and other calcium-containing tissues. The removal of calcium from bones is primarily done via osteoclasts (i.e. bone-releasing = breakdown bone).

A: calcitonin is secreted by the thyroid gland as the antagonist to parathyroid hormone. Calcitonin reduces blood $[Ca^{2+}]$ by depositing it onto bone via osteoblasts (i.e. build bone).

C: aldosterone, secreted by the adrenal cortex, increases Na^+ reabsorption in the kidneys.

D: glucagon (α cells of the islets of Langerhans) from the pancreas raises blood glucose.

E: antidiuretic (ADH) is also known as vasopressin; its function is to increase water reabsorption in the collecting tubules of the kidneys.

33. D is correct. *Antagonistic* refers to an action which is opposite, while *agonist* means to mimic the effect. An antagonistic relationship exists between insulin and glucagon. Both hormones are synthesized by the pancreas but insulin lowers glucose blood levels, while glucagon raises glucose blood levels.

A: ACTH and TSH are unrelated in their actions. ACTH (adrenocorticotropic hormone) is secreted by the anterior pituitary gland and stimulates the adrenal gland. TSH (thyroid stimulating hormone) activates the thyroid to produce thyroxine.

B: oxytocin and prolactin are both *agonists* (i.e. same effect) of the same process, specifically the production of milk by the mammary gland. Oxytocin is secreted by the *posterior* pituitary gland and stimulates *release* of milk during lactation. Prolactin is secreted by the *anterior* pituitary gland and stimulates *production* of milk during lactation.

C: vitamin D and parathyroid hormone (PTH) are both *agonists* of the same process because both increase blood calcium levels.

34. E is correct.

35. D is correct. See explanations for questions **32** and **54**.

36. B is correct. Aldosterone is a steroid hormone produced by the outer adrenal cortex in the adrenal gland. It is produced by the adrenal gland by low blood pressure in response to *low sodium or elevated potassium*. Aldosterone increases blood Na^+ levels and decreases blood K^+ concentration.

It plays a central role in the regulation of blood pressure mainly by acting on the distal tubules and collecting ducts of the nephron. It increases the reabsorption of ions and water in the kidney, causing the conservation of Na^+, secretion of K^+, increased H_2O retention and increased blood pressure.

37. E is correct.

38. C is correct. Glucagon is a peptide hormone secreted by the pancreas that raises blood glucose levels. Its effect is opposite that of insulin, which lowers blood glucose levels. The pancreas releases glucagon when blood glucose levels are too low. Glucagon causes the liver to convert stored glycogen into glucose, which is released into the bloodstream. Conversely, high blood glucose levels stimulate the release of insulin, which allows glucose to be taken up and used by insulin-dependent tissues. Thus, glucagon and insulin are part of a feedback system that keeps blood glucose levels within a narrow range.

Glucagon generally elevates the amount of glucose in the blood by promoting gluconeogenesis and glycogenolysis. Glucose is stored in the liver in the form of glycogen, which is a polymer of glucose molecules. Liver cells (hepatocytes) have glucagon receptors. When glucagon binds to the glucagon receptors, the liver cells convert the glycogen polymer into individual glucose molecules, and release them into the bloodstream. This process is known as glycogenolysis. As these stores become depleted, glucagon then signals the liver and kidney to synthesize additional glucose by gluconeogenesis. Glucagon turns off glycolysis in the liver, causing glycolytic intermediates to be shuttled to gluconeogenesis.

A: calcitonin is secreted by the thyroid gland and lowers blood Ca^{2+} levels.

B: estrogen is secreted by the ovaries, and maintains secondary sex characteristics (i.e. menstrual cycles in females) and is responsible for maintenance of the endometrium.

D: oxytocin is produced by the hypothalamus and is secreted by the posterior pituitary to stimulate uterine contractions during labor and milk secretion during lactation.

E: thyrotropin (also known as TSH) is an anterior pituitary hormone that stimulates the thyroid to release thyroxine and triiodothyronine that increase metabolic rate.

39. A is correct. The main function of hormone-sensitive lipase (HSL) is to mobilize stored fats. When the body needs to mobilize energy stores (e.g. during fasting) HSL is activated through a positive response to catecholamines (i.e. epinephrine, norepinephrine) in the absence of insulin. Insulin inhibits HSL. Insulin is a peptide hormone produced by beta cells of the pancreas. It is central to regulating carbohydrate and fat metabolism. When plasma glucose levels are low, the decreased insulin levels signal the adipocytes to activate hormone-sensitive lipase. HSL converts – via hydrolysis – the triglycerides into free fatty acids to be utilized as energy by the cells of the body.

B: epinephrine (also known as adrenaline) is a hormone and a neurotransmitter. Epinephrine has many functions in the body, such as regulating heart rate, blood vessel and air passage diameters, and metabolism. Adrenaline release is a crucial component of the fight-or-flight response of the sympathetic nervous system. Epinephrine is one of a group of monoamines called the catecholamines. It is produced in some neurons of the central nervous system and also in the adrenal medulla (inner region) from the amino acids phenylalanine and tyrosine.

C: estrogen promotes female secondary sex characteristics. Estrogens, in females, are produced primarily by the ovaries and by the placenta during pregnancy. Follicle-stimulating hormone (FSH) stimulates the ovarian production of estrogens by the ovarian follicles and corpus luteum. Some estrogens are also produced in smaller amounts by other tissues such as the liver, adrenal glands, breasts and fat cells. These secondary sources of estrogens are especially important in postmenopausal women.

D: glucagon is a peptide hormone secreted by the pancreas and is one of four hormones (e.g. epinephrine, cortisol and growth hormone) that raise blood glucose levels. Its effect is opposite that of insulin, which lowers blood glucose levels. The pancreas releases glucagon when blood sugar (glucose) levels fall too low. Glucagon causes the liver to convert stored glycogen into glucose, which is released into the bloodstream. High blood glucose levels stimulate the release of insulin. Insulin allows glucose to be taken up and used by insulin-dependent tissues. Thus, glucagon and insulin are part of a feedback system (i.e. antagonists) that keeps blood glucose levels stable.

E: see explanation for question **29**.

40. C is correct. See explanation for question **3**.

41. A is correct. Melatonin is an important and multifunctional hormone secreted by the pineal gland that has numerous biological effects triggered by the activation of melatonin receptors. Also, it is a pervasive and powerful antioxidant, with a particular role in the protection of nuclear and mitochondrial DNA. Melatonin affects circadian rhythms, mood, timing of puberty, aging and other biological processes.

42. C is correct. The endocrine islets of Langerhans account for only 1–2% of the pancreas and secrete insulin (to lower blood sugar), glucagon (to raise blood sugar) and somatostatin (to inhibit the release of both insulin and glucagon).

Hormones produced in the islets of Langerhans are secreted by five types of cells.

- Alpha cells produce glucagon (15–20% of total islet cells)

- Beta cells produce insulin (65–80%)

- Delta cells produce somatostatin (3–10%)

- Gamma cells produce pancreatic polypeptide (3–5%)

- Epsilon cells produce ghrelin – hunger stimulating peptide (<1%)

A: cortisol is a hormone released (via bloodstream) by the adrenal cortex in response to stress.

B: trypsin is an exocrine (via duct) pancreatic product necessary for digestion of protein.

D: pepsin is secreted by chief cells in the stomach to digest protein.

43. D is correct.

44. E is correct. Epinephrine (also known as adrenaline) is a peptide hormone released by the adrenal medulla and is a physiological stimulant (i.e. Epipen for anaphylaxis) that aids in *fight or flight* responses.

A: epinephrine causes bronchial dilation for deeper respiration and increased O_2 intake.

B: epinephrine release is stimulated by the sympathetic nervous system.

C: epinephrine is a peptide hormone synthesized by the adrenal medulla, while the adrenal cortex synthesizes steroid hormones (e.g. aldosterone and cortisol).

45. C is correct. Growth-hormone-releasing hormone (GHRH), also known as growth-hormone-releasing factor, is a 44-amino acid peptide hormone produced in the hypothalamus and carried to the anterior pituitary gland where it stimulated the secretion of the growth hormone (GH). The hypothalamus controls all secretions that come out of the pituitary. Hypothalamic hormones are referred to as *releasing* and *inhibiting* hormones because of their influence on the anterior pituitary.

The eight hormones of the anterior pituitary:

• *Adrenocorticotropic hormone* (ACTH) targets the adrenal gland and results in secretions of glucocorticoid, mineralocorticoid and androgens.

• *Beta-endorphin* targets opioid receptor and inhibits the perception of pain.

• *Thyroid-stimulating hormone* (TSH) targets the thyroid gland and results in secretions of thyroid hormones.

• *Follicle-stimulating hormone* (FSH) targets the gonads and results in growth of the reproductive system.

• *Luteinizing hormone* (LH) targets gonads to trigger ovulation, maintain corpus luteum and secrete progesterone (in females) or to stimulate testosterone secretion (for males).

• *Growth hormone* (aka somatotropin; GH) targets the liver and adipose tissue and promotes growth, lipid and carbohydrate metabolism

• *Prolactin* (PRL) targets the ovaries and mammary glands and results in secretions of estrogens/progesterone; it also stimulates milk production.

• *Leptin* targets corticotrophic and thyrotrophic cells (both are cell types in the anterior pituitary) and results in secretions of TSH and ACTH.

46. A is correct.

47. B is correct.

48. D is correct. Hormones are classified as steroids (i.e., cholesterol derived) and amino acid hormones. Amino acid hormones include amines (derived from a single amino acid, either tryptophan or tyrosine), peptide hormones (short chains of amino acids) and protein hormones (longer chains of amino acids).

Steroid hormones can be grouped into two classes: corticosteroids (mineralocorticoids and glucocorticoids) and sex steroids (androgens, estrogens and progestogens).

49. E is correct. Cyclic AMP (cAMP) is a second messenger derived from ATP and functions for intracellular signal transduction in many organisms by transferring into cells the effects of hormones like glucagon and adrenaline, which cannot pass through the plasma membrane. cAMP is involved in the activation of protein kinases, binds to and regulates the function of ion channels (e.g. HCN channels) and some other cyclic nucleotide-binding proteins. cAMP and its associated kinases function in several biochemical processes, including the regulation of glycogen, sugar and lipid metabolism.

50. A is correct. Exocrine glands release their secretions through ducts, while endocrine glands release their secretions into the bloodstream. The exocrine secretions of the pancreas are protease, lipase and amylase (which aid in the digestion of food), and bicarbonate ions (which buffer the pH of the chyme) coming from the stomach.

Glucagon is a pancreatic endocrine secretion in response to low blood glucose that increases glucose levels through the degradation of glycogen and by decreasing the uptake of glucose by muscles.

51. A is correct. Insulin is produced by beta cells of the pancreas and decreases blood glucose by simulating muscle cells to uptake glucose from the blood. Aldosterone is synthesized by the adrenal gland and increases the reabsorption of Na^+ and H_2O by the distal convoluted tubules of the nephron.

52. D is correct.

53. D is correct.

54. C is correct. The parathyroid gland (via PTH) and the thyroid (via calcitonin) are antagonist hormones that regulate blood calcium concentration. The thyroid gland is located in the middle of the lower neck, below the larynx (i.e. voice box) and just above the clavicles (i.e. collarbone). Humans usually have four parathyroid glands, which are usually located on the posterior surface of the thyroid gland.

Parathyroid hormone (PTH) is a polypeptide hormone containing 84 amino acids secreted by the chief cells of the parathyroid glands. PTH acts to increase blood $[Ca^{2+}]$, whereas calcitonin (by the C cells of the thyroid gland) decreases $[Ca^{2+}]$. PTH increases $[Ca^{2+}]$ in the blood by acting upon the parathyroid hormone 1 receptor (high levels in bone and kidney) and the parathyroid hormone 2 receptor (high levels in the central nervous system, pancreas, testis and placenta).

55. E is correct. Thyroxine is one of the thyroid hormones secreted by the thyroid gland and plays an important role in regulating metabolism. In adults, thyroid deficiency (i.e. hypothyroidism) results in a decreased rate of metabolism, which produces symptoms such as weight gain, fatigue, intolerance to cold and a swelling of the thyroid (i.e. goiter). A decreased metabolic rate means that the body is using less energy per day and fewer dietary calories are required.

Thus, unless a hypothyroid patient changes her diet, she will gain weight because excess calories are converted and stored in adipose tissue as fat. In response to low levels of thyroid hormone, the pituitary secretes thyroid-stimulating hormone, which increases thyroid hormone production.

However, if the thyroid is unable to increase its hormone synthesis, it will simply hypertrophy (i.e. increase in mass) and this results in the formation of a goiter (i.e. swollen pouch in the throat region).

A: estrogen is not the cause of the patient's symptoms.

B: cortisol deficiency is not the cause of the patient's symptoms.

D: aldosterone deficiency is not the cause of the patient's symptoms.

56. B is correct. Aldosterone is a steroid hormone secreted from the adrenal gland involved in the regulation of blood pressure by acting on the distal tubules and collecting ducts of the nephron. It increases reabsorption of ions and water in the kidney to cause the conservation of sodium, secretion of potassium, increased water retention and increased blood pressure.

The renin-angiotensin system regulates blood pressure and water (i.e. fluid) balance. When blood volume is low, juxtaglomerular cells (in kidneys) activate their prorenin and secrete renin into circulation. Plasma renin then carries out the conversion of angiotensinogen (released by the liver) to angiotensin I. Angiotensin I is subsequently converted to angiotensin II by the angiotensin converting enzyme (located in the lungs). Angiotensin II is a potent vaso-active peptide that causes blood vessels to constrict, resulting in increased blood pressure. Angiotensin II also stimulates the secretion of aldosterone.

57. B is correct.

58. A is correct. The radio-labeled hormone enters the nucleus of the liver cells, which means the hormone is soluble in the plasma membrane. Steroid hormones are hydrophobic and therefore pass through the plasma membrane and enter the target cells where the hormone binds to an intracellular (i.e. cytoplasmic) receptor and then, bound to the receptor, the hormone/receptor complex enters the nucleus and directly binds to the DNA (i.e. transcription factor) and therefore influences the transcription of mRNA.

Peptide hormones (i.e. ligand) bind to a receptor on the surface of the target cell's membranes. The ligand-receptor complex may trigger the release of a second messenger (e.g. G-protein or cAMP) or the ligand-receptor complex may be carried into the cytoplasm by receptor-mediated endocytosis. In the event of a second messenger, a series of (i.e. second messenger) events within the cell are responsible for the hormone's activity (i.e. phosphorylation or dephosphorylation to change the shape/activity of enzymes within the cell). Thus, peptide hormones do not influence the transcription of mRNA directly, because peptide hormones do not enter the nucleus of their target cells and activate transcription.

B: steroid hormones are often derivatives of cholesterol (i.e. hydrophobic).

D: amino acids contain hydrophilic residues and cannot cross the hydrophobic lipid bilayer of the plasma membrane or the nuclear membrane.

Chapter 4.3: Nervous System

1. C is correct. Astrocytes, which collectively form astroglia, are star-shaped glial cells in the brain and spinal cord. They are involved in multiple processes and carry out many important functions. Schwann cells are cells that produce the myelin sheath around neuronal axons in the peripheral nervous system.

2. A is correct. Sensation is the awareness of changes in the internal and external environments. For the sensation to occur, the stimulus energy must match the specificity of the receptor, the stimulus must be applied within a sensory receptor's receptive field (the smaller the receptive field, the greater the brain's ability to correctly localize the stimulus site) and the stimulus energy must be converted into energy of a graded potential (i.e., transduction). The graded potential can be depolarizing or hyperpolarizing.

3. C is correct. The parasympathetic nervous system maintains homeostasis (*rest and digest*) that increases gut motility, modulates heart rate, constricts the bronchi and the pupils.

The sympathetic nervous system prepares the body for action (*fight or flight*), which decreases digestive tract activity, increases heart rate, dilates the pupils and relaxes the bronchi.

4. D is correct. The autonomic nervous system (ANS) consists of motor neurons that: innervate smooth and cardiac muscle and glands, make adjustments to ensure optimal support for body activities, operate via subconscious control (involuntary nervous system or general visceral motor system) and have viscera as most of their effectors. The effectors of the ANS are cardiac muscle, smooth muscle and glands.

The ANS uses a 2-neuron chain to its effectors. Cell body of first neuron (i.e., preganglion neuron) resides in the brain or spinal cord. The axon of the first neuron (i.e., preganglionic axon) synapses with the second motor neuron (i.e., ganlionic neuron) in an autonomic ganglion outside of the central nervous system. The axon of the ganglionic neuron (i.e., postsynaptic axon) extends to the effector organ.

5. B is correct. Retinal (retinaldehyde) is one of the many forms of vitamin A. Retinal bound to opsin proteins is the chemical basis of vision. Vision begins with the photoisomerization of retinal. When the 11-*cis*-retinal chromophore absorbs a photon, it isomerizes from the 11-*cis* state to the all-*trans* state. The absorbance spectrum of the chromophore depends on its interactions with the opsin protein to which it is bound; different opsins produce different absorbance spectra.

6. E is correct.

---

7. B is correct. Myopia (i.e. nearsightedness) is a condition of the eye where light does not focus directly on the retina but focuses in front. The image of a distant object is out of focus, while the image of a closer object is in focus. This condition is generally due to the shape of the eye being too long.

8. E is correct. Bipolar neurons are neurons that have two extensions. They are specialized sensory neurons for the transmission of special senses (e.g. sight, smell, taste, hearing and vestibular functions). Common examples of bipolar neurons are the bipolar cell of the retina and the ganglia of the vestibulocochlear nerve. Bipolar neurons are also extensively utilized to transmit efferent (motor) signals to control muscles. Additionally, they are found in the spinal ganglia when the cells are in an embryonic condition.

9. B is correct.

10. A is correct. The blood-brain barrier helps block harmful substances (e.g., bacteria, toxins, metabolic waste products) from entering the brain. It does allow for the passage of oxygen, glucose, amino acids, alcohol and anesthetics.

11. C is correct. The medulla controls many vital functions such as breathing, heart rate and gastrointestinal activity.

A: cerebrum (i.e. cerebral cortex) processes and integrates sensory input and motor responses, and is also involved in memory and creativity.

B: cerebellum is important in coordinating muscles and aids in balance by receiving input from the inner ear, hand-eye coordination and the timing of rapid movements.

D: see explanation for question **45**.

E: pituitary gland (along with hypothalamus) is important for controlling the endocrine system.

12. B is correct. Acetylcholine (ACh) is the neurotransmitter found at neuromuscular junctions for motor and memory functions. It is also found at synapses in the ganglia of the visceral motor system. ACh carries signals at all neuromuscular (nerve-to-skeletal muscle) connections.

13. E is correct. In anatomy, the term decussation (i.e. crossing) means the same thing as chiasma (or chiasm). In genetics, the chiasma is the point where the two chromatids are interwoven, as observed between the S phase and anaphase.

14. A is correct.

15. D is correct. The autonomic nervous system is composed of sympathetic and parasympathetic divisions. Both divisions rely on a two-neuron motor pathway away from the spinal cord and a two-neuron sensory pathway toward the spinal cord.

The ANS is unique in that it requires a sequential two-neuron efferent pathway; the preganglionic (first) neuron must first synapse onto a postganglionic (second) neuron before innervating the target organ. The preganglionic neuron begins at the *outflow* and synapses at the postganglionic neuron's cell body. The postganglionic neuron then synapses at the target organ.

The sympathetic nervous system has a short preganglionic neuron and a long postganglionic neuron.

16. D is correct. Voltage-gated ion channels are a type of transmembrane proteins activated by changes in membrane potential. They are essential to the initiation and propagation of action potentials in neurons.

17. E is correct. The cerebellum, along with the pons and medulla oblongata, is located in the hindbrain. All higher brain sensory neurons and motor neurons pass through the hindbrain. The main function of the cerebellum is coordinating unconscious movement. Hand-eye coordination, posture and balance are all controlled by the cerebellum. Therefore, damage to the cerebellum would most likely result in loss of muscle coordination.

The cerebrum is divided into left and right hemispheres that are further subdivided into four lobes. The cerebrum is responsible for the coordination of most voluntary activities, sensation and *higher functions* (including speech and cognition). The extremities are also controlled by the spinal cord.

18. A is correct. The somatic nervous system controls skeletal muscles and consists of sensory and motor nerves.

19. A is correct. The fovea is located in the center of the macula region of the retina. The fovea is responsible for sharp central vision necessary for reading, driving and for visual detail. Cone cells are one of the two types of photoreceptor cells in the retina and are responsible for color vision and eye color sensitivity. Cone cells function best in relatively bright light, as opposed to rod cells, which function better in dim light. Cone cells are densely packed in the fovea (0.3 mm in diameter, rod-free area), but quickly reduce in number towards the periphery of the retina.

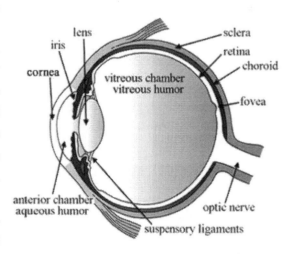

The six to seven million cones in a human eye are concentrated towards the macula.

C: rod cells are photoreceptor cells in the retina of the eye that function in less intense light than cone cells. Rods are concentrated at the outer edges of the retina and are used in peripheral vision. There are approximately 125 million rod cells in the human retina. Rod cells are more sensitive than cone cells and are almost entirely responsible for night vision.

20. D is correct.

21. C is correct. The cell bodies of somatic sensory neurons are not located in the CNS. They are located in dorsal root ganglia just behind the spinal cord all along the length of the spinal cord.

The PNS is divided into somatic and visceral parts. The somatic part consists of the nerves that innervate the skin, joints and muscles. The cell bodies of somatic sensory neurons lie in dorsal root ganglia of the spinal cord. The visceral part (also known as the autonomic nervous system) contains neurons that innervate the internal organs, blood vessels and glands. The autonomic nervous system itself consists of two parts: the sympathetic nervous system and the parasympathetic nervous system.

22. B is correct.

23. E is correct. The medulla oblongata monitors CO_2 levels and pH of the blood. It regulates breathing, temperature, heart rate and reflex activities such as sneezing, coughing and swallowing.

A: the reticular activating system (RAS) is a network of neurons in the brainstem that receives input from all sensory systems and sends non-specific information to the brain.

B: see explanation for question **45**.

C: cerebellum controls muscle coordination and tone, and maintains posture.

D: cerebral cortex controls vision, hearing, smell, voluntary movement and memory.

24. B is correct.

25. D is correct. The eye takes approximately 20–30 minutes to fully adapt from bright sunlight to complete darkness, becoming ten thousand to one million times more sensitive than at full daylight. In this process, the eye's perception of color changes as well. However, it takes approximately five minutes for the eye to adapt to bright sunlight from darkness. This is due to cones obtaining more sensitivity during the first five minutes when entering the light, but the rods take over after five or more minutes.

Rhodopsin is a biological pigment in the photoreceptors of the retina that immediately photobleaches in response to light. Rods are more sensitive to light and take longer to fully adapt to the change in light. Rods' photopigments regenerate more slowly, and need about 30 min to reach their maximum sensitivity. However, their sensitivity improves considerably within 5–10 minutes in the dark. Cones take approximately 9–10 minutes to adapt to the dark. Sensitivity to light is modulated by changes in intracellular calcium ions and cyclic guanosine monophosphate.

26. A is correct.

27. C is correct. Parasympathetic stimulation elicits *rest and digest* responses (i.e. not stress or immediate survival).

A: piloerection is mediated by skeletal muscles that do not receive autonomic innervation.

B: contractions of abdominal muscles during exercise do not receive autonomic innervation.

D: increased rate of heart contractions is a sympathetic response.

28. D is correct. Graded potentials are changes in membrane potential that vary in size. In the absence of action potentials, a graded potential decays as it travels away from the site of origin.

29. E is correct. Acetylcholine (ACh) is a neurotransmitter that causes depolarization of the postsynaptic membrane when released by the presynaptic terminal of another neuron. ACh is removed from the synapse by the enzyme acetylcholinesterase, which catalyzes the hydrolysis of acetylcholine into choline and acetate.

Insecticides, such as Diazinon, are anti-cholinesterases, because they block the activity of acetylcholinesterase, and acetylcholine is not degraded. This increases the concentration of acetylcholine in the synapse, while the presynaptic membrane continues to secrete acetylcholine in response to presynaptic action potentials.

C: acetylcholinesterase inactivity would increase (not *decrease*) postsynaptic depolarization, because if acetylcholine is not degraded, it will continuously bind to its receptors on the postsynaptic membrane and depolarize it.

D: anti-cholinesterase does not affect the activity of other neurotransmitters (e.g. epinephrine or norepinephrine), and therefore all synaptic nervous transmission does not cease.

30. A is correct.

31. D is correct. The cornea is the transparent front part of the eye that covers the iris, pupil and anterior chamber. The cornea, with the anterior chamber and lens, refracts light, with the cornea accounting for approximately two-thirds of the eye's total optical power (43 dioptres). While the cornea contributes most of the eye's focusing power, its focus is fixed. The curvature of the lens, by comparison, can be adjusted to "tune" the focus depending upon the object's distance.

The lens is a transparent, biconvex structure in the eye that (along with the cornea) helps to refract light to be focused on the retina. In humans, the refractive power of the lens is about 18 dioptres (1/3 of the total refractive power). The lens, by changing shape, functions to adjust the focal distance of the eye to focus on objects at various distances. This allows a sharp real image of the object to be formed on the retina. This adjustment of the lens is known as accommodation. The lens is more flat on its anterior side than on its posterior side.

A dioptre is a unit of measurement of the optical power of a lens or curved mirror. It is equal to the reciprocal of the focal length measured in meters (i.e. 1/meters) and therefore is a unit of reciprocal length. For example, a 3-dioptre lens brings parallel rays of light to focus at 0.33 meters.

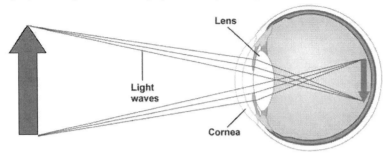

32. B is correct.

33. E is correct. Epinephrine is part of the sympathetic nervous system, and its release prompts the *fight or flight* response.

A: aldosterone is involved in sodium reabsorption by the kidney.

B: dopamine is important for motor control, arousal, motivation, cognition, and reward, as well as a number of basic lower-level functions including nausea, lactation and sexual gratification.

C: acetylcholine is a neurotransmitter. In the heart, acetylcholine neurotransmission has an inhibitory effect, which lowers heart rate. However, at neuromuscular junctions in skeletal muscle, acetylcholine also acts as an excitatory neurotransmitter.

D: insulin causes cells to take up glucose and is not part of the sympathetic nervous system.

34. D is correct. The cerebral cortex is the outer layer of the brain, made up of gray matter. White matter is a different type of nerve fibers found in the brain's inner layer.

35. B is correct. The cerebellum is part of the hindbrain (i.e. posterior part of the brain) and consists of the pons and medulla oblongata. The cerebellum receives sensory information from the visual and auditory systems, as well as information about the orientation of joints and muscles.

The cerebellum's main function is hand-eye coordination. It also receives information about the motor signals being initiated by the cerebrum and integrates the inputs to produce balance and unconscious coordinated movement. Damage to the cerebellum could damage any one of these functions, while total destruction would eliminate all and would seriously impair coordinated movement.

A: thermoregulation is a function of the hypothalamus (a part of the cerebrum).

C: sense of smell (i.e. olfaction) is a function of the cerebrum.

D: urine formation is the primary function of the kidneys, along with some hormonal regulation.

36. C is correct.

37. E is correct.

38. C is correct.

39. B is correct. Pressure waves (i.e. sound) are converted into neural signals by hair cells in the organ of Corti of the cochlea. The organ of Corti (found only in mammals) is part of the cochlea of the inner ear and has hair cells (i.e. auditory sensory cells). The organ of Corti is the structure that transduces pressure waves into electrical signals as action potentials.

40. A is correct.

41. C is correct. The nervous system is divided into the peripheral nervous system and the central nervous system. The peripheral nervous system (PNS) is divided into the autonomic nervous system (ANS) and somatic nervous system (SNS). The autonomic nervous system regulates the internal environment by way of involuntary nervous pathways. The ANS innervates smooth muscle in blood vessels and the digestive tract, and innervates the heart, the respiratory system, the endocrine system, the excretory system and the reproductive system.

The autonomic nervous system is divided into sympathetic, parasympathetic and enteric nervous systems. The functions of ANS can be divided into sensory and motor subsystems. The sensory subsystem consists of those receptors and neurons that transmit signals to the central nervous system. The motor subsystem transmits signals from the central nervous system to effectors.

The sympathetic division innervates those pathways that prepare the body for immediate action known as the "fight-or-flight" response. Heart rate and blood pressure increase, blood vessels in the skin constrict and those in the heart dilate (i.e. vasoconstriction and vasodilation). Pathways innervating the digestive tract are inhibited, and epinephrine (i.e. adrenaline) is secreted by the adrenal medulla, increasing the conversion of glycogen into glucose and therefore increasing blood glucose concentration.

A: the central nervous system (CNS) consists of the brain and the spinal cord.

B: the somatic system innervates skeletal muscle, and its nervous pathways are typically under voluntary control.

D: the parasympathetic system innervates nervous pathways that return the body to homeostatic conditions following exertion. Heart rate, blood pressure and blood glucose concentration decrease, blood vessels in the skin dilate and those in the heart constrict, and the digestive process is no longer inhibited.

42. E is correct. Endorphin is a neurotransmitter that works as a chemical process to control and hinder pain perception. When endorphins are released, fewer pain impulses are sent to the brain.

43. C is correct. A photoreceptor cell is a specialized type of neuron found in the retina that is capable of phototransduction (i.e. light is converted into electrical signals). Photoreceptors convert light (visible electromagnetic radiation) into signals that can stimulate biological processes by triggering a change in the cell's membrane potential. Activation of rods and cones involves hyperpolarization.

Rods and cones, when not stimulated, depolarize and release glutamate (neurotransmitter). In the dark, cells have a relatively high concentration of cGMP (cyclic guanosine monophosphate), which opens ion channels (Na^+ and Ca^{2+}). The positively charged ions that enter the cell change the cell's membrane potential, cause depolarization, and lead to the release of glutamate. Glutamate depolarizes some neurons and hyperpolarizes others.

44. C is correct. The brain stem is the posterior part of the brain that joins with the spinal cord. It consists of three parts: the pons, medulla and midbrain. The midbrain is in charge of vision, hearing, motor control, alertness and temperature regulation. The pons contains neural pathways that carry signals from and to the brain. The medulla is associated with breathing, heart rate and blood pressure functions.

45. B is correct. The hypothalamus controls hunger, thirst, sleep, water balance, blood pressure, temperature regulation and libido. It also has an important role in modulating the endocrine system.

E: substantia nigra is a small midbrain area that forms a component of the basal ganglia and is involved in Parkinson's disease.

46. D is correct. Cerebrospinal fluid is a clear fluid in the brain and spinal cord. It acts as a buffer for the brain and provides mechanical and immunological protection. Red blood cells are usually not present in the cerebrospinal fluid.

47. E is correct. Otitis is a general term for inflammation or infection of the ear.

A: glaucoma is an increase of pressure in the aqueous humor from blockage of aqueous humor outflow.

B: amblyopia, also called "lazy eye," is the most common cause of vision impairment in children, when the vision in one of the eyes is reduced because the eye and the brain are not working together properly. The eye itself looks normal, but it is not being used normally, because the brain is favoring the other eye.

C: myopia (nearsightedness) causes the image to form in front of the retina.

D: hyperopia (i.e. hypermetropia or farsightedness) causes the image to form behind the retina.

48. A is correct.

49. B is correct. A reflex arc requires one muscle group to contract, while the antagonistic muscle group must relax (i.e. inhibitory signal). This excitation/inhibitory response prevents conflicting contractions by antagonistic muscle groups.

A: reflex arcs often are confined to a three-neuron network (i.e. afferent, interneuron and efferent) with the spinal cord. The interneuron (within the spinal cord) integrates the communication between the afferent and efferent neurons.

C: motor neurons exit the ventral side of the spinal cord.

D: reflex arcs do not require processing from the brain (i.e. no conscious thought before responding) so the cerebral cortex is not involved.

50. D is correct.

51. A is correct. A reflex arc is a stimulus coupled with a rapid motor response and is meant for quickness and protection. Examples include recoiling a finger away from a hot stove, whereby the reflex arc begins when the stove is touched, which stimulates a sensory nerve. The sensory nerve directs a signal towards the CNS and synapses with an interneuron, which connects the sensory and motor neurons within the spinal cord. The interneuron synapses with a motor neuron that delivers the response signal to an arm and causes movement away from the stove. The reflex arc requires no processing (or input) from the brain.

There are two types of reflex arc: autonomic reflex arc which affects internal organs and somatic reflex arc which affects muscles.

B: a very brief delay occurs at the two synaptic junctions, because time is necessary for neurotransmitters to diffuse.

C: a reflex arc may consist of only two neurons (one sensory and one motor neuron) and is referred to as *monosynaptic* as it has a single chemical synapse. In peripheral muscle reflexes (e.g., patellar reflex, achilles reflex), brief stimulation of the muscle spindle causes contraction of the agonist or effector muscle. In *polysynaptic* reflex pathways, one or more interneurons connect afferent (i.e., sensory) and efferent (i.e., motor) signals. With exception of the simplest reflexes, all reflex arcs are polysynaptic.

D and E: sensory neurons can synapse in the brain, but reflex arc neurons synapse in the spinal cord.

52. A is correct. White matter consists of bundles of myelinated nerve fibers. Myelin is a white fatty substance that wraps around nerve fibers and increases the speed of neural communication.

53. D is correct. Blood vessels can be directly visualized only during examination of the eyes, specifically in the retina. The condition of these blood vessels indicates the health of small blood vessels in the nervous system (including signs of atherosclerosis or of diabetic vascular disease that can affect peripheral nerves and the brain).

54. A is correct. Broca's area is located in the frontal lobe of the brain (in the dominant hemisphere – usually the left) and is responsible for speech production. Damage to Broca's area is commonly associated with speech made up of content vocabulary where the content of the information is correct, but the sentence structure and fluidity is missing.

55. E is correct. The organ of Corti and the semicircular canals contain hair cells with microcilia (i.e. small hair) projecting from the apical surface of the cell into the surrounding fluid.

Movement of the fluid around the hair cells in the organ of Corti detects sound.

Movement of the fluid around the hair cells in the semicircular canals detects changes in body orientation.

I: hair on skin is different and is not formed by "hair cells," but by dead epithelial cells.

56. C is correct.

57. B is correct. The sympathetic nervous system is involved in the *fight or flight* response. Stimulation of this branch of the autonomic nervous system increases in the amount of adrenaline secretion and is characterized by dilation of the pupils, an increase in heart rate, an increase in respiratory rate, dilation of the bronchi, increased blood flow to the skeletal muscles and less blood flow to the digestive organs.

The parasympathetic nervous system produces *rest and digest* responses that are consistent with the other answer choices.

58. C is correct. Afferent nerves are sensory neurons that carry impulses from sensory stimuli towards the central nervous system, while efferent nerves carry neural impulses away from the central nervous system towards muscles.

59. A is correct.

60. C is correct.

Chapter 4.4: Circulatory System

1. B is correct.

2. D is correct. Hemoglobin does exhibit positive cooperative binding (i.e. sigmoid shape O_2 binding curve). When O_2 binds to the first of four hemoglobin subunits, this initial binding of O_2 changes the shape of Hb and increases the affinity of the three remaining subunits for O_2. Conversely, as the first O_2 molecule is unloaded, the other O_2 molecules dissociate more easily. Cooperative binding is necessary for hemoglobin transport of O_2 to the tissues.

B: fetal hemoglobin acquires O_2 from the maternal circulation, so fetal hemoglobin has a higher O_2 affinity than adult hemoglobin.

C: CO poisoning occurs because this odorless gas has a very high hemoglobin affinity and binds tightly to hemoglobin, preventing oxygen from binding to the occupied site of the hemoglobin.

3. E is correct. At the arteriole end of the capillary bed, the hydrostatic pressure is approximately 30-36 mmHg, while the opposing osmotic pressure is approximately 25-27 mmHg. The larger hydrostatic pressure forces fluid out of the capillaries and into the interstitial space.

At the venule end of the capillary bed, the osmotic pressure is greater than the hydrostatic pressure, which has dropped to 12-15 mmHg. The larger osmotic pressure draws fluid into the capillaries from the interstitial space and permits the liquid to return to the heart.

Most of the fluid (blood) is forced from the capillaries to the interstitial space at the arteriole end (via hydrostatic pressure) and is reabsorbed by the capillaries at the venule end (via osmotic pressure).

4. B is correct.

5. A is correct.

6. A is correct. O_2 binding to hemoglobin exhibits a positive cooperativity with a Hill coefficient greater than 1.

Positive cooperativity is due to the first (of four) O_2 binding weakly to hemoglobin. The binding of first O_2 increases the affinity of the hemoglobin for additional O_2 binding. The binding curve has a sigmoid shape.

Negative cooperativity occurs when the ligand binding to the first site decreases the affinity for additional ligands to bind.

7. E is correct. Veins are thin-walled inelastic vessels that usually transport deoxygenated blood (pulmonary vein is the exception) *toward* the heart. Veins do not have a strong pulse because of the fluid exchange that occurs in the capillaries. Because of their low pressure, blood flow along veins depends on compression by skeletal muscles, rather than on the elastic smooth muscles that line the arteries.

A: ventricles are the pumping chambers of the heart while the atria are collecting chambers.

B: arteries transport blood *away* from the heart and are thick walled, muscular elastic circulatory vessels. Blood in the arteries is usually oxygenated. The pulmonary arteries are the exception whereby deoxygenated blood from the heart, returning from the body, is pumped to the lungs.

C: mammals have a four chambered heart.

D: lymphatic system is an open circulatory system that transports excess interstitial fluid (lymph) to the cardiovascular system and maintains constant fluid levels. Lymph enters the bloodstream at the thoracic duct that connects to the superior vena cava.

8. D is correct.

9. C is correct. Blood has three main roles in the body: transport, protection and regulation.

In its transport role, blood transports the following substances:

- Gases (oxygen and carbon dioxide) between the lungs and tissues
- Nutrients from the digestive tract and storage sites to the cells
- Waste products to be removed by the liver and kidneys
- Hormones from where they are secreted to their target cells

Blood's protective role includes:

- Leukocytes (i.e., white blood cells) destroy invading microorganisms and cancer cells
- Antibodies and other proteins destroy pathogenic substances
- Platelet factors initiate blood clotting and minimize blood loss

Blood helps regulate:

- pH by interacting with acids and bases
- Water balance by transferring water to and from tissues

10. B is correct. Myoglobin accepts O_2 from hemoglobin, because myoglobin has a higher affinity for O_2 than hemoglobin. Myoglobin releases O_2 into the cytochrome c oxidase system, because cytochrome c oxidase has a higher affinity for O_2 than myoglobin or hemoglobin.

Cytochrome c oxidase is the last enzyme in the respiratory electron transport chain located in the mitochondrial (or bacterial) membrane. It receives an electron from each of four cytochrome c molecules, and transfers them to one O_2 molecule, converting molecular O_2 to two molecules of H_2O. In the process, it binds four protons from the inner aqueous phase to make H_2O, and also translocates four H^+ across the membrane to establish a transmembrane difference of proton electrochemical potential that the ATP synthase then uses to synthesize ATP via oxidative phosphorylation.

11. E is correct. Platelets are cell fragments that lack nuclei and are involved in clot formation. Platelets (i.e. thrombocytes) are small, disk shaped clear cell fragments (i.e. do not have a nucleus), 2–3 μm in diameter, which are derived from fragmentation of precursor megakaryocytes. The average lifespan of a platelet is normally just 5 to 9 days.

Platelets are a natural source of growth factors circulating in the blood and are involved in hemostasis (formation of blood clots to stop bleeding). A low platelet count may result in excessive bleeding, while too high of a number may result in the formation of blood clots (thrombosis), which may obstruct blood vessels and result in stroke, myocardial infarction, pulmonary embolism or the blockage of blood vessels, for example, in the extremities of the arms or legs.

A: erythrocytes (i.e. red blood cells) are the oxygen carrying components of the blood and contain hemoglobin that binds up to four molecules of oxygen.

B: macrophages perform phagocytosis of foreign particles and bacteria by engulfing them, digesting the material and then presenting the fragments on their cell surface.

C: B cells mature into memory cells, or antibody producing cells, during immune responses.

D: T cells lyse virally infected cells or secrete proteins that stimulate the development of B cells or other types of T cells.

12. A is correct. A person with AB blood has neither anti-A nor anti-B antibodies, so there is no agglutination reaction when any type of blood is transfused into an AB person

13. E is correct.

14. C is correct. Myoglobin is a single polypeptide chain and has tertiary (3°) protein structure.

A: primary (1°) structure refers to the linear sequence of amino acids in the polypeptide.

B: secondary (2°) protein structure has alpha (α) helixes and beta (β) sheets.

D: quaternary (4°) structure refers to protein consisting of 2 or more peptide chains. Hemoglobin is the classic example of a 4° structure protein with 2 α and 2 β chains.

E: the highest level of protein structure organization is quaternary (4°). 5° does not exist.

15. C is correct. CO, like O_2, binds to hemoglobin, but more strongly than O_2, and its binding is almost irreversible. CO is formed by incomplete combustion of fuels (i.e. hydrocarbons), which results from faulty space heaters or from barbecue grills.

A: CO is not irritating, as it is odorless and colorless.

B: CO does not affect the cytochrome chain.

D: CO does not form complexes in blood.

E: CO does not affect the Na^+/K^+ pump.

16. B is correct. Arterial hydrostatic pressure (i.e. blood pressure) is higher than the hydrostatic pressure at the venous side of the capillaries. Hydrostatic pressure tends to drive fluids out of the blood and into the interstitial tissues on the arterial side.

Osmotic pressure is greater in the plasma than in interstitial fluid, because plasma has a much higher protein (i.e. albumin) concentration. Therefore, at the venule end, osmotic pressure draws fluids into the blood from the interstitial tissues, because osmotic pressure is higher than the hydrostatic pressure at the venule.

17. A is correct. D: erythroblast is a type of red blood cell that still retains a cell nucleus and is the immediate precursor of an erythrocyte.

18. B is correct. Severe combined immunodeficiency (SCID) causes abnormalities in T and B-cells. B-cells are produced in the bone marrow. T-cells are produced in the bone marrow and mature in the thymus. SCID is a genetic disorder characterized by the absence of functional T-lymphocytes, which results in a defective antibody response due to either direct involvement with B lymphocytes or through improper B lymphocyte activation due to non-functional T-helper cells. Consequently, both B cells and T cells are impaired due to a genetic defect.

19. D is correct. Capillaries have a higher hydrostatic pressure at the arteriole end and a lower hydrostatic pressure at the venule end. As blood flows from arterioles to capillaries, blood pressure gradually drops due to friction between the blood and the walls of the vessels, and the increase in cross-sectional areas provided by numerous capillary beds.

Blood plasma in the capillaries has a higher osmotic pressure than the pressure in interstitial fluid. This is a result of the greater amount of dissolved solutes in the blood plasma of the capillaries.

Hydrostatic pressure is defined as the force per area that blood exerts on the walls of the blood vessels. The pumping force of the heart through the blood vessels creates hydrostatic pressure.

20. D is correct. pH is represented as a log scale: a pH change of 1 unit (e.g. pH 4 *vs.* 5) = 10× difference.

A pH difference of 2 units (e.g. pH 3 *vs.* 5): 10 × 10 = 100× difference.

A pH difference of 3 units (e.g. pH 2 *vs.* 5): 10 × 10 × 10 = 1,000× difference.

21. B is correct. Albumin is the main protein produced in the liver and functions to maintain osmotic pressure. Osmotic pressure is exerted in a blood vessel's plasma by pulling water into the circulatory system.

22. E is correct. Angiotensin is a peptide hormone derived from the precursor angiotensinogen produced in the liver, which causes vasoconstriction and a subsequent increase in blood pressure. It is part of the renin-angiotensin system that is a major target for drugs that lower blood pressure.

The kidney secretes renin when blood volume is low, which cleaves angiotensinogen into angiotensin I that is converted into angiotensin II by the angiotensin-converting enzyme (ACE). Angiotensin II stimulates the release of aldosterone from the adrenal cortex. Aldosterone promotes sodium retention in the distal nephron (in the kidney), which increases blood pressure.

C: parathyroid hormone (PTH) increases plasma $[Ca^{2+}]$.

D: calcitonin (hormone produced by the thyroid gland) acts to decrease $[Ca^{2+}]$.

23. C is correct. In an adult, all haematopoiesis (i.e. formation of blood cellular components) occurs in the bone marrow. In the fetus, haematopoiesis occurs in the fetal liver. The spleen acts as a reservoir for red blood cells and filters the blood.

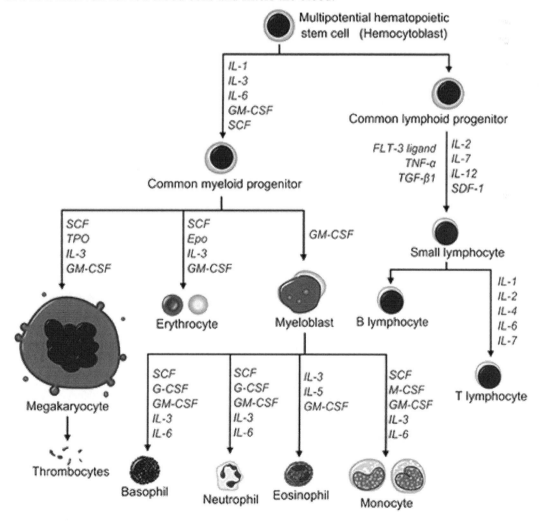

A: erythrocytes (mature red blood cells) are not nucleated to create more space for hemoglobin.

B: blood platelets are crucial for the clotting of blood.

D: leukocytes (white blood cells) such as macrophages and neutrophils engulf foreign matter. Neutrophils are the most abundant white blood cells and account for approximately 50-70% of all white blood cells.

E: erythrocytes are produced in bone marrow to replace erythrocytes during their 120 day life span.

24. C is correct. Blood returns to the heart via the inferior and superior vena cava (i.e. hollow veins) and enters the right atrium of the heart according to the following flow:
Superior/inferior vena cava → right atrium → right ventricle → pulmonary artery → lungs → pulmonary vein

25. A is correct. Monocytes are a type of white blood cells and have granules that are very fine, thus making the structures invisible under a light microscope.

26. E is correct. Surfactants are compounds that lower the surface tension and form micelles. Micelle formation requires numerous hydrophobic interactions and minimizes exposure to hydrophilic surfaces. Surfactants are a lipid-based compound. Micelle formation is necessary during the absorption of lipids (i.e. dietary fats) by the small intestine. A lack of micelle formation results in lipid-soluble vitamin (e.g. A, D, E and K) deficiency.

27. D is correct. Blood traveling from the left ventricle enters the aorta and then flows to all of the body areas (i.e. systemic circulation) except the lungs (i.e. pulmonary circulation). Examples of systemic circulation include blood in the brachiocephalic artery (travels to the head and the shoulders) and blood in the renal artery (travels to the kidney to be filtered).

Superior/inferior vena cava → right atrium → right ventricle → pulmonary artery (*pulmonary circulation*) → lungs → pulmonary vein → left atrium → left ventricle → aorta (*systemic circulation*) → superior/inferior vena cava.

For systemic circulation, blood moves from arteries into arterioles, then capillaries, where nutrients, waste, and energy are exchanged. Blood then enters venules, collects in veins, is transported to the superior and inferior venae cavae and then enters the right atrium and flows into the right ventricle.

For pulmonary circulation, blood is transported to the lungs via the pulmonary artery where capillary beds around the alveoli exchange gas (O_2 into the blood and CO_2 from the blood) in the lungs. Then the pulmonary veins bring blood back to the left atrium to start the systemic circulation.

28. A is correct. O blood contains neither A nor B antigens, so there is no agglutination reaction when O type blood is infused into the patient.

29. D is correct.

30. B is correct. Surfactants are compounds that lower the surface tension (or interfacial tension) between two liquids or between a liquid and a solid. Surfactants may act as detergents, emulsifiers and foaming agents. Pulmonary surfactants are also naturally secreted by type II cells of the lung alveoli in mammals to form a layer over the alveolar surface, which reduces alveolar collapse by decreasing surface tension within the alveoli.

31. A is correct. Alveoli are thin air sacs and are the sites of gas (e.g. O_2 and CO_2) exchange via passive diffusion in the lungs between the air and blood.

B: pleura is the outer lining of the lungs filled with pleural fluid that lubricates the lungs.

C: bronchi are the two main branches of the air intake pathway, with one bronchus for each lung.

D: bronchioles are smaller subdivisions of the bronchi.

E: trachea (i.e. windpipe) is the region of the air intake pathway between the glottis and the bronchi.

32. E is correct. The right atrium is the upper chamber of the right side of the heart. The blood that is returned to the right atrium by the superior and inferior vena cava is deoxygenated (low in O_2) and passed into the right ventricle to be pumped through the pulmonary artery to the lungs for O_2 and removal of CO_2.

The left atrium receives newly oxygenated blood from the lungs by the pulmonary vein. The blood is passed into the strong left ventricle to be pumped through the aorta to the different organs of the body.

In the normal heart, the right ventricle pumps blood through the pulmonary arteries to the lungs during pulmonary circulation. The left ventricle pumps blood through the aorta to all parts of the body for systemic circulation. If blood mixes between the ventricles, some of the blood in the right ventricle (deoxygenated) is mixed with the blood in the left ventricle (oxygenated) for the systemic circulation. This mixing lowers the O_2 supplied to the tissues.

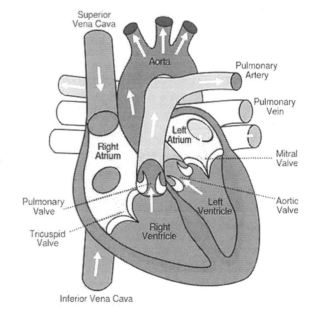

33. A is correct. Blood plasma is the liquid part of blood, consisting primarily of water (up to 95%). It holds blood cells in suspension acting as the extracellular matrix.

34. B is correct. Antibodies are proteins; therefore plasma cells have a well-developed rough endoplasmic reticulum.

C: smooth ER is responsible for steroid synthesis (e.g. Leydig cells for testosterone synthesis).

D: mitochondria function for ATP synthesis via Krebs cycle and the electron transport chain.

35. A is correct. The aorta carries oxygenated blood from the heart through the systemic (i.e. body) circulation and (like the pulmonary vein) has the greatest partial pressure of O_2.

B: coronary veins carry deoxygenated blood to the right side of the heart.

C: superior vena cava carries deoxygenated blood from the upper regions of the body back to the heart.

D: pulmonary arteries carry deoxygenated blood *away* from the heart towards the lungs. The pulmonary artery is the only artery in the adult carrying deoxygenated blood.

E: the inferior vena cava carries deoxygenated blood from the lower half of the body into the right atrium of the heart.

36. C is correct.

37. C is correct.

38. E is correct. The major histocompatibility complex (MHC) is a set of cell surface molecules that mediate interactions of leukocytes (i.e. white blood cells), which are immune cells, with other cells and leukocytes. MHC determines compatibility of donors for an organ transplant and susceptibility to an autoimmune disease (e.g. type I diabetes, lupus, celiac disease). Protein molecules—either of the host's own phenotype or of other biologic entities—are continually synthesized and degraded in a cell. MHC molecules display, on the cell surface, a molecular fraction (i.e. epitope) of a protein. The presented antigen (i.e. epitope from the degraded protein) can be either *self* or *nonself*.

MHC class II can be conditionally expressed by all cell types, but normally occurs only on antigen-presenting cells (APC): macrophages, B cells and especially dendritic cells (DCs). An APC uptakes an antigen, processes it and returns a molecular fraction (i.e. epitope) to the APC's surface.

Dendritic cells are found in tissues that are in contact with the external environment such as skin (i.e. Langerhans cell) or the inner lining of the nose, lungs, stomach and intestines. Dendritic cells are also in an immature state in the blood. Once activated, they migrate to the lymph nodes to interact with T and B cells to initiate an adaptive immune response.

A: neutrophils are part of innate immunity and destroy tissue invaders in a non-specific manner.

C: erythrocytes are not part of the immune system.

D: macrophages are part of innate immunity and destroy tissue invaders in a non-specific manner.

39. C is correct. Deoxygenated blood returns from the systemic circulation and drains into the right atrium from both the inferior and superior vena cava. From the right atrium, blood is pumped through the tricuspid valve into the right ventricle, which then pumps blood to the lungs via the pulmonary arteries. CO_2 is exchanged for O_2 in the alveoli of the lungs. Oxygenated blood is returned to the left atrium via the pulmonary veins. From the left atrium the blood is pumped through the mitral valve into the left ventricle, which ejects it into the aorta for systemic circulation.

If a tracer substance were injected into the superior vena cava, it would take the longest amount of time to reach the left ventricle.

A: the left atrium receives oxygenated blood via the pulmonary veins.

B: blood with the tracer would travel from the right ventricle to the lungs via the pulmonary arteries and return through the pulmonary veins to the left atrium.

D: from systemic circulation, the tracer first enters the right atrium and then passes through the tricuspid valve as it is pumped into the right ventricle.

E: the tricuspid valve separates the right atrium from the right ventricle and prevents the backflow of blood into the right atrium when the right ventricle contracts, forcing blood into the pulmonary artery.

40. D is correct. Semi-lunar valves prevent the backflow of blood as it leaves the heart via the aorta (systemic circulation) and pulmonary arteries (pulmonary circulation), because the ventricles have negative pressure once they start to relax before receiving more blood from the atria.

41. C is correct. Kidneys produce a hormone called erythropoietin (EPO), which promotes the formation of red blood cells by the bone marrow in response to low oxygen levels.

42. B is correct. The atrioventricular (AV) node is located at the junction between the atria and the ventricles and functions to delay the conduction impulse to the ventricles for a fraction of a second. A wave of excitation spreads out from the sinoatrial (SA) node through the atria along specialized conduction channels.

This activates the AV node, which delays conduction impulses by approximately 0.12s. This delay in the cardiac pulse ensures that the atria have ejected their blood into the ventricles before the ventricles contract.

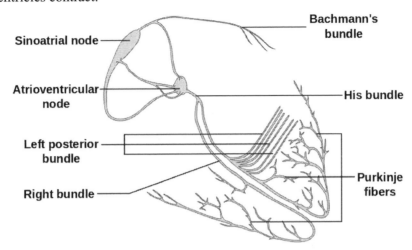

43. D is correct. The mitral valve is located between the left atrium and the left ventricle and prevents the backflow of blood into the left atrium during the contraction of the left ventricle. A patient suffering from stenosis of the mitral valve has impeded blood flow from the left atrium into the left ventricle. Therefore, the effect is an increase in blood volume in the left atrium and a reduced net movement of blood from the left atrium into the left ventricle, which results in increased left atrial pressure and decreased left ventricular pressure.

A: the right atrium receives deoxygenated blood by venous circulation and is too far removed for the circulatory process to be affected by mitral valve stenosis.

B: the left ventricle ejects blood into the aorta as it begins systemic circulation. Since volume/pressure in the left ventricle is reduced due to impeded flow from the left atrium, the patient experiences decreased pressure in the aorta.

44. C is correct.

45. E is correct. Hemorrhage is a loss of blood from the circulatory system. Hemorrhage can be external (when blood is lost to the outside of the body through a wound or a natural opening) or internal (when blood escapes from the blood vessels into other structures within the body). A moderate blood loss results in a decrease in cardiac output (volume of blood pumped by the heart each minute).

46. C is correct. The cardiac conduction pathway begins with the SA node located in the right atrium. The conduction signal then travels to the AV node located between the atria and the ventricles before it moves down the bundle of His, which then splits into the right/left Purkinje fibers.

The Purkinje fibers spread the conduction signal to the two ventricles, which contract simultaneously to eject blood. The blood from the right ventricle enters the pulmonary artery, while the blood from the left ventricle enters the aorta.

SA node → AV node → bundle of His → Purkinje fibers

47. B is correct. Blood in the left ventricle has just returned via the pulmonary vein from the lungs where it was oxygenated. The pulmonary vein is the only vein in the adult that carries oxygenated blood.

A: the right atrium receives blood via the inferior and superior venae cavae returning from systemic circulation. Blood returning from the body has low O_2 content as it is transported to the lungs.

C: pulmonary artery blood is pumped to the lungs to be oxygenated and is the least oxygenated blood in the circulatory system.

D: lymph vessels, such as the thoracic duct, return lymphatic fluid back to the venous circulation. The lymph fluid has a very low O_2 partial pressure.

E: the superior vena cava returns deoxygenated blood to the heart from the upper regions of the body.

48. C is correct. See explanation to question **21** of *Chapter 4.10*.

49. B is correct.

50. E is correct. The availability of O_2 needed to sustain mental and physical alertness decreases with altitude. A person remaining at high altitudes acclimates to these environmental conditions. Availability of O_2 decreases as the air density (i.e. the number of molecules of O_2 per given volume) drops with increasing altitude. Dehydration results from the higher rate of H_2O vapor loss from the lungs that contributes to the symptoms of altitude sickness.

A person at high altitude needs increased blood flow to deliver more blood to the tissues, because the blood carries less O_2.

2,3-BPG is an allosteric effector present in human red blood cells. It interacts with deoxygenated hemoglobin β subunits by decreasing their affinity for O_2, so it allosterically promotes the release of the remaining O_2 molecules bound to the hemoglobin, thus enhancing the ability of RBCs to release O_2 near tissues that need O_2 most.

51. C is correct. The blood from the pulmonary artery is pumped to the lungs to be oxygenated and is the least oxygenated blood in the circulatory system.

A: lymph vessels, such as the thoracic duct, return lymphatic fluid back to the venous circulation. The lymph fluid has a very low O_2 partial pressure.

B: left ventricle blood has just returned via the pulmonary vein from the lungs where it was oxygenated. The pulmonary vein is the only vein in the adult that carries oxygenated blood.

D: the inferior vena cava returns deoxygenated blood to the right atrium from the lower regions of the body.

E: the right atrium receives blood via the inferior and superior venae cavae returning from systemic circulation. Blood returning from the body has low O_2 content as transported to the lungs.

52. E is correct. The afferent arteriole supplies blood to the kidney at the glomerulus. Protein does not normally enter the kidney in the filtrate (fluid destined to become excreted as urine) and the loss of blood plasma into the kidney increases the protein concentration (relative to plasma). Presence of proteins in the filtrate is a clinical indication of kidney abnormalities.

D: vasa recta form a series of straight capillaries of the kidney in the medulla and are parallel to the loop of Henle. These vessels branch off the efferent arterioles of juxtamedullary nephrons (nephrons closest to the medulla), enter the medulla, and surround the loop of Henle. Each vasa recta has a hairpin turn in the medulla and carries blood at a very slow rate, crucial for the support of countercurrent exchange that maintains the concentration gradients established in the renal medulla.

The maintenance of this concentration gradient is one of the components responsible for the kidney's ability to produce concentrated urine. On the descending portion of the vasa recta, NaCl and urea are reabsorbed into the blood, while H_2O is secreted. On the ascending portion of the vasa recta, NaCl and urea are secreted into the interstitium while H_2O is reabsorbed.

53. B is correct. Atrioventricular (AV) valves prevent backflow of blood from the ventricles into the atria of the heart. They are anchored to the walls of the ventricles by chordae tendineae.

54. C is correct. Perspiration is the secretion of water, salts and urea from the sweat (i.e. sudoriferous) pores of the skin. As the sweat comes into contact with air, it evaporates and cools the skin. Thus, perspiration is a thermoregulatory mechanism involved in heat dissipation, *not* heat conservation. Eccrine sweat glands are distributed almost all over the body, though their density varies from region to region. Humans utilize eccrine sweat glands as a primary form of cooling. Apocrine sweat glands are larger, have a different secretion mechanism, and are mostly limited to the axilla (i.e. armpits) in humans.

A: constriction of selective blood vessels at the skin's surface reroutes blood flow from the skin to deeper tissues, which minimizes the loss of body heat.

B: shivering is a thermoregulatory mechanism that conserves body heat by intensive, rhythmic involuntary contraction of muscle tissue that increases internal heat production.

D: piloerection (i.e. *goose bumps*) is a reflex contraction of the small muscles found at the base of skin hairs, which causes the hairs to stand erect, forming an insulating layer that minimizes the loss of body heat.

55. D is correct. Platelets are small pieces of large megakaryocyte cells. When a wound occurs, platelets are activated to stick to the edge of the wound and release their contents that stimulate the clotting reaction. Platelets release several growth factors including platelet-derived growth factor (PDGF), a potent chemotactic agent (direct movement from chemical stimuli) and TGF beta, which stimulates deposits of extracellular matrix. Both growth factors play a significant role in the repair and regeneration of connective tissues.

A: leukocytes (i.e. white blood cells) are immune system cells involved in defending against infectious disease. Five different types of leukocytes exist. All leukocytes are produced and derived from haematopoietic stem cells in the bone marrow. They live for about three to four days in the average human body and are found throughout the body, including the blood and lymphatic system.

The name "white blood cell" derives from the physical appearance of a blood sample after centrifugation. White cells are found in the *buffy coat* (a thin, typically white layer of nucleated cells between the sedimented red blood cells and the blood plasma). All leukocytes have common traits, but each is distinct in form and function. A major distinction is the presence or absence of granules (i.e. granulocytes or agranulocytes).

Granulocytes (polymorphonuclear leukocytes) are characterized by the presence of differently staining granules in their cytoplasm. These granules (lysozymes) are membrane-bound enzymes that act primarily in the digestion of endocytosed particles. There are three types of granulocytes: *neutrophils*, *basophils* and *eosinophils*, which are named according to their staining properties. Agranulocytes (mononuclear leukocytes) are characterized by the absence of granules in their cytoplasm. The cells include *lymphocytes*, *monocytes* and *macrophages*.

B: erythrocytes (red blood cells) are the most common type of blood cell and the principal means of delivering oxygen (O_2) to the body.

C: lymphocytes are a type of white blood cell in the immune system, a landmark of the adaptive immune system. Lymphocytes can be categorized into either large or small lymphocytes. Large granular lymphocytes include natural killer cells (NK cells), while small lymphocytes consist of T cells and B cells.

56. A is correct. The hepatic portal system directs blood from parts of the gastrointestinal tract to the liver. The inferior vena cava is a large vein that is responsible for carrying deoxygenated blood into the heart.

57. E is correct. Increased pulmonary vascular resistance leads to right-to-left circulation through the foramen ovale and the ductus arteriosus – as in newborns suffering from persistent fetal circulation. The foramen ovale is a fetal opening that connects the right atrium with the left atrium. Ductus arteriosus is a fetal vessel that allows blood to flow from the pulmonary artery directly to the aorta. The foramen ovale and the ductus arteriosus normally close at birth due to pressure changes. However, in newborns with persistent fetal circulation, the foramen ovale and the ductus arteriosus remain open, which causes blood flows from the right atrium directly into the left atrium and from the pulmonary artery directly into the aorta.

58. A is correct. In humans, CO_2 is carried through the venous system and is expired (breathed out) through the lungs. Therefore, the CO_2 content in the body is high in the venous system, and decreases in the respiratory system, resulting in lower levels along any arterial system.

CO_2 is carried in blood in three different ways (exact % vary depending whether it is arterial or venous blood):

1) most (about 70% to 80%) is converted to bicarbonate ions HCO_3^- by carbonic anhydrase in the red blood cells by the reaction $CO_2 + H_2O \rightarrow H_2CO_3 \rightarrow H^+ + HCO_3^-$

2) 5% – 10% is dissolved in the plasma (liquid portion minus cells) *vs.* serum (liquid portion minus cells and clotting factors)

3) 5% – 10% is bound to hemoglobin as carbamino compounds (i.e. CO_2 added to amino group of hemoglobin protein).

Hemoglobin is the main O_2-carrying molecule in erythrocytes that transports O_2 and CO_2. However, the CO_2 bound to hemoglobin does not bind to the same site as O_2 but combines with the N-terminal groups on the four globin chains. Because of allosteric effects on the hemoglobin molecule, the binding of CO_2 decreases the amount of O_2 that is bound for a given partial pressure of O_2.

59. B is correct. The movement of fluids occurs between the capillaries and the tissues. Blood enters at the arterial end of the capillary bed as it passes to the venule end. In the capillary bed, gases, nutrients, wastes and fluids are exchanged with tissue cells in the interstitial space.

Two opposing forces affect the movement of fluids between the capillaries and the interstitial space. At the arteriole end, hydrostatic pressure (i.e. blood pressure) is greater than osmotic pressure, which forces fluids out of the capillaries and into the interstitial spaces. At the venule end, osmotic pressure is greater than hydrostatic pressure, which forces fluids back into the capillaries and returns the interstitial fluid into the circulatory system.

Starling's hypothesis proposes that not all of the fluid can be returned to the capillary and some fluid travels through the interstitial spaces towards the lymphatic system, where it is filtered by the lymph nodes and then, via the thoracic duct, returns to the circulatory system.

Chapter 4.5: Lymphatic and Immune Systems

1. B is correct. The thymus is a specialized organ of the adaptive immune system. It is different from other organs (e.g. heart, liver, kidney), because thymus is at its largest size during childhood.

The thymus provides an inductive environment for development of T lymphocytes from hematopoietic progenitor cells. The thymus is largest and most active during the neonatal and pre-adolescent periods. By the early teens, the thymus begins to atrophy, and thymic stroma (i.e. stroma is connective tissue cells of any organ) is mostly replaced by adipose (fat) tissue.

The thymus is the only lymphoid organ that does not *directly* fight antigens. It is a maturation site for T lymphocyte precursors, because these precursors must be isolated from foreign antigens to prevent their premature activation.

Histologically, each lobe of the thymus can be divided into a central medulla and a peripheral cortex, which is surrounded by an outer capsule. The cortex and medulla play different roles in the development of T-cells. Cells in the thymus can be divided into stromal cells and cells of hematopoietic origin (derived from bone marrow resident hematopoietic stem cells). Developing T-cells are referred to as thymocytes. Stromal cells include thymic cortical epithelial cells, thymic medullary epithelial cells and dendritic cells.

2. E is correct. Prions are misfolded proteins that cause disease, but do not elicit an immune response (e.g. mad cow disease). Erythrocytes deliver O_2 to cells and are not part of the immune system. While they are not an indicator of prion exposure, E is the only choice that is not indicative of an immune response.

A: plasma cells are mature B-cells that secrete antibodies. B-cells develop in the bone marrow and are part of the humoral immune response.

C: T-cells develop in bone marrow and mature in the thymus to produce cell-mediated immunity.

D: monocytes (like macrophages) are phagocytic cells and are part of innate (i.e. non-specific) immunity.

3. B is correct.

4. D is correct. Gamma globulins, also known as immunoglobulins (Ig), are Y-shaped protein antibodies. Gamma globulins are produced by B cells and used by the immune system to identify and neutralize foreign objects, such as bacteria and viruses. Antibodies are secreted by white blood cells called plasma cells and recognize a unique part of the foreign target (i.e. antigen).

5. B is correct. The small intestine absorbs nutrients and has a large surface area due to the villi and microvilli facing the lumen of the gastrointestinal tract.

6. E is correct.

7. C is correct. Humoral immunity is also called the antibody-mediated system because it is mediated by macromolecules (as opposed to cell-mediated immunity) found in extracellular fluids (e.g. secreted antibodies, complement proteins and certain antimicrobial peptides). Humoral immunity is so named because it involves substances found in the humours (i.e. body fluids). The immune system is divided into a more primitive innate immune system and adaptive immune system (i.e. acquired), each of which contains humoral and cellular components (e.g. phagocytes, antigen-specific cytotoxic T-lymphocytes, and cytokines).

Immunoglobulins are glycoproteins that function as antibodies. The terms *antibody* and *immunoglobulin* are used interchangeably. Antibodies are synthesized and secreted by plasma cells that are derived from the B cells of the immune system. They are found in the blood, tissue fluids and in many secretions. In structure, they are large Y-shaped globular proteins. In mammals there are five types of antibodies: IgA, IgD, IgE, IgG, and IgM. Each immunoglobulin class differs in its biological properties and has evolved to deal with different antigens.

An antibody is used by the acquired immune system to identify and neutralize foreign objects like bacteria and viruses. Each antibody recognizes a specific antigen unique to its target and, by binding, antibodies can cause agglutination and precipitation of antibody-antigen products, prime for phagocytosis by macrophages and other cells, block viral receptors, and stimulate other immune responses (i.e. complement pathway).

B: cytotoxic T cells (also known as cytotoxic T lymphocytes, T-killer cells, cytolytic T cells, CD8 or killer T cells) function in cell mediated immunity. Cytotoxic T cells are a type of white blood cell that destroys cancer cells, virally infected cells, or damaged cells. Most cytotoxic T cells express T-cell receptors (TCRs) that recognize specific antigen molecules capable of stimulating an immune response that are often produced by viruses or cancer cells.

8. D is correct.

9. C is correct. Antibodies are produced in response to antigens (e.g. viral coat proteins, bacterial cell walls) that are detected by T cells. T cells stimulate B cells to become plasma cells and secrete antibodies. B cells are also stimulated to become memory cells, which remain dormant in the interstitial fluid and lymphatics until the same antigens are detected, and provide a rapid humoral response.

A: neurons are not involved in the immune response.

B: T cells are the immune response organizers and can become cytotoxic cells (i.e. kill invading cells), helper T cells (i.e. recruit other T and B cells) and suppressor T cells (i.e. inactivate the immune response when the antigen has been cleared).

D: macrophages are phagocytic cells that engulf and digest bacterial cells and foreign material.

E: natural killer cells are immune system cells that destroy invading cells.

10. D is correct. Plasma cells are white blood cells that secrete antibodies which target and bind to antigens (foreign substances).

11. E is correct. Phagocytosis and pinocytosis are movements *into* the cell and are forms of endocytosis. Phagocytosis is the process by which either foreign particles invading the body or minute food particles are engulfed and broken down. The cell membrane of the phagocyte invaginates to capture the particle and forms a vacuole with the enclosed material to join a lysosome that contains enzymes that hydrolyze the enclosed material. Pinocytosis is a similar process, but describes the injections of a droplet of liquid.

A: apoptosis is programmed cell death.

B: necrosis is a form of cell injury that results in the premature cell death in living tissue by autolysis.

C: exocytosis is the transport of membrane-enclosed material *out* of the cell.

D: endocytosis is ATP-dependent transport of material into the cell.

12. D is correct.

13. C is correct. From the diagram, lymph flow increases (to a point) in proportion to the increase in pressure of the interstitial fluid. To decide whether an increase in interstitial fluid protein increases lymph flow, determine whether an increase in interstitial fluid protein increases interstitial fluid pressure.

When proteins move out of the capillaries and into the interstitial space, the solute concentration in the interstitial space increases. Fluid then flows out of the capillaries and into the interstitial space, which increases both the interstitial fluid volume and fluid pressure.

A: fluid movement from the interstitial spaces into the capillaries would decrease interstitial fluid pressure, but an increase of proteins in the interstitial causes fluid movement out of the capillaries.

D: increase in interstitial fluid pressure increases lymph flow (until point 0 – see graph).

14. B is correct. Phagocytes are cells that function to eliminate foreign invaders (e.g., particles, microorganisms, dying cells) in the body by phagocyting (i.e., digesting) them. Macrophages and neutrophils play an important role in the inflammatory response by releasing proteins and other substances that control infection, but can also damage the host tissue.

15. A is correct.

16. A is correct. Monocytes and macrophages are non-specific immune system cells involved in phagocytosis of foreign matter.

B: platelets are produced by the bone marrow and are part of the clotting cascade.

C: T-cells are part of the cell-mediated immune response and do not phagocytize other cells.

D: B-cells are part of adaptive immunity. They secrete antibodies and do not phagocytize other cells.

17. D is correct. B cells (also called B lymphocytes) are a type of lymphocyte. They participate in the humoral immunity of the adaptive immune system. B cells mature in the bone marrow. Activated, B cells go through a two-step differentiation process producing short-lived plasmablasts (for immediate immune response) and long-lived plasma cells and memory B cells (for persistent protection). Plasma cells release into the blood and lymph antibody molecules closely modelled after the receptors of the precursor B cell. When released, they bind to the target antigen (i.e., foreign substance) and initiate its destruction.

18. E is correct. Erythroblastosis fetalis (known as Rh incompatibility) occurs when an Rh^+ fetus develops within an Rh^- mother and requires an Rh^+ father. The Rh^- mother, if exposed to the Rh+ antigens, produces antibodies against the Rh proteins of the fetus. This occurs for subsequent children between an Rh^- mother and an Rh^+ father, because the first child sensitizes the mother to Rh protein and stimulates the production of antibodies to Rh^+ antigens. These anti-Rh antibodies are present when the second Rh^+ fetus develops. Normally, the maternal and fetal blood systems are completely separate, but the incompatibility may occur late in the pregnancy because of mixing of maternal and fetal.

19. B is correct.

20. A is correct. The bone marrow synthesizes both B and T cell lymphocytes. B cells mature in the bone marrow, while T cells migrate to the thymus for maturation.

T cells have three functions: 1) activating B-cells to respond to antigens, 2) stimulating the growth of phagocytic macrophages and 3) destroying antigens (i.e. foreign invaders) and abnormal tissue (e.g. cancer).

D: thymus is important for immunity in both children and adults.

E: erythrocytes develop in the bone marrow and circulate for about 120 days before their components are recycled by macrophages.

21. D is correct. Peyer's patches are organized lymphoid follicles, named after the 17th-century Swiss scientist Johann Conrad Peyer. They are a part of the gut associated lymphoid tissue usually found in the small intestine, mainly in the distal jejunum and the ileum.

Peyer's patches play an important role in the immune response within the gut. When pathogenic microorganisms and other antigens enter the GI tract, they encounter macrophages, dendritic cells, B-lymphocytes and T-lymphocytes found in Peyer's patches.

22. A is correct. Erythrocytes have antigens expressed on their cell surfaces and are classified by the type of antibody elicited. The four major blood types are A, B, AB, and O. Three alleles determine these four groups, because alleles for A and B are codominant, while the allele O is recessive.

The A allele encodes for the A antigen on the erythrocyte, the B allele encodes for the B antigen, and the O allele does not encode for any antigen. A person with type A blood has the genotype AA or AO, a person with type B blood is either BB or BO, a person with type AB blood is AB and a person with type O blood is OO.

The presence (or absence) of certain antibodies determines blood groups. The blood serum does not contain antibodies to endogenous blood type antigens but produces antibodies to the other blood antigens. For example, a patient with type A blood has antibodies to the B antigen, a patient with type B blood has antibodies to the A antigen, and a patient with type AB blood has neither anti-A nor anti-B antibodies since their red blood cells carry both antigens. Patients with type O blood have both anti-A and anti-B antibodies since their red blood cells have neither the A nor B antigens.

Agglutination (i.e. clumping of the blood) occurs when anti-A or anti-B antibodies react with the antigens on the blood cell. Therefore, when type A blood is mixed with another blood type (except AB), antibodies from the other blood type will cause agglutination. Precipitation does not occur in the test tube that contains type AB blood.

23. D is correct.

24. A is correct.

25. B is correct. The innate immune system responds to any and every foreign invader with the white blood cells (i.e. granulocytes) and with inflammation. Granulocytes include neutrophils, eosinophils and basophils because of the presence of granules in their cytoplasm. They are also referred to as polymorphonuclear cells (PMNs) due to their distinctive lobed nuclei.

Neutrophil granules contain a variety of toxic substances that kill or inhibit growth of bacteria and fungi. Similar to macrophages, neutrophils attack pathogens by activating a respiratory burst of strong oxidizing agents (e.g. hydrogen peroxide, free oxygen radicals and hypochlorite). Neutrophils are the most abundant type of phagocyte (i.e. 50 to 60% of the total circulating leukocytes) and are usually the first cells to arrive at the site of an infection.

Innate immunity does not involve humoral immunity (B cell) or cell mediated immunity (T cell).

26. A is correct. The thymus is a lymphoid organ within the immune system where T-cells and lymphocytes mature. The thymus produces thymosin, which is the hormone responsible for creating T-cells. Its structure includes a cortex and medulla, which help the development of T-cells.

27. E is correct. The type A and type B alleles are codominant to the O allele. Thus, a person with genotype AA or AO has type A blood. a person with genotype BB or BO has type B blood, and a person with genotype OO has type O blood.

A person with type AB blood has the genotype AB and produces gametes with either the A allele or the B allele. A person with type O blood has the genotype OO and produces gametes with the O allele. The only two genotypes for a mating of AB x OO are AO or BO (type A or type B blood).

28. A is correct. The islets of Langerhans are the regions of the pancreas that contain endocrine cells (alpha, beta, delta, PP and epsilon cells) and are separated from the surrounding pancreatic tissue by a thin fibrous connective tissue capsule. See explanation for **42** of *Chapter 4.2* for more information.

29. B is correct. CD4 T helper cells are white blood cells that are an essential part of the human immune system. They are called helper cells because one of their main roles is to send signals to other types of immune cells, including CD8 killer cells that destroy the infection or virus. If CD4 cells become depleted (for example, in an untreated HIV infection, or following immune suppression prior to a transplant), the body is left vulnerable to a wide range of infections that it would otherwise be able to fight.

The immune system can be subdivided into non-specific (innate) and adaptive (i.e. acquired) immunity. The innate immune system consists of physical barriers (e.g. skin and mucous membranes) and cells that non-specifically remove invaders (e.g. mast cells, macrophages, neutrophils, etc.). The adaptive immune system synthesizes cells directed against specific pathogens and has the capacity to respond rapidly if a similar pathogen is encountered. The innate system always responds with the same mechanisms (and efficiency) to every infection.

T and B cells are part of the acquired immunity system. B cells are part of the humoral division, while T cells are part of the cell-mediated division. Acquired immunodeficiency syndrome (AIDS) infects and kills T-cells, which results in a loss of cell-mediated immunity. Destruction of B cells would cause a loss of humoral immunity.

30. D is correct.

31. B is correct. Lymph nodes are linked by lymph vessels and contain phagocytic cells (leukocytes) that filter the lymph, remove and destroy foreign particles/pathogens. Lymphatic fluid absorbs fat and fat soluble vitamins.

A: the liver breaks down hemoglobin and uses its components to produce bile salts.

C: amino acids are not absorbed from the digestive tract by the lymphatic system or lymph nodes.

D: glucagon is produced by the pancreas and increases the blood concentration of glucose.

E: O_2 is carried to the tissues for cellular respiration by the hemoglobin protein in red blood cells.

32. D is correct. Antigens are any foreign substance (e.g., bacteria) that enters the body and induces an immune response, specifically the production of antibodies to destroy the antigen.

33. A is correct. Lymph is the fluid that circulates throughout the lymphatic system. The lymph is formed when the interstitial fluid is collected through lymph capillaries. The lymph is derived from the interstitial fluid and its composition constantly changes as the blood and the surrounding cells continuously exchange substances with the interstitial fluid. Lymph returns protein and excess interstitial fluid to the circulation. Lymph collects bacteria and brings them to lymph nodes where they are destroyed.

34. E is correct. The effectiveness of a vaccine depends on the organism's immune system. Plasma cells (a differentiated form of a B cell) secrete antibodies (i.e. immunoglobulin). The acquired immune system consists of humoral (i.e. antibodies) and cell-mediated (i.e. T cells) responses. The humoral response is mediated by B cells that are synthesized and matured in the bone marrow. When stimulated by antigens, B cells differentiate into several types of cells including memory B cells and plasma B cells that secrete antibodies.

A: cell-mediated immunity involves T cells that originate in the bone marrow (as B cells do) but then migrate to and mature in the thymus. T-helper cells regulate the activity of other T and B cells.

C: macrophages are non-specific immune cells that are released by the bone marrow and are involved in phagocytosis but do not synthesize or release immunoglobulin (antibodies).

35. D is correct.

36. B is correct. The small intestine is lined with villi (i.e. finger-like projections) that increase the absorptive surface area. Within the core of the villi are capillaries and lacteals (i.e. lymphatic vessels). Amino acids and glucose pass through the walls of the villi into the capillary. Fatty acids and glycerol pass into the lacteals and are reconverted into lipids. Lymph and the absorbed lipids empty into the central circulatory system via the thoracic ducts that join the superior vena cava at the junction of the left subclavian vein and left jugular vein (near the shoulders and below the clavicle).

37. E is correct.

38. B is correct. Anemia is a decreased red blood cell count. The spleen is similar in structure to a large lymph node and acts primarily as a blood filter. It removes old red blood cells, recycles iron and holds a reserve of blood, which can be valuable in case of hemorrhagic shock. The spleen synthesizes antibodies in its white pulp and removes antibody-coated bacteria and antibody-coated blood cells via the blood and lymph node circulation. Anemia results from abnormalities of fragile red blood cells that make them more susceptible to rupture by the spleen. Normally, the spleen destroys red blood cells about every 120 days.

A: adenoids (also known as a pharyngeal tonsil or nasopharyngeal tonsil) are masses of lymphatic tissue situated posterior to the nasal cavity, in the roof of the nasopharynx where the nose blends into the throat.

39. D is correct. T cells can be categorized into helper, cytotoxic (killer), memory, regulatory (suppressor) and natural killer T cells.

40. A is correct. The lymphatic system transports excess fluid and proteins from the interstitial space into the circulatory system. The lymphatic system does not drain excess fluid from deeper peripheral nerves, the central nervous system, superficial portions of the skin or from bone.

B: macrophages in lymph nodes engulf bacteria and other foreign particles.

C: the lymphatic system functions to remove proteins from interstitial spaces.

D: the lymphatic system is also important in the absorption of nutrients (particularly lipids) from the small intestine via lacteals (i.e. specialized lymph vessels).

41. A is correct.

The liver is involved in several key metabolic roles. Red blood cell synthesis occurs in the bone marrow and is not associated with the liver.

B: the liver deaminates amino acids by removing the amino terminus to allow the amino acid remnants to enter the metabolic pathways to provide energy.

C: the liver is involved in fat metabolism. If glycogen stores are high, glucose is converted into fatty acids and stored as triglycerides.

D: the liver stores carbohydrates as glycogen when carbohydrates and energy are abundant.

42. B is correct.

B-lymphocytes (B-cells) are a type of white blood cell that creates antibodies. They develop from stem cells in the bone marrow.

43. B is correct. Antibodies bind to antigens through interactions between the antibody's variable region and the antigen. The fragment antigen-binding (Fab fragment) is a region on an antibody that binds to antigens. It is composed of one constant and one variable domain of each of the heavy and the light chain, at the amino terminal end of the monomer. The two variable domains bind the epitope on their specific antigens.

A: antibodies do not assist in phagocytosis of cells.

D: antibodies are produced by plasma cells derived from stem cells in the bone marrow.

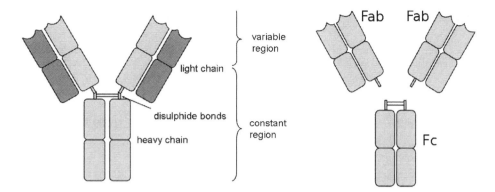

44. D is correct. Tissue repair after an injury is not part of the inflammatory response, it is the restoration of tissue structure and function via two separate processes: regeneration and replacement. For regeneration, the cell type that was destroyed must be able to replicate. Replacement is when tissue is replaced with a scar tissue.

45. B is correct. Agglutinins are the antibodies that recognize the blood types and are named because the antibodies cause agglutination (i.e. clumping or sticking) of blood. Blood type is determined by the presence of antibodies (i.e. agglutinins) to A or B antigens. There are two types of antigens on the surface of red blood cells. A person with the A antigens produces agglutinins to type B blood (anti-B antibodies). A person with the B antigens produces agglutinins to type A blood (anti-A antibodies).

Type O blood has neither A nor B antigens on the surface of the red blood cells but has antibodies (i.e. agglutinins) to both type A and type B blood.

C: Rh factor (named after the rhesus monkey it was discovered in) is another red blood cell antigen but is independent of the ABO blood types.

D: type B blood has B antigens and makes anti-A agglutinins. Type AB blood has both antigens and makes no agglutinins (otherwise the blood would coagulate itself).

46. E is correct.

47. C is correct. Fluid from lymphatic tissues is returned to the circulation at the right and left lymphatic ducts that feed into veins in the upper portion of the chest. The lymphatic system is part of the circulatory system, comprising a network of conduits called lymphatic vessels that carry a clear fluid called lymph towards the heart. The lymph system is not a closed system.

The circulatory system processes about 20 liters of blood per day through capillary filtration, which removes plasma while leaving the blood cells. Roughly 17 liters of the filtered plasma actually get reabsorbed directly into the blood vessels, while the remaining 3 liters are left behind in the interstitial fluid. The primary function of the lymph system is to provide an accessory route for these excess 3 liters per day to get returned to the blood. Lymph is essentially recycled blood plasma.

Lymphatic organs play an important part in the immune system, having a considerable overlap with the lymphoid system. Lymphoid tissue is found in many organs, particularly the lymph nodes, and in the lymphoid follicles associated with the digestive system such as the tonsils. Lymphoid tissues contain lymphocytes, but also contain other types of cells for support. The lymphatic system also includes all the structures dedicated to the circulation and production of lymphocytes (the primary cellular component of lymph), which includes the bone marrow thymus, spleen and the lymphoid tissue associated with the digestive system.

48. D is correct. Each lymphocyte has membrane-bound immunoglobulin receptors specific for a particular antigen. When the receptor is engaged, proliferation of the cell occurs resulting in a clone of antibody-producing cells (plasma cell). This takes place in secondary lymphoid organs (i.e., spleen and the lymph nodes). The theory of clonal selection explains the mechanism for the generation of diversity of antibody specificity.

49. C is correct. B lymphocytes (B cells), when exposed to a specific antigen, differentiate and divide into either memory or plasma cells. The plasma cells produce antibodies which recognize and bind the specific antigen that activated the precursor B lymphocytes.

A: cytotoxic T cells originate from T cells (T lymphocytes), not B lymphocytes, and bind to and destroy the antigen directly.

B: lymphokines originate from T cells (T lymphocytes), not B lymphocytes, and are secreted by helper T cells that activate other T cells, B cells and macrophages.

D: macrophages are highly specialized in removal of dying/dead cells and cellular debris. They do not originate from lymphocytes. When a monocyte enters damaged tissue through the endothelium of a blood vessel (i.e. leukocyte extravasation), it undergoes a series of changes to become a macrophage. Monocytes are attracted to a damaged site by chemotaxis (i.e. chemical signals) triggered by stimuli such as damaged cells, pathogens and cytokines released by macrophages already at the site.

50. E is correct.

51. B is correct.

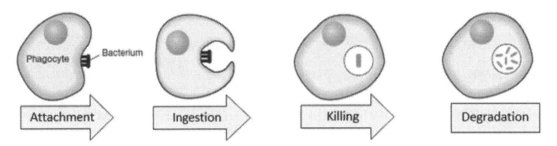

Stages of phagocytosis from attachment to degradation

52. C is correct. Memory B cells are formed after a primary infection and differentiate into plasma cells (i.e. B lymphocytes) when exposed to a specific antigen.

A: phagocytosis is an example of nonspecific (innate) immunity. Phagocytes include monocytes (large white blood cells), macrophages (which stimulate lymphocytes and other immune cells to respond to pathogens), neutrophils (one of the first-responders of inflammatory cells to migrate towards the site of inflammation), mast cells (involved in wound healing and defense against pathogens – best known for their role in allergy and anaphylaxis) and dendritic cells (antigen-presenting cells that function as a bridge between the innate and adaptive immune systems).

B: stomach acid is an example of innate (i.e. nonspecific) immunity.

D: skin is an example of innate (i.e. nonspecific) immunity.

53. A is correct.

54. E is correct. Herbivores cannot digest plant cell walls (i.e. cellulose), but have long digestive systems, so microorganisms can thrive to digest cellulose and provide the herbivores with the nutrients (i.e. cellulose is composed of glucose monomers) from degradation products.

55. A is correct.

56. B is correct. Humans have approximately 500-600 lymph nodes distributed throughout the body, with clusters found in the underarms, groin, neck, chest and abdomen. Enlargement of the lymph nodes is recognized as a common sign of infectious, malignancy or autoimmune disease.

C: inflammatory responses draw fluid into the affected area.

57. C is correct. Interferons are signaling proteins that develop in response to pathogens (i.e., viruses, bacteria, parasites, tumor cells). Interferons are named for their ability to interfere with the replication of RNA and DNA in a virus.

58. D is correct. CD4 cells are T helper cells that assist other white blood cells in immunologic processes. These cells express the CD4 glycoprotein on their surfaces. Helper T cells become activated when they are presented with peptide antigens by MHC class II molecules.

CD8 cells are cytotoxic T cells that destroy virus-infected cells and tumor cells, and are also involved in organ transplant rejection. These cells express the CD8 glycoprotein on their surfaces. They recognize their targets by binding to antigen associated with MHC class I molecules, which are present on the surface of all nucleated cells.

59. E is correct.

Blood types (also called blood groups) are classified by the presence, or absence, of inherited antigens on the surface of erythrocytes. These antigens include proteins, carbohydrates, glycolipids or glycoproteins. There are no O antigens. Type A⁻ (A negative) blood has A antigens present on the erythrocyte, but no Rh factor (Rh⁻) and no B antigens. Type A⁻ blood produces anti-B antibodies that only bind to B antigens and antibodies to the Rh antigen.

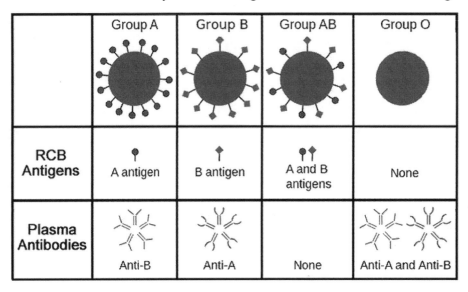

	Group A	Group B	Group AB	Group O
RCB Antigens	A antigen	B antigen	A and B antigens	None
Plasma Antibodies	Anti-B	Anti-A	None	Anti-A and Anti-B

Blood type is determined by the ABO blood group antigens present on red blood cells.

60. D is correct.

Within the first few minutes after the injury, platelets (i.e. thrombocytes) aggregate at the injury site to form a fibrin clot that reduces active bleeding. The speed of wound healing depends on bloodstream levels of platelets, fibrin, and hormones such as oxytocin.

During the inflammation phase, bacteria and cell debris are removed via phagocytosis by white blood cells.

The proliferation phase is characterized by the release of blood factors into the wound that cause the migration and division of cells, angiogenesis, collagen deposition, formation of granulation tissue and wound contraction.

Fibrin is a fibrous protein involved in the clotting of blood. It is formed from fibrinogen by the protease thrombin. Fibrin form a "mesh" that plugs or clots (in conjunction with platelets) over a wound site. Fibrin is involved in signal transduction, blood coagulation, and platelet activation.

Chapter 4.6: Digestive System

1. C is correct. The liver is a multi-function organ and is necessary for proper physiology within humans. The liver does not manufacture red blood cells in the adult, although it is a source of red blood cells in the developing fetus. In the adult, red blood cells are formed primarily in the red bone marrow.

A: glycogen is stored primarily in the liver and also in muscle cells for release when blood glucose levels are low.

B: liver detoxifies many compounds such as alcohol, drugs and other metabolites. For example, the liver hydrolyzes pharmaceutical compounds and prepares them for excretion by the kidneys.

D: liver forms urea from the metabolism of excess dietary amino acids. Deamination removes amino groups that are combined with CO_2 via reactions to form urea as a moderately toxic waste.

E: liver also regulates blood sugar levels by removing glucose from blood and storing it as glycogen. Conversely, the liver releases glycogen to increase plasma glucose when levels are low.

2. E is correct. Insulin decreases blood sugar levels in several ways, including the inhibition of glycogenolysis.

B: Glucagon is a peptide hormone secreted by the pancreas that raises blood glucose levels as an antagonist to the actions of insulin. The pancreas releases glucagon when blood glucose levels fall too low. Glucagon causes the liver to convert stored glycogen into glucose, which is released into the bloodstream.

C: adrenaline (i.e. epinephrine) is involved in the sympathetic response of *fight or flight*.

3. B is correct.

4. D is correct. The epiglottis is the cartilaginous structure that diverts food into the esophagus and prevents food from entering the trachea. The epiglottis closes off the respiratory tract (i.e. trachea) and covers the glottis (i.e. opening at the top of the trachea).

A: tongue is a muscle that manipulates food for mastication, but does not close off the trachea.

B: larynx is the voice box and is located below the glottis.

C: glottis is the opening at the top of the trachea (i.e. windpipe), between the vocal cords, and is covered by the epiglottis.

E: esophageal sphincter (also known as cardiac sphincter) separates the esophagus from the stomach.

5. D is correct. The process of digestion is aided by a variety of mechanical and chemical stimuli. Mechanoreceptors and chemoreceptors embedded in the lining of the GI tract respond to stimuli (e.g., stretching of the organ, osmolarity, pH, presence of substrates and end products of digestion) and initiate reflexes. These reflexes may activate or inhibit glands to secrete substances (e.g., digestive juices, hormones) blood or mix contents and move them along the length of the GI tract.

6. C is correct. Chylomicrons are lipoprotein particles that consist of triglycerides (85-92%), phospholipids (6-12%), cholesterol (1-3%) and proteins (1-2%). Chylomicrons are one of the five major groups of lipoproteins (chylomicrons, VLDL, IDL, LDL and HDL) that enable cholesterol and fats to move through the hydrophilic blood plasma. They transport dietary lipids from the intestines to liver, adipose, cardiac and skeletal muscle tissue, where their triglyceride components are unloaded by the activity of lipoprotein lipase. Chylomicron remnants are taken up by the liver.

Lipids are hydrophobic and require carrier proteins for transport within the hydrophilic blood plasma. The chylomicrons are the carrier molecules, but they are too large to pass into the small arterioles. Therefore, absorbed lipids are initially passed into lacteals (in the small intestine) of the lymphatic system, which then transports lipids into large veins of the cardiovascular system.

7. B is correct. The esophagus is a muscular tube through which food (i.e. chyme) passes from the pharynx to the stomach.

A: glottis is the opening at the top of the trachea (i.e. windpipe), between the vocal cords, and is covered by the epiglottis.

C: trachea (i.e. windpipe) connects the pharynx and larynx to the lungs, allowing the passage of air.

D: larynx (i.e. voice box) manipulates pitch and volume and is located below the glottis.

E: fundus is the upper part of the stomach, which forms a bulge higher than the opening of the esophagus (farthest from the pylorus).

8. E is correct.

9. D is correct. Lipid production and deposition depend on both hormones and genetic predisposition. The liver is the largest organ (i.e. with glandular functions) and is involved in the detoxification (e.g. drugs and alcohol), production of bile for emulsification of dietary lipids, storage of fat-soluble vitamins (A, D, E & K), synthesis of lipids, metabolism of cholesterol and production of urea.

10. C is correct. The liver produces bile that is stored in the gallbladder before its release into the duodenum of the small intestine. Bile contains no enzymes but emulsifies fats, breaking down large globules into small fat droplets.

A: the large intestine absorbs water and vitamins K, B_1 (i.e. thiamine), B_2 (i.e. riboflavin) and B_{12} (i.e. cobalamin).

B: the small intestine completes chemical digestion and is where individual nutrients (i.e. monomers of proteins, carbohydrates and lipids) are absorbed.

D: the gallbladder receives bile from the liver and stores it for release into the small intestine. Bile is released upon stimulation from the hormone cholecystokinin (CCK), which is synthesized by I-cells in the mucosal epithelium of the small intestine.

E: pancreas produces digestive enzymes (e.g. amylase, trypsin, and lipase) and hormones (e.g. insulin and glucagon).

11. A is correct.

12. D is correct. Type 1 diabetes can occur from a deficiency of insulin produced by beta cells of the pancreas. With a deficiency/absence of insulin, plasma glucose rises to dangerously high levels and patients excrete excess glucose in the urine.

A: insulin is not stimulated by decreased blood glucose levels.

B: insulin is not produced by the thyroid gland.

C: decreased levels of circulating erythrocytes (i.e. anemia) are not related to diabetes.

E: hematocrit is the volume percentage (%) of red blood cells in blood, normally about 40% for women and 45% for men. It is considered an integral part of a person's complete blood count results, along with hemoglobin concentration, white blood cell count, and platelet count.

13. B is correct. The large intestine functions to absorb water from the remaining indigestible food matter, and then passes waste material from the body. Some salts and minerals are also reabsorbed with this water, and bacteria within the large intestine also produce vitamin K.

The large intestine consists of the cecum, appendix, colon, rectum and anal canal. The large intestine takes about 16 hours to finish the digestion of food. It removes water and any remaining absorbable nutrients from the food before sending the indigestible matter to the rectum. The colon absorbs vitamins created by the colonic bacteria, such as vitamin K (especially important, as the daily ingestion of vitamin K is not normally enough to maintain adequate blood coagulation), vitamin B_{12}, thiamine and riboflavin. The large intestine also secretes K^+ and Cl^-. Chloride secretion increases in cystic fibrosis. Recycling of various nutrients takes place in colon, such as fermentation of carbohydrates, short chain fatty acids and urea cycling.

A: duodenum is the anterior section of the small intestine, which connects to the posterior end of the stomach and precedes the jejunum and ileum. It is the shortest part of the small intestine, where most chemical digestion takes place. The duodenum is largely responsible for the enzymatic breakdown of food in the small intestine. The villi of the duodenum have a leafy-looking appearance and secrete mucus (found only in the duodenum).

The duodenum also regulates the rate of emptying of the stomach via hormonal pathways. Secretin and cholecystokinin are released from cells in the duodenal epithelium in response to acidic and fatty stimuli present there when the pylorus (from the stomach) opens and releases gastric chyme (i.e. food products) into the duodenum for further digestion. These cause the liver and gallbladder to release bile, and the pancreas to release bicarbonate and digestive enzymes such as trypsin, amylase and lipase into the duodenum.

C: jejunum is the second section of the small intestine, which connects to the duodenum (anterior end) and the ileum (posterior end). The lining (i.e. enterocytes) of the jejunum is specialized for the absorption of small nutrient particles, which have been previously digested by enzymes in the duodenum. Once absorbed, nutrients (with the exception of fat, which goes to the lymph) pass from the enterocytes into the enterohepatic circulation and enter the liver via the hepatic portal vein, where the blood is processed.

D: ileum mainly absorbs vitamin B_{12} and bile salts and whatever products of digestion that were not absorbed by the jejunum. The wall itself is made up of folds, each of which has many tiny finger-like projections, known as villi, on its surface. In turn, the epithelial cells that line these villi possess even larger numbers of microvilli. The lining of the ileum secretes the proteases and carbohydrases responsible for the final stages of protein and carbohydrate digestion.

E: the mouth does not absorb water, but masticates (chews) and moistens food. It also converts a small amount of starch into maltose using the salivary amylase enzyme.

14. E is correct. See explanations for questions **45**, **54** and **57**.

15. D is correct. Proteins are denser than lipids, and lipoproteins with higher protein content are even more dense. Lipoproteins with low density transport mostly lipids and contain little protein (i.e. VLDL transport more lipids/less protein than LDL). Chylomicrons are formed by the small intestine and have the lowest density and the highest lipid/lowest protein content of all lipoproteins.

16. D is correct. The oral cavity begins mechanical (i.e. mastication) and chemical digestion (i.e. salivary amylase and lingual lipase) of food (i.e. bolus). Mechanical digestion breaks down large particles into smaller particles through the searing action of teeth, thus increasing the total surface area of the ingested food.

Chemical digestion begins in the mouth when the salivary glands secrete saliva that contains salivary amylase, which hydrolyzes starch into simple sugars. Saliva also lubricates the bolus to facilitate swallowing and provides a solvent for food particles. The muscular tongue manipulates the food during chewing and pushes the bolus into the pharynx.

17. B is correct. Bile does not contain digestive enzymes. It is an emulsifying agent released by the gallbladder that acts to increase the surface area of dietary fats, allowing more contact with lipase to break down fats into smaller particles.

18. C is correct. Salts (e.g. K^+Cl^- and Na^+Br^-) and acids (e.g. H_2SO_4 dissociates into H^+ and HSO_4^-) are electrolytes that dissociate into ions. Cations are positive charges, and anions are negative charges. Molecules with ionic bonds (i.e. differences in electronegativity) dissociate in aqueous solutions such as blood. Glucose, sucrose and fructose have covalently bonded carbon backbones and are not electrolytes.

19. E is correct. Sodium bicarbonate ($NaHCO_3$) buffers the blood to maintain a slightly basic pH (7.35) of the blood. $NaHCO_3$ is the conjugate base of H_2CO_3 (i.e. carbonic acid) that results from the combination of carbon dioxide and water: $CO_2 + H_2O \leftrightarrow H_2CO_3 \leftrightarrow H^+ + HCO_3^-$

20. B is correct.

21. A is correct. The digestive tract is subdivided by muscular sphincters: the esophagus-stomach (lower esophageal sphincter), stomach-duodenum (pyloric sphincter) and ileum-colon (ileocecal sphincter).

B: the small intestine is the only organ that has a high concentration of villi.

D: peristalsis is the process of wave-like contractions that move food along the GI tract. It takes place along the entire length of the digestive system.

E: Peyer's patches are aggregations of lymphoid tissue in the lowest portion of the small intestine (i.e. ileum). They differentiate the ileum from the duodenum and jejunum. The duodenum can be identified by Brunner's glands. The jejunum has neither Brunner's glands nor Peyer's Patches.

22. C is correct. The low pH of the stomach (~2 to 3) is essential for the function of the protease enzymes (e.g. pepsin) that hydrolyze proteins into their amino acids.

A: peristalsis propels food (i.e. chyme) and indigestible waste (i.e. feces) through the system.

B: amylases, lipases and bicarbonate are released through the pancreatic duct.

D: glucose and amino acids are absorbed into the blood, while dietary lipids, separated into glycerol and free fatty acids, are absorbed by lacteals (special vessels within the villi of the small intestine) that transport the ingested components into the lymphatic system.

E: see explanations for questions **45**, **54** and **57**.

23. A is correct. Hydrochloric acid released in the stomach denatures the proteins, unfolding them out of their three-dimensional shapes. The enzyme pepsin hydrolyzes proteins into fragments of various sizes called peptides.

24. B is correct. Parietal cells are epithelial cells of the stomach that secrete gastric acid (HCl) and intrinsic factor. Tagamet is an H_2-receptor antagonist that inhibits the parietal cells from releasing HCl, and the pH of the stomach increases.

D: chief cells produce pepsin, which digests protein.

25. E is correct. All carbohydrates absorbed into the bloodstream via the small intestine must be hydrolyzed to monosaccharides prior to absorption. The digestion of starch (i.e. plant polymer of glucose) begins with the action of salivary alpha-amylase/ptyalin (its activity is minimal compared with pancreatic amylase in the small intestine). Amylase hydrolyzes starch to alpha-dextrin that is digested to maltose. The products of digestion of alpha-amylase, along with dietary disaccharides are hydrolyzed to their corresponding monosaccharides by enzymes (e.g. maltase, isomaltase, sucrase and lactase) present in the brush border of small intestine.

Maltose is the disaccharide produced when amylase breaks down starch.

26. C is correct. Gastrin is a digestive hormone responsible for the stimulation of acid secretions in the stomach in response to the presence of peptides and proteins.

27. B is correct. For the same molecular weight, more of the positively charged compounds and less of the negatively charged compounds are filtered compared to neutral compounds. This occurs because the *negative charge* on the cellulose filtration membrane attracts positively charged and repels negatively charged compounds.

28. B is correct. Lactase is an enzyme released into the small intestine that hydrolyzes lactose (i.e. a disaccharide) into its monomer sugars of glucose and galactose.

A: kinase is an enzyme that phosphorylates (i.e. adds a phosphate) its substrate.

C: lipase hydrolyzes lipids into glycerol and free fatty acids.

D: zymogen (i.e. ending in ~ogen) is the inactive enzyme that is cleaved under certain physiological conditions into its active form. Examples are pepsinogen, trypsinogen, and chymotrypsinogen, which are cleaved in the small intestine into pepsin, trypsin, and chymotrypsin, respectively).

E: phosphatase removes a phosphate from its substrate (i.e. opposite action to kinase).

29. E is correct.

30. C is correct. The gastrointestinal tract is lined with involuntary smooth muscle that is controlled by the autonomic nervous system. The esophagus is the only exception, with the upper 1/3 consisting of voluntary skeletal muscle (e.g. swallowing), middle 1/3 consisting of a mixture of skeletal and smooth muscle and the lower 1/3 consisting of smooth muscle.

31. C is correct. Lipids are the primary means of food storage in animals, and lipids release more energy per gram (9 kcal/gram) than carbohydrates (4 kcal/gram) or proteins (4 kcal/gram). Lipids also provide insulation and protection against injury as the major component of adipose (fat) tissue.

A: proteins are mostly composed of amino acids with the elements C, H, O, and N, but may also contain S (i.e. sulfur in cysteine).

B: α helices and β pleated sheets are secondary structures of proteins.

D: the C:H:O ratio of carbohydrates is 1:2:1 ($C_nH_{2n}O_n$).

E: maltose is a disaccharide composed of two molecules of glucose.

32. A is correct. Hepatocytes make up the main tissue of the liver and are 70-85% of the liver's cytoplasm. They are involved in protein synthesis, protein storage, modification of carbohydrates, synthesis of cholesterol, bile salts and phospholipids, detoxification, modification, and excretion of substances. The hepatocyte also initiates formation and secretion of bile.

33. E is correct. Acinar cells synthesize and secrete pancreatic enzymes.

Pepsinogen and pancreatic proteases (e.g. trypsinogen) are secreted as zymogens (i.e. inactive precursors). Proteases must be inactive while being secreted, because they would digest the pancreas and the GI tract before reaching their target location.

A: protease is a general term for enzymes that digest proteins.

B: salivary amylase digests carbohydrates in the mouth and is secreted in its active form.

D: bicarbonate is not an enzyme but buffers the pH of the gastrointestinal system and blood. The centroacinar cells of the exocrine pancreas secrete bicarbonate, secretin and mucin.

34. D is correct. The small intestine is where enzymatic digestion is completed and where the food monomers (i.e. amino acids, sugars, glycerol and fatty acids) are absorbed. The small intestine is highly adapted for absorption because of its large surface area formed by villi (i.e. finger-like projections). Amino acids and monosaccharides pass through the villi walls and enter into the capillary system, while dietary lipids, after hydrolysis into glycerol and free fatty acids, are absorbed by lacteals within the villi of the small intestine that connect with the lymphatic system.

A: stomach is a large muscular organ that stores, mixes and partially digests dietary proteins.

B: gallbladder stores bile prior to its release in the small intestine.

C: the large intestine functions in the absorption of salts and water.

E: the rectum provides for transient storage of feces prior to elimination through the anus.

35. C is correct.

36. E is correct. Bicarbonate (HCO_3^-) is not a digestive enzyme, but is an important compound that buffers (i.e. resists changes in pH) both the blood and fluids of the gastrointestinal tract. HCO_3^- is released into the duodenum as a weak base that increases the pH of the chyme entering the small intestine.

If the small intestine is too acidic, the protein lining of the small intestine (i.e. mucus of the stomach lining is absent in the small intestine) would be digested, pancreatic enzymes would denature and be unable to digest the food in the small intestine.

37. A is correct. Carbohydrates (e.g. glucose, fructose, lactose and maltose) and proteins both provide 4 calories per gram (i.e.4 kcal/gram). Fats (lipids) are energy dense and provide 9 calories per gram.

38. C is correct.

B: Goblet cells are glandular simple columnar epithelial cells that secrete mucin, which forms mucus when dissolved in water. They use both apocrine (i.e. bud secretions) and merocrine (i.e. exocytosis via a duct) methods for secretion. They are found scattered among the epithelial lining of organs (i.e. intestinal and respiratory tracts) and inside the trachea, bronchus, and larger bronchioles in the respiratory tract, small intestines, the colon, and conjunctiva in the upper eyelid (i.e. goblet cells supply the mucus tears).

39. B is correct. The proper digestion of macromolecules is required for adequate absorption of nutrients out of the small intestine. Carbohydrates must be broken down into monosaccharides like glucose, fructose, and galactose. Lactose is a disaccharide of glucose and galactose; sucrose is a disaccharide of glucose and fructose; maltose is disaccharide of two units of glucose. Disaccharide digestion into monomers occurs at the intestinal brush border of the small intestine via enzymes like lactase, sucrase and maltase.

A: amino acids are the monomers of proteins. Proteins must be hydrolyzed into mono-peptides, di-peptides or tri-peptides for absorption in the duodenum of the small intestine.

D: lipids are degraded into free fatty acids and glycerol for absorption in the small intestine.

40. D is correct. Vitamin K is used by the liver to produce prothrombin, which is a clotting protein involved in the cascade for the formation of a fibrin clot. A deficiency of dietary vitamin K results in hemorrhagic diseases. Vitamins A, D, E, and K are fat-soluble vitamins. *Essential* means that the nutrient cannot be produced by the body and must be included in the diet, while *necessary* means that it is required (without a reference to its source – dietary or able to be synthesized by the body from precursors).

A: vitamin A is necessary for growth of skin, hair and mucous membranes and needed for night vision and bone growth.

B: vitamin B_{12} is an essential (i.e. must be ingested from food) water-soluble vitamin for red blood cell formation and also for proper functioning of the nervous system.

C: vitamin D is necessary for proper bone and tooth development and also for absorption of dietary calcium and phosphate.

E: vitamin E is an antioxidant that protects cell membranes and prevents degradation of vitamin A.

41. A is correct. The large intestine (i.e. colon) absorbs vitamins created by bacteria, such as vitamin K, vitamin B_{12}, thiamine and riboflavin.

C: while bacteria in the large intestine do produce gas, this role is not essential.

D: Bilirubin is the yellow breakdown product of normal heme catabolism. Heme is found in hemoglobin as the principal component of red blood cells; bilirubin is excreted in bile and urine, and elevated levels may indicate certain diseases. It is responsible for the yellow color of bruises, the background straw-yellow color of urine, the brown color of feces and the yellow discoloration in jaundice (i.e. resulting from liver disease).

42. B is correct. Saliva is a complex mixture of water, mucus, electrolytes, enzymes and antibacterial compounds. Saliva lubricates bolus (food ingested into the mouth) and buffers the oral cavity pH to 6.2-7.4. Saliva contains lingual lipase and salivary amylase enzymes that digest lipids and carbohydrates, respectively. These enzyme classes are also produced in the pancreas and secreted into the duodenum of the small intestine.

A: chymotrypsin (along with trypsin) is secreted by the pancreas to digest proteins.

C: pepsin is produced by gastric chief cells to initiate the breakdown of protein.

D: trypsin (along with chymotrypsin) is secreted by the pancreas to digest proteins.

E: secretin helps regulate the pH within the duodenum through inhibiting gastric acid secretion by the parietal cells of the stomach, and through stimulating bicarbonate production by the centroacinar cells and intercalated ducts of the pancreas.

43. D is correct. The cardiac sphincter (also known as the lower esophageal sphincter) regulates the flow of material (i.e. bolus) from the esophagus into the stomach. The food substance that was termed bolus in the mouth and esophagus is, upon entry into the stomach, referred to as chyme.

A: gallbladder is a small storage organ for bile (digestive secretion of the liver) before it is released into the duodenum of the small intestine, via CCK, for the emulsification of dietary fats.

B: pyloric sphincter regulates the flow of chyme from the stomach into the small intestine.

C: epiglottis is the structure that blocks the opening to the trachea (i.e. windpipe) during swallowing.

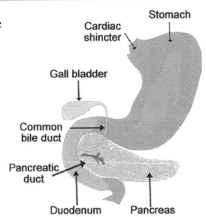

44. B is correct.

45. E is correct. CCK stimulates the release of enzymes needed to digest fat (i.e. pancreatic lipase) and protein (i.e. trypsin and chymotrypsin). Salivary amylase is produced in the mouth and digests carbohydrates. CCK is released by I cells in the mucosal epithelium of the duodenum and proximal jejunum and is not related to salivary amylase.

D: bile is synthesized by the liver, and released from the gallbladder as an emulsifying agent to increase the surface area of lipids and facilitate their digestion in the duodenum.

46. C is correct. A patient develops a peptic ulcer either when her gastric mucosa overproduces HCl or when mucosal defenses are inadequate to protect the stomach mucosa lining from the normal concentration of HCl. The pH in the stomach is about 2 because of the secretion of HCl, and this acidic environment is needed by the gastric (stomach) enzyme pepsin that functions optimally at pH 2. Peptic ulcers can be treated with antacids, which neutralize the HCl and raise the pH. By increasing pH, pepsin becomes nonfunctional and therefore, pepsin activity is the most affected by an overdose of antacid.

The small intestine has a pH of between 6 and 7 and is relatively alkaline due to the secretion of bicarbonate. Trypsin, procarboxypeptidase, and lipase are all enzyme products of the exocrine pancreas secreted into the small intestine, and function best in the alkaline environment.

A: procarboxypeptidase is an enzyme that hydrolyses proteins.

B: trypsin is an enzyme that hydrolyses protein.

D: lipase is an enzyme that hydrolyses lipids.

47. A is correct.

48. E is correct. Intrinsic factor is secreted as a glycoprotein produced by the parietal cells of the stomach and is required for the absorption of vitamin B_{12} (cobalamin) in the small intestine. The stomach also secretes two other hormones: gastrin is a peptide hormone that stimulates secretion of gastric acid (HCl) by the parietal cells of the stomach; pepsin is produced by the chief cells in the stomach to digest dietary proteins. The stomach also absorbs caffeine and ethanol (i.e. alcohol).

49. B is correct. Continuous muscle contraction depletes all available O_2 and causes the muscle fiber to depend on anaerobic respiration for energy (ATP) production. Anaerobic respiration converts pyruvate into lactic acid to regenerate NAD^+ as a necessary component for glycolysis in the absence of O_2.

50. D is correct. See explanations for questions **45, 54** and **57**.

51. A is correct. Intrinsic factor is secreted by parietal cells along with HCl. Intrinsic factor is essential for the absorption of vitamin B_{12} in the ileum.

C: chief cells release pepsinogen (i.e. zymogen), which is converted into pepsin.

D: mucus cells protect the stomach lining from damage by pepsin and HCl. Erosion of the mucous cells by the gram-negative *Helicobacter pylori* (*H. pylori*) is the primary cause of gastric ulcers.

52. E is correct. Bile is an emulsifying agent that increases the surface area of dietary lipids (i.e. fats), allowing an increase in contact with lipase (i.e. enzyme) to dissociate fats into smaller particles. Bile is not an enzyme and does not catalyze a chemical change of fats. The lipids are just separated into smaller micelles after interaction with bile. Bile is made up of bile salts that are cholesterol derivatives and pigments from the breakdown of hemoglobin.

A: proteases are enzymes that hydrolyze proteins into amino acids.

B: proteins are large polymers of amino acids linked together by peptide bonds.

C: enzymes catalyze chemical reactions by lowering the energy of activation.

D: hormones are chemical messengers released into the blood that signal target cells.

53. A is correct.

Pepsin functions at an optimum pH of about 2.0, but becomes denatured in the small intestine, where pH has been increased to between 6 and 7 by bicarbonate ions.

B: pancreatic amylase (i.e. small intestine) and salivary amylase (i.e. mouth) digest starch.

D: pepsinogen is a zymogen converted into pepsin by HCl in the stomach.

54. C is correct.

Cholecystokinin is a hormone released by the wall of the small intestine due to the presence of acidic chyme from the stomach. CCK stimulates the gallbladder to release bile and the pancreas to release digestive enzymes into the duodenum. See explanations for questions **45** and **57**.

55. B is correct. Most chemical (i.e. enzymatic) digestion occurs in the small intestine. Within the small intestine, most digestion of starch/ disaccharide, all lipid (fat) digestion, most protein digestion and all absorption of monomers occurs. Prior to the small intestine, some starch is digested in the mouth by salivary amylase into the disaccharide maltose. There is also within the stomach a small amount of protein digestion, in which the stomach enzyme pepsin splits proteins into smaller chains of amino acids (i.e. peptides). However, the digestion occurring before the small intestine is incomplete, and the majority of the digestive process occurs within the small intestine.

A: liver is not part of the alimentary (i.e. gastrointestinal) canal, and food does not pass through the liver. The liver does produce bile (i.e. an emulsifying agent), which increases the surface area of fats and mixes them within the watery enzyme environment of the small intestine.

E: pancreas produces several digestive enzymes, including the proteases (e.g. trypsin and chymotrypsin) and fat-digesting enzymes (e.g. pancreatic lipase and pancreatic amylase). Pancreatic enzymes function within the small intestine and not within the pancreas itself.

56. B is correct. The liver is the largest and most metabolically complex organ in the body. The liver performs detoxification by hydrolyzing and then excreting harmful materials (e.g. alcohol, toxins) or decomposing materials destined for excretion (e.g. hemoglobin) from the blood.

The liver also regulates blood sugar, lipids and amino acids. Additionally, it synthesizes about 700 mg/day of endogenous cholesterol and also serves as a site for blood, glycogen and vitamin storage.

57. A is correct. The pancreatic duct leads from the pancreas to the common bile duct into the second portion of the duodenum. It carries pancreatic enzymes (e.g. pancreatic amylase and lipase), and some proteases (e.g. trypsin, elastin, amylase and chymotrypsin) from the pancreas to the duodenum. The pancreatic secretions contain bicarbonate in response to the hormones cholecystokinin and secretin (i.e. released by the walls of the duodenum in response to food). Bicarbonate neutralizes the acidic stomach contents (i.e. chyme), as they enter the duodenum.

The pancreas is both an endocrine gland that produces hormones (e.g. insulin, glucagon and somatostatin) and an exocrine gland that produces enzymes for digestion of proteins, lipids and fats in the duodenum. The pancreatic hormones do not pass through the pancreatic ducts, but are secreted directly into the bloodstream.

I: diabetic crisis does not result from the obstruction of the pancreatic duct, because the release of the pancreatic hormones (insulin or glucagon) is not affected.

II: acromegaly results from excessive secretion of growth hormone in an adult and causes excessive bone growth of some facial bones, resulting in a distorted facial appearance. There is no connection between growth hormone from the anterior pituitary and blockage of the pancreatic duct.

58. E is correct.

59. A is correct. Bile is formed in the liver and released by the gallbladder to emulsify fats by increasing their surface area to assist lipase in breaking them down. Bile salt anions are hydrophilic on one side and hydrophobic on the other side. They aggregate around droplets of fat (triglycerides and phospholipids) to form micelles, with the hydrophobic sides towards the fat and hydrophilic sides facing outwards. The hydrophilic sides are negatively charged, and this charge prevents fat droplets coated with bile from re-aggregating into larger fat particles.

60. B is correct. Bacteria of the large intestine produce benefits for the host (e.g. vitamin production), but when bacteria invade other tissues they can cause disease.

61. E is correct. Pepsinogen (zymogen of pepsin) is released by the chief cells in the stomach. Pepsin and many other proteases are secreted as *zymogens* that are biologically inactive until they reach their proper extracellular environment. Pepsin is secreted as pepsinogen (i.e. inactive form) that is cleaved by HCl (with the removal of 44 amino acids from the zymogen) in the acidic conditions of the stomach to form pepsin as an active protease.

Stomach is coated by mucus which provides physical barrier between gastric juices containing hydrolytic enzymes and the stomach walls.

Chapter 4.7: Excretory System

1. A is correct. Renin regulates the body's mean arterial blood pressure because it mediates the extracellular volume (i.e. blood plasma, lymph and interstitial fluid), and arterial vasoconstriction.

2. C is correct. The individual collecting ducts in the kidney empty into much larger ducts called the minor calyces, which join to form a major calyx. The major calyces join to form the renal pelvis. Each of the two ureters originates from the renal pelvis of the kidney and connects to the urinary bladder that stores urine. A single urethra transports urine from the bladder for discharge from the body.

3. E is correct. Aldosterone is produced by the adrenal cortex and stimulates both the reabsorption of Na^+ from the collecting duct, and the secretion of K^+, because aldosterone activates the Na^+/K^+ pumps at the distal convoluted tubule. Na^+ reabsorption draws H_2O with it, increasing blood volume, increasing blood pressure and producing concentrated urine. Aldosterone release is stimulated by angiotensin II, which is influenced by renin (i.e. renin-angiotensin system).

Vasopressin (ADH; antidiuretic hormone) is secreted by the hypothalamus and stored in the posterior pituitary. Vasopressin increases H_2O reabsorption by opening water channels in the collecting ducts of the nephron (compared to indirect action of aldosterone that increases salt reabsorption).

4. C is correct.

5. B is correct. An obstruction of urine discharge from the nephron results in increased hydrostatic pressure (i.e. pressure of the liquid) in Bowman's capsule. Fluid accumulates and decreases renal function because it interferes with the filtration of blood. The increase in hydrostatic pressure in the kidney impedes the passage of blood into Bowman's capsule as putative filtrate.

Humans have two kidneys, but a kidney stone in the urethra (single tube out of the bladder) results in a fluid backup in the urinary bladder, the ureter and effectively to both kidneys.

6. E is correct. Water reabsorption in the kidneys is directly proportional to the relative osmolarity of the interstitial tissue compared to the osmolarity of the filtrate. When the osmolarity of the filtrate is higher than the osmolarity of the kidney tissue, water diffuses *into* the nephron. When the osmolarity of the filtrate is lower than the osmolarity of the kidney tissue, water diffuses *out* of the nephron.

Infusing the nephron of a healthy person with a concentrated NaCl solution increases the filtrate osmolarity and water diffuses *into* the nephron.

Since the volume of urine excreted is inversely proportional to the amount of water reabsorption, there is an increase in urine volume.

7. D is correct. ADH (antidiuretic hormone) is also known as vasopressin. Its two primary functions are to retain water within the body and to constrict blood vessels. ADH regulates the retention of water by acting to increase water absorption in the collecting ducts of the nephron.

8. B is correct. Glucose, amino acids and phosphate are reabsorbed in the proximal convoluted tubule through secondary active transport. Urea is a waste product that may (or may not) be reabsorbed by the kidney but does not involve active transport.

Reabsorption Site	Reabsorbed nutrient	Notes
Early proximal tubule	Glucose (100%), amino acids (100%), bicarbonate (90%), Na^+ (65%), Cl^-, phosphate and H_2O (65%)	PTH inhibits phosphate excretion AT II stimulates Na^+, H_2O and HCO_3^- reabsorption
Thin descending loop of Henle	H_2O	Reabsorbs via medullary hypertonicity and makes urine hypertonic
Thick ascending loop of Henle	Na^+ (10–20%), K^+, Cl^-; indirectly induces paracellular reabsorption of Mg^{2+}, Ca^{2+}	This region is impermeable to H_2O and the urine becomes less concentrated as it ascends
Early distal convoluted tubule	Na^+, Cl^-	PTH causes Ca^{2+} reabsorption
Collecting tubules	Na^+(3–5%), H_2O	Na^+ is reabsorbed in exchange for K^+, and H^+, which is regulated by aldosterone ADH acts on the V2 receptor and inserts aquaporins on the luminal side

9. D is correct. See explanation for question **36**.

10. A is correct.

11. C is correct. Hydrostatic (i.e. blood) pressure assists diffusion in the process of waste removal by the Bowman's capsule.

12. B is correct.

13. E is correct. The glomerulus is the capillary portion of the nephron where glucose, water, amino acids, ions, and urea pass through the capillary bed and enter Bowman's capsule, while larger plasma proteins and cells remain within the glomerulus. Glucose and amino acids pass into the filtrate (i.e. the urine), but are completely reabsorbed into the blood. Urea is filtered and excreted in the urine, and Na^+ and other salts are filtered but partially reabsorbed.

14. B is correct.

A: ureter is innervated by both parasympathetic and sympathetic nerves

C: the urinary bladder has a transitional epithelium and does not produce mucus

D: urine exits the bladder when both the autonomically controlled internal sphincter and the voluntarily controlled external sphincter open. Incontinence results from problems with these muscles.

E: detrusor muscle is a layer of the urinary bladder wall made of smooth muscle

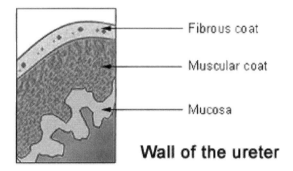

Wall of the ureter

15. E is correct. Glucose is normally completely reabsorbed from the filtrate and is not excreted in the urine. The presence of glucose in the urine signifies that glucose transporters in the proximal convoluted tubule (not loop of Henle) are unable to reabsorb glucose from the filtrate.

16. A is correct.

17. C is correct. Hypertonic refers to a greater solute concentration. A hypertonic solution has a higher concentration of solutes outside the cell than inside the cell. When a cell is placed into a hypertonic solution, water flows out of the cell to balance the concentration of the solutes.

Hypotonic refers to a lesser solute concentration. A hypotonic solution has a lower concentration of solutes outside the cell than inside the cell. When a cell is placed into a hypotonic solution, water flows into the cell, causing it to swell and possibly burst (i.e. plasmolysis).

Depending on their environment, organisms have different needs to maintain their H_2O balance. Evolution results in various renal structures for osmoregulation based on habitat. For example, fish living in saltwater maintain hypotonic body fluids, because the hypertonic environment in which they live tends to draw H_2O out of them by osmosis. They obtain H_2O by ingesting ocean water but this raises their internal osmolarity, so they excrete salt to lower osmolarity.

18. C is correct.

19. C is correct. Gout results when uric acid is not properly eliminated. The kidneys are the main organs that secrete, eliminate, filter and reabsorb compounds in order to maintain homeostasis. Thus, gout may be caused by a renal (kidney) problem.

A: the spleen is a lymphoid organ that filters blood for pathogens.

B: the large intestine reabsorbs water and is the site of vitamin K synthesis by symbiotic bacteria. The distal end of the colon is the portion that stores indigestible material as feces.

D: the liver synthesizes blood proteins (e.g. albumins, fibrinogen, some globulins), produces bile, detoxifies metabolites and stores fat soluble vitamins (A, D & E). Vitamin K is fat soluble, but is excreted from the body without being stored.

20. A is correct. The proximal tubule of the nephron is divided into an initial convoluted portion and a straight (descending) portion. The proximal convoluted tubule reabsorbs into the peritubular capillaries approximately two-thirds of the salts (i.e. electrolytes), water and all filtered organic solutes (e.g. glucose and amino acids).

21. A is correct. The solute concentration varies depending on the region of the kidney that filtrate is traveling.

The cortex (i.e. as a descriptive term) is the outer region of the kidney and has the lowest solute concentration. Filtrate that enters the nephron moves through the proximal convoluted tubule, loop of Henle, distal convoluted tubule, collecting duct, renal pelvis and then out of the kidney and into the bladder. The convoluted tubules are within the cortex, while the loop of Henle, the collecting duct and pelvis are in the medulla (i.e. middle region). As filtrate travels from the cortex to the medulla, it is subject to increasing concentration gradients that reabsorb water to concentrate the urine.

B: nephron is the functional unit of the kidney.

C: pelvis (in the medulla) has a very high concentration gradient and reabsorbs H_2O and salts.

D: medulla has a very high concentration gradient to produce concentrated urine.

E: epithelium does not refer to any region of the kidney.

22. E is correct.

23. C is correct.

24. B is correct.

25. C is correct. Passive diffusion of Na^+ occurs at the loop of Henle. The descending limb is permeable to H_2O and less impermeable to salt, and thus only indirectly contributes to the concentration of the interstitium. As the filtrate descends deeper into the medulla, the hypertonic interstitium osmosis causes H_2O to flow out of the descending limb until the osmolarity of the filtrate and interstitium equilibrate.

Longer descending limbs allow more time for H_2O to flow out of the filtrate, so a longer loop of Henle creates a more hypertonic filtrate than shorter loops. The thin ascending loop of Henle is impermeable to H_2O.

A: the thick segment of the ascending loop of Henle uses active transport for Na^+.

B: the distal convoluted tubule uses active transport for Na^+.

D: the proximal convoluted tubule uses active transport for Na^+.

26. D is correct. Angiotensin II acts at the Na^+/H^+ exchanger in the proximal tubules of the kidney to stimulate Na^+ reabsorption and H^+ excretion, which is coupled with bicarbonate reabsorption. This increases blood volume, pressure and pH. ACE (i.e. angiotensin converting enzyme) inhibitors are major anti-hypertensive drugs.

27. A is correct. The sodium-hydrogen exchange carrier moves Na^+ ions out of the urine and into epithelial cells and H^+ ions, as an anti-port carrier (i.e. in the opposite direction) into the urine. Because H^+ ions are moving out of epithelial cells, the pH inside the cells would be basic (~7.35) and the expected pH of the urine would be acidic. The pH of urine can vary between 4.6 and 8, with 7 being normal.

28. B is correct. Polyuria is a condition defined as excessive production of urine (i.e. greater than 2.5L/day). The most common cause of polyuria is uncontrolled diabetes mellitus (i.e. lack of insulin) that causes osmotic diuresis. In the absence of *diabetes mellitus*, the common causes are primary polydipsia (intake of excessive fluids), *central diabetes insipidus* (i.e. deficiency of vasopressin) and *nephrogenic diabetes insipidus* (i.e. improper response of the kidneys to ADH).

29. E is correct. The metabolism of amino acids by the liver involves deamination (i.e. removal of the amino group). Depending on the organism, nitrogenous wastes are excreted as ammonia, urea or uric acid. Organisms where H_2O is abundant (e.g. fish) excrete a dilute urine in the form of ammonia, while organisms in arid conditions excrete uric acid (e.g. some birds desiccate urine so it becomes a pellet).

30. D is correct.

31. C is correct. The kidneys and the respiratory system work together to maintain the body fluids near a pH of 7.35. Each system alters a different variable. The respiratory system regulates the level of CO_2, while the kidneys control the $[HCO_3^-]$ (concentration of bicarbonate).

$$CO_2 + H_2O \leftrightarrow H_2CO_3 \leftrightarrow H^+ + HCO_3^-$$

When CO_2 levels increase, the reaction shifts to the right, increasing the $[H^+]$ in the blood. An increase in $[H^+]$ decreases the pH. The lungs serve to release (via expiration) excess CO_2 to maintain the pH of the blood at about 7.35.

32. B is correct.

33. A is correct. The urethra is a tube that connects the urinary bladder to the male or female genitals for the removal of urine. During ejaculation, sperm travels from the epididymis, through the vas deferens, and through the urethra that opens to the outside from the tip of the penis. In males, the urethra functions in both the reproductive and excretory systems. In females, the reproductive and excretory systems do not share a common (i.e. urethral) pathway. Sperm enter the vagina and travel through the cervix, uterus, and fallopian tubes. Urine leaves the body through the urethra. In females, the vagina and urethra never connect because they are separate openings.

B: ureters serve the same excretory function in both males and females. The ureter is the duct connecting the kidney to the bladder. Urine is formed in the kidneys, travels to the bladder via the ureters and is stored in the bladder until excreted from the body through the urethra.

C: the prostate gland (found only in males) contributes the majority of ejaculated fluid to semen. It secretes prostatic fluid, which is alkaline, and neutralizes the acidity of any residual urine in the urethra. It also protects the sperm from the acidic conditions in the female reproductive tract.

D: vas deferens (also known as ductus deferens) transports sperm from the epididymis to the ejaculatory duct in anticipation of ejaculation.

E: the epididymis is a collection of coiled tubes on top of the seminiferous tubules in the male reproductive tract. Sperm are produced in the seminiferous tubules. They mature and acquire motility in the epididymis, where they are stored until ejaculation.

34. E is correct. In secondary active transport (i.e. coupled transport or co-transport), energy is used to transport molecules across a membrane. In contrast to primary active transport, there is no direct coupling of ATP. Secondary active transport uses the electrochemical potential difference, created by pumping ions out of the cell, to drive the process. The movement of an ion (or molecule) from the side, where it is more concentrated to where it is less concentrated, increases entropy and serves as a source of energy for metabolism (e.g. H^+ gradient in the electron transport chain for oxidative phosphorylation via the ATP synthase).

35. C is correct. Vasopressin (also known as anti-diuretic hormone; ADH) is released from the posterior pituitary in response to reductions in plasma volume or in response to increased osmolarity.

Aldosterone is a mineral corticoid released by the adrenal cortex in response to low blood pressure. It regulates blood pressure mainly by acting on the distal tubules and collecting ducts of the nephron.

36. B is correct. It is the high osmolarity of the medulla that produces hypertonic urine. The greater the osmolarity within the medulla (i.e. longer loop of Henle), the more concentrated the urine.

A: animals living in an arid environment produce hypertonic urine.

C: greater hydrostatic pressure does not affect urine osmolarity.

D: increasing the rate of filtrate does not affect urine osmolarity.

37. C is correct.

B: water is reabsorbed in the descending loop, while electrolytes are actively reabsorbed in the ascending loop.

38. E is correct. The nephron is the functional renal unit and is composed of the proximal convoluted tubule, the descending and ascending loops of Henle, the distal convoluted tubule and the collecting duct. A nephron eliminates wastes from the body, regulates blood volume and blood pressure, controls levels of electrolytes and metabolites, and regulates blood pH.

The proximal convoluted tubule reabsorbs 2/3 of all H_2O entering the nephron; the descending loop of Henle passively reabsorbs H_2O; the distal convoluted tubule only reabsorbs H_2O when stimulated by aldosterone; the collecting duct reabsorbs H_2O only when stimulated by antidiuretic hormone.

Unlike the descending limb, the thin ascending loop of Henle is impermeable to H_2O and is a critical feature of the countercurrent exchange mechanism that concentrates urine.

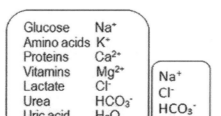

Secretion and reabsorption of various solutes throughout the nephron.

39. B is correct. Renal clearance is the maximum concentration for a solute in the plasma, when its concentration is greater than the capacity for the nephron to reabsorb the solute, so it is excreted in the urine. For example, the renal clearance threshold for glucose is about 180mg/100 ml.

40. B is correct. Plasma osmolarity levels are detected by the osmoreceptors of the hypothalamus that signal the pituitary gland to secrete ADH (also known as vasopressin). ADH acts on the nephron to increase permeability to H_2O, thereby increasing H_2O reabsorption that decreases plasma osmolarity. Therefore, increased ADH secretion lowers plasma osmolarity.

A: dehydration results from an excess excretion of dilute urine (i.e. diuresis) that may result from an excessive loss of fluid or an increased plasma osmotic pressure. Therefore, dehydration increases plasma osmolarity but is not a likely result of increased plasma osmolarity.

C: decreased H_2O permeability in the nephron increases plasma osmolarity, which leads to increased ADH secretion. ADH increases H_2O reabsorption by increasing H_2O permeability of the nephron. Therefore, decreasing water permeability increases plasma osmotic pressure.

D: excretion of dilute urine results from either low plasma osmolarity or inhibition of ADH secretion (e.g. by alcohol or caffeine). A low solute concentration (i.e. osmolarity) of blood decreases pituitary secretion of ADH. Low ADH levels reduce H_2O reabsorption and lead to the excretion of dilute urine, which raises plasma osmolarity because of the loss of H_2O from the blood.

41. A is correct. Renal clearance is the amount of liquid filtered out of the blood or the amount of substance cleared per time (i.e. volume/time). However, the kidney does not completely remove a substance from the renal plasma flow. In physiology, clearance is the rate at which waste substances are cleared from the blood (e.g. renal plasma clearance). Substances have a specific clearance rate that depends on its filtration characteristics (e.g. glomerular filtration, secretion from the peritubular capillaries and reabsorption from the nephron). A constant fraction of the substance is eliminated per unit time, but the overall clearance is variable, because the amount of substance eliminated per unit time changes with the concentration of the substance in the blood.

42. C is correct. For females, the anatomy of the excretory system from the urethra and into the body is: urethra → bladder → opening to the ureter → ureter → renal pelvis of the kidney

43. D is correct. The adrenal cortex synthesizes cortisol and aldosterone. A deficiency of aldosterone reduces reabsorption of Na^+ and H_2O from urine, which increases the volume of urine excreted.

A: sex steroids are synthesized primarily by the gonads (i.e. ovaries and testes).

C: decreased resistance to stress results from a deficiency of cortisol.

44. E is correct. For males, the anatomy of the excretory system from the urethra and into the body is: urethra → prostate → ejaculatory duct → vas deferens → epididymis → seminiferous tubules

45. B is correct. Renin is part of the renin-angiotensin system that influences blood pressure. Renin is formed by the juxtoglomerular apparatus near the distal convoluted tubule of the nephron and acts indirectly on the adrenal cortex but does not act on the pituitary gland.

When blood pressure falls (e.g. heavy bleeding), renin is released by the kidneys and converts the zymogen of angiotensinogen to angiotensin I by the angiotensin converting enzyme (ACE).

Angiotensin II causes vasoconstriction (to increase blood pressure) and aldosterone release from the adrenal cortex that opens ion channels in convoluted tubules (to increase H_2O reabsorption).

46. C is correct.

47. A is correct. Antidiuretic hormone (ADH) is released from the posterior pituitary and stimulates the collecting duct in the kidneys to reabsorb H_2O to concentrate the urine. Persons with *diabetes insipidus* do not respond to ADH despite having normal plasma concentrations of ADH. Their urine remains diluted because the collecting ducts fail to reabsorb H_2O.

D: another disease, *diabetes mellitus*, affects the pancreas and elevates blood glucose levels.

48. C is correct.

49. D is correct. Glucose is reabsorbed in the proximal convoluted tubule. H_2O is reabsorbed everywhere along the nephron except the ascending limb where only salt (e.g. Na^+, K^+ and Cl^-) is reabsorbed.

A: K^+ is secreted into the filtrate (i.e. solution destined to be urine) and reabsorbed in the proximal convoluted tubule and the ascending loop of Henle.

B and C: Cl^- and Na^+ are reabsorbed along the nephron, except in the descending limb and collecting ducts.

E: amino acids are reabsorbed in the proximal convoluted tubule.

50. E is correct. The brush border increases the surface area to reabsorb solutes from the filtrate. The brush border is composed of villi and does not, unlike cilia, influence the direction or rate of fluid movement.

51. B is correct. See explanation for question **3**.

52. A is correct. When the glomeruli are unable to filter adequate amounts of fluid, the major physiological effect is retention of salt and H_2O, since these can no longer be excreted by the kidneys.

The blood travels via the interlobular artery to the afferent arteriole and then to the glomerulus through the network of capillaries. Blood flows out of the glomerulus via the efferent arteriole. Bowman's capsule is part of the kidney and is a round, double-walled structure. The plasma (liquid component of blood) is filtered by passing through the glomerular membrane and the epithelial layer of Bowman's capsule, and drains into the efferent arteriole. The filtered blood leaves through the tubule at the top and rejoins the interlobular vein.

B: urinary output decreases significantly to less than 500 mL/day compared to the average volume of 1-2 L/day.

C: glomerulus inflammation results in the inability of urea to be excreted and causes an excessive accumulation of urea in the blood (i.e. uremia).

D: the quantity of salt and H_2O in the extracellular fluid increases drastically and decreases in the filtrate (i.e. urine), which leads to an excess of extracellular fluid in body tissue (i.e. edema).

53. E is correct.

54. A is correct. Renin secretion catalyzes the conversion of angiotensin I to angiotensin II, which increases the secretion of aldosterone. If renin is blocked, aldosterone cannot cause increased synthesis of Na^+ absorbing proteins, and Na^+ absorption decreases. Without renin secretion, production of angiotensin II decreases.

B: blood pressure decreases, which reduces the amount of blood entering Bowman's capsule.

D: blood pressure would decrease.

E: platelets are involved in blood clotting and are not related to the production of angiotensin II.

55. B is correct.

56. E is correct. Antidiuretic hormone (ADH) is released by the posterior pituitary and causes the collecting tubule to become more permeable to H_2O. Therefore, more H_2O is reabsorbed and the volume of urine decreases while it becomes more concentrated.

A: filtration only occurs at the interface of the glomerulus and Bowman's capsule.

B: salts, glucose and amino acids are reabsorbed via active transport, while H_2O is reabsorbed via diffusion.

C: ammonia is converted into urea in the liver and then is excreted by the kidney.

D: in the nephron, the glomerulus, Bowman's capsule, the proximal and the distal convoluted tubules are in the cortex, while the descending and descending tubules, as well as the collecting duct, span the cortex, outer medulla and inner medulla. The loop of Henle is in the inner medulla.

57. C is correct. Vasopressin (ADH) and aldosterone both result in concentrated urine to prevent H_2O loss. Without food, the patient's plasma glucose levels are low and, therefore, her insulin is also low, because insulin is stimulated by high plasma glucose levels as a trigger to increase glucose uptake by cells.

58. D is correct. Long loops of Henle on juxtamedullary nephrons allow for a greater concentration of urine via a net loss of solute to the medulla. This process is critical to the function of other parts of the nephron. A high concentration of solutes in the medulla allows for passive absorption of H_2O from the filtrate in other areas of the nephron.

59. C is correct. Insulin is a hormone secreted by the beta cells of the pancreas in response to high blood glucose levels. It decreases blood glucose levels by stimulating cells to uptake glucose and by stimulating the conversion of glucose into its storage form (glycogen) in liver cells and muscle cells. An overdose of insulin can (and often does) lead to convulsions because of the resulting sharp decrease in blood glucose concentration.

A: diabetics would become dehydrated, if untreated (with proper insulin dosage), because the excess glucose in the nephrons causes water (as filtrate) to diffuse into the nephrons. With high glucose levels, the net result is an increase in urine excretion. Insulin decreases urine excretion by decreasing blood glucose levels, and glucose does not join the filtrate.

B: insulin increases the conversion of glucose into glycogen. Therefore, an overdose of insulin would not cause an increase in the conversion of glycogen back into glucose. Glucagon, also secreted by the pancreas, stimulates the conversion of glycogen into glucose.

D: the presence of glucose is a clinical manifestation of diabetes because of high blood glucose concentration that arises from an insufficiency (or complete lack of) insulin production, or because insulin-specific receptors are insensitive to insulin. In untreated diabetes, there is a high blood glucose concentration, along with a high urine glucose concentration, since the kidneys are unable to reabsorb glucose because of its excess in the blood. Controlled doses of insulin are used to stimulate glucose uptake by muscle and adipose cells, and the conversion of excess glucose into glycogen. Insulin alleviates the symptoms of diabetes and would not cause an increased glucose concentration in the urine.

Chapter 4.8: Muscle System

1. C is correct. The sarcolemma (also called the myolemma) is the cell membrane of a striated muscle fiber cell. It consists of a lipid bi-layer plasma membrane and an outer coat of a thin layer of polysaccharide material (glycocalyx). Glycocalyx contacts the basement membrane that contains thin collagen fibrils and specialized proteins (e.g. laminin) to act as a scaffold for the muscle fiber to adhere.

The neuromuscular junction connects the nervous system to the muscular system via synapses between efferent nerve fibers and muscle fibers (i.e. muscle cells). When an action potential reaches the end of a motor neuron, voltage-dependent calcium channels open and calcium enters the neuron. Calcium binds to sensor for vesicle fusion with the plasma membrane and subsequent release of the neurotransmitter (e.g. acetylcholine) from the motor neuron into the synaptic cleft.

On the post-synaptic membrane surface, the sarcolemma has invaginations (i.e. postjunctional folds), which increases the surface area of the post-synaptic membrane at the boundary of the synaptic cleft. These postjunctional folds form the motor end-plate that has numerous acetylcholine receptors. The binding of acetylcholine to the post-synaptic receptor on the muscle cells depolarizes the muscle fiber and causes a cascade that eventually results in muscle contraction.

2. A is correct. Cardiac muscle exhibits cross striations formed by alternating segments of myosin thick and actin thin protein filaments. Like skeletal muscle, the primary structural proteins of cardiac muscle are actin and myosin. However, in contrast to skeletal muscle, cardiac muscle cells may be branched, instead of linear and longitudinal.

E: histologically, T-tubules in cardiac muscle, compared to skeletal muscle, are larger and broader and run along the Z-discs; there are fewer T-tubules in cardiac muscle than in skeletal muscle.

3. D is correct. Phosphocreatine and ATP are energy storage molecules located in muscle fibers that release energy used for contraction.

A: lactose (i.e. milk sugar) is a disaccharide composed of glucose and galactose.

B: ADP (adenosine diphosphate) is a lower energy form of ATP (adenosine triphosphate).

C: lactic acid is a product of anaerobic respiration and is not a direct energy source.

E: cAMP is a second messenger found in target cells of peptide hormones.

4. E is correct.

5. A is correct. Neither the actin thin filament nor the myosin thick filament change in length during muscle contraction. Muscle contraction occurs via the *sliding filament model* by increasing overlap of actin and myosin filaments.

6. A is correct. The sarcolemma (also called myolemma) is the cell membrane of a striated muscle fiber cell. The motor end plate is a large terminal formation at the point of juxtaposition of the axon of a motor neuron and the striated muscle cell that it connects to.

7. D is correct. Slow twitch fibers produce 10 to 30 contractions per second, while fast twitch fibers produce 30 to 70 contractions per second.

8. E is correct. Permanent sequestering of Ca^{2+} in the sarcoplasmic reticulum (its normal storage compartment in the muscle) prevents Ca^{2+} from binding to the troponin that causes the conformational change in tropomyosin to move and expose the myosin binding sites on the actin thin filament.

A: Ca^{2+} depletion does not cause depolymerization of actin thin filaments because such an event would be a repetitive problem every time Ca^{2+} is sequestered into the sarcoplasmic reticulum as contraction is complete and the muscle relaxes.

B: permanent contraction of the muscle occurs if ATP is absent because ATP hydrolysis is necessary for the release of the myosin cross-bridge from the actin thin filament.

C: resorption of Ca^{2+} from bone (via osteoclast activity) results in decreased bone density.

9. C is correct.

Sliding filament model of skeletal muscle

10. A is correct.

D: *treppe* stimulus is the gradual increase in muscular contraction following rapidly repeated stimulation.

11. A is correct. The vagus nerve is part of the parasympathetic nervous system. Heart rate is set by the SA node innervated by the vagus nerve (i.e. parasympathetic). The rhythmic pace of the SA nerve is faster than the rate of normal heart beats, but the vagus nerve slows the contractions of the heart to its observed (resting) pace. Without inhibition from the vagus, the heart would normally beat between 100 – 120 beats per minute.

12. C is correct. During strenuous activity, continuous muscle contractions deplete all available O_2 (i.e. stored on myoglobin) and cause the muscle fiber to rely on anaerobic respiration for energy. Reducing pyruvate into lactic acid regenerates NAD^+ that is needed for glycolysis to continue in the absence of O_2 (anaerobic conditions).

13. D is correct. A neuromuscular junction is a synapse between the motor neuron and the muscle fiber.

14. A is correct.

15. E is correct. The relaxation of smooth muscles that line the circulatory system results in dilation of blood vessels.

16. A is correct.

17. B is correct.

18. A is correct. Shivering is a protective mechanism of increases in muscle contractions, which raises core body temperature.

19. C is correct.

20. E is correct.

21. A is correct.

22. B is correct.

23. C is correct. The ventricles of the heart require long contraction periods to pump the viscous blood.

B: gap junctions permit rapid communication that result in adjacent cardiac muscle cells contracting simultaneously.

D: Na^+ is outside the resting neuron and flows into the cell during the action potential.

E: K^+ is inside the resting neuron and flows out of the cell during the action potential.

24. C is correct. Intercalated discs are microscopic identifying features of cardiac muscle. Cardiac muscle consists of individual heart muscle cells (cardiomyocytes) connected by intercalated discs to work as a single functional organ or syncytium. Intercalated discs support synchronized contraction of cardiac tissue and occur at the Z line of the sarcomere.

25. A is correct. After death, cellular respiration ceases to occur, depleting the body of oxygen used in the making of ATP. The corpse hardens and becomes stiff (in humans, it commences after 3-4 hrs, reaches maximum stiffness after 12 hours and gradually dissipates by approximately 24 hrs after death). Unlike in normal muscular contraction during life of the organism, after death, the body is unable to complete the cycle and release the coupling between the myosin and actin, creating a state of muscular contraction until the breakdown of muscle tissue by enzymes during decomposition.

26. E is correct. Muscles move by contracting, which brings the origin and insertion closer together. This process often moves the insertion, while the origin remains fixed in position.

27. A is correct.

C: titin is important for contractions of striated muscle tissues. It connects the Z line to the M line in the sarcomere and contributes to force at the Z line and resting tension in the I band region.

28. D is correct.

29. D is correct. During muscle contraction, Ca^{2+} binds to troponin, which causes the tropomyosin strands to move and expose the myosin-binding sites on the actin thin filaments.

30. E is correct.

31. C is correct.

32. D is correct. Antagonistic muscles move bones in opposite directions relative to a joint and require one muscle to relax while the other muscle contracts.

33. E is correct.

34. D is correct.

35. E is correct. Tendons connect muscle to bone, while ligaments connect bone-to-bone. Tendons are not cartilage.

36. C is correct.

37. A is correct. See explanations for questions **8** and **41**.

38. B is correct. Peristalsis is a contractile process of smooth muscle that is innervated by the autonomic nervous system.

A: diaphragm contracts, the volume of the thoracic cavity increases, and air is drawn into the lungs.

C: cardiac muscle action potentials are conducted by gap junctions.

D: a knee jerk reflex contracts skeletal muscles innervated by the somatic nervous system.

39. A is correct. See explanation for question **41**.

40. B is correct. Unlike skeletal and cardiac muscle, smooth muscle does not contain the calcium-binding protein troponin, and calmodulin has a regulatory role. In skeletal muscle, calcium is stored in the sarcoplasmic reticulum within the cell, while in smooth muscle, calcium is stored when bound to calmodulin within the cell. Contraction for smooth muscle is initiated by a calcium-regulated phosphorylation of myosin, rather than a calcium-activated troponin system.

41. D is correct. Skeletal muscle has multinucleated fibers with a striated appearance from the repeating motifs of actin and myosin filaments. These filaments slide past each other and shorten during contraction. The process of contraction does not require energy but ATP is needed for release for the cross-bridges between the actin and myosin during the next contraction process. Each muscle fiber is innervated by neurons from the somatic division of the PNS. The axon releases an action potential to each muscle fiber, but this action potential cannot pass from one muscle fiber to another. When an action potential reaches the muscle fiber, it causes the release of Ca^{2+} from the sarcoplasmic reticulum to permit the sliding of the actin and myosin filaments.

42. D is correct.

43. B is correct. Smooth muscle contains thin actin filaments and thick myosin filaments (albeit not striated) and requires Ca^{2+} for contraction.

44. C is correct.

45. E is correct.

46. C is correct. The diaphragm is a skeletal muscle innervated by the phrenic nerve of the somatic nervous system. Smooth muscle and cardiac muscle are involuntary because they are innervated by the autonomic nervous system.

47. A is correct.

48. E is correct.

49. B is correct. Characteristics of smooth muscle: involuntary, without T-tubules, lacking troponin and tropomyosin, without repeating striations (Z disks) of sarcomeres, single unit with the whole muscle contracting/relaxing.

Smooth muscle cells undergo involuntary contractions and are located 1) in blood vessels for regulating blood pressure, 2) in the gastrointestinal tract to propel food during digestion and 3) in the bladder to discharge urine.

50. A is correct.

51. A is correct.

52. E is correct. Cardiac muscle, unlike the other two muscle types, is capable of spontaneous depolarization resulting in contraction. Both smooth and skeletal muscle require stimulation (e.g. neurotransmitter) to initiate depolarization and contraction.

A: cardiac muscle has only one or two centrally located nuclei.

B: cardiac muscle is striated.

D: cardiac muscle is innervated by the autonomic nervous system. The somatic motor system innervates skeletal muscle and controls voluntary actions (e.g. walking, standing).

53. A is correct.

54. D is correct.

55. E is correct. Myoglobin is an iron- and oxygen-binding protein found in the muscle tissue of vertebrates and almost all mammals. It has a similar function to hemoglobin, which is the iron- and oxygen-binding protein in red blood cells. Myoglobin is the primary oxygen-carrying pigment of muscle tissues. High concentrations of myoglobin allow organisms to hold their breath for longer periods (e.g. diving animals).

56. C is correct. A muscle is a bundle of parallel fibers. Each fiber is a multinucleated cell created by the fusion of several mononucleated embryonic cells. Skeletal muscle is responsible for voluntary movement and is innervated by the somatic nervous system. Skeletal muscle is made up of individual components known as myocytes (i.e. muscle cells). These long cylindrical multinucleated cells are also called myofibers and are composed of myofibrils. The myofibrils are composed of actin and myosin filaments repeated in units called a sarcomere – the basic functional unit of the muscle fiber. The sarcomere is responsible for skeletal muscle's striated appearance and forms the basic machinery necessary for muscle contraction.

A: cardiac muscle composes the muscle tissue of the heart. These muscle fibers possess characteristics of both skeletal and smooth muscle fibers. As in skeletal muscle, the cardiac muscle has a striated appearance, but cardiac muscle cells generally have only one or two centrally located nuclei. Cardiac muscle is innervated by the autonomic nervous system that modulates the rate of heartbeats.

B: smooth muscle, found in the digestive tract, bladder, uterus, and blood vessel walls, is responsible for involuntary action and is innervated by the autonomic nervous system. It does not appear as striated and has one centrally located nucleus.

57. D is correct.

58. D is correct. Cardiac muscle contains actin/myosin filaments as in striated skeletal muscle. A syncytium is a multinucleated cell that results from the fusion of several individual cells. Cardiac muscle is not a true syncytium but adjoining cells are linked by gap junctions that communicate action potentials directly from the cytoplasm of one myocardial cell to another. The sympathetic nervous system increases heart rate, while parasympathetic stimulation (via the vagus nerve) decreases heart rate.

59. B is correct.

60. D is correct. From the propagation of an action potential to the axon terminus, ACh is the neurotransmitter released into the neuromuscular junction when vesicles fuse with the presynaptic membrane and release ACh via exocytosis. ACh then diffuses across the synapse and binds to specific receptors on the postsynaptic membrane. Acetylcholinesterase is an enzyme that degrades ACh in the synaptic cleft to prevent constant stimulation by ACh of the postsynaptic membrane.

Chapter 4.9: Skeletal System

1. C is correct.

2. D is correct. A tendon connects bone to muscle, while a ligament connects bone to bone.

C: aponeuroses are layers of flat broad tendons.

3. C is correct. Synovial fluid functions as a lubricant that decreases friction between the ends of bones as they move relative to each other.

A: bone cells (i.e. osteocytes) receive blood circulation for adequate nutrition and hydration, but synovial fluid is not involved in these functions.

4. B is correct.

5. E is correct. Osteocytes are derived from osteoprogenitors, some of which differentiate into active osteoblasts that do not divide and have an average half life of 25 years. In mature bone, osteocytes reside in lacunae spaces, while their processes reside inside canaliculi spaces.

When osteoblasts become trapped in the matrix that they secrete, they become osteocytes. Osteocytes are networked to each other via long cytoplasmic extensions that occupy tiny canals (i.e. canaliculi) and exchange nutrients and waste through gap junctions. Osteocytes occupy the lacuna space. Although osteocytes have reduced synthetic activity and (like osteoblasts) are not capable of mitotic division, they are actively involved in the dynamic turnover of bone matrix.

6. D is correct. There are five types of bones: long, short, flat, irregular and sesamoid. Short bones are those bones that are as wide as they are long. Their primary function is to provide support and stability with little to no movement (e.g. tarsals in the foot and carpals in the hand).

Cancellous bone (i.e., trabecular or spongy bone) is one of two types of bone tissue. The other type of bone tissue is cortical bone (i.e., compact bone).

7. A is correct.

8. E is correct.

9. D is correct. Cartilage does not contain nerves. Unlike other connective tissues, cartilage does not contain blood vessels. Cells of the cartilage (chondrocytes) produce a large amount of extracellular matrix composed of collagen fibers, elastin fibers and abundant ground substance rich in proteoglycan. Cartilage is classified as *elastic cartilage*, *hyaline cartilage* and *fibrocartilage*, which differ in the relative amounts of the three main components.

10. A is correct.

11. C is correct.

12. C is correct. The middle ear contains three bones: the malleus, incus and stapes, which amplify the vibrations of the tympanic membrane (i.e. ear drum) and transmit it to the oval window that leads to the inner ear and receptors on the auditory nerve.

The larynx, nose, outer ear and skeletal joints are composed of cartilage. Cartilage is neither innervated nor vascularized.

13. D is correct. The two types of bone marrow are red marrow (which consists mainly of hematopoietic tissue) and yellow marrow (which is mainly made up of fat cells). The red marrow gives rise to red blood cells, platelets and most white blood cells. Both types of bone marrow contain numerous blood vessels and capillaries. At birth, all bone marrow is red, but, with age, more of it is converted to yellow marrow until about half of adult bone marrow is yellow.

Red marrow is found mainly in the flat bones (e.g. sternum, pelvis, vertebrae, cranium, ribs and scapulae) and in the cancellous (i.e. spongy bone) material at the epiphyseal ends of long bones (e.g. femur and humerus). Yellow marrow is found in the medullary cavity (i.e. hollow interior of the middle portion of long bones). In cases of severe blood loss, the body can convert yellow marrow back to red marrow to increase blood cell production.

14. B is correct. Parathyroid hormone (PTH) increases plasma Ca^{2+} by increasing the activity of osteoclast cells.

D: osseous tissue forms the rigid part of the bones making up the skeletal system.

15. E is correct.

16. B is correct.

17. A is correct.

18. A is correct.

19. E is correct. Ligaments attach bone to bone.

A: an osteocyte is the most common cell in mature bone and is derived from osteoprogenitors, some of which differentiate into active osteoblasts. In mature bone, osteocytes and their processes reside inside spaces called lacunae and canaliculi, respectively. Osteocytes contain a nucleus and a thin ring piece of cytoplasm. When an osteoblast becomes trapped in the matrix that it secretes, it becomes an osteocyte.

B: tendons attach muscle to bones.

D: periosteum is the membrane that covers the surface of all bones.

20. B is correct.

21. A is correct.

22. C is correct. Thyroid hormones (e.g. thyroxine) have important general metabolic effects, but are not involved in bone remodeling. They do increase the basal metabolic rate, affect protein synthesis, help regulate long bone growth and neural maturation. They also increase the body's sensitivity to catecholamines (e.g. adrenaline). The thyroid hormones are essential to proper development and differentiation of all cells of the human body. These hormones also regulate protein, fat and carbohydrate metabolism, affecting how cells use energetic compounds. They also stimulate vitamin metabolism.

Calcitonin, vitamin D and PTH are involved in Ca^{2+} metabolism and bone remodeling.

23. A is correct.

24. E is correct. Bones store calcium and phosphate, support and protect the body, produce blood cells and store fat. Bones do not regulate the temperature of blood.

25. B is correct.

26. C is correct. Osteon is the functional unit of much compact bone. Osteons are roughly cylindrical structures that are typically several millimeters long and around 0.2mm in diameter. Each osteon consists of concentric layers, or lamellae, of compact bone tissue that surrounds a central canal (the Haversian canal) which contains the bone's nerve and blood supplies. The boundary of an osteon is the cement line.

A: periosteum membrane covers the outer surface of all bones, except at the joints of long bones.

B: endosteum lines the inner surface of all bones.

D: trabeculae are tissue elements in the form of a small strut or rod (e.g. head of the femur), generally having a mechanical function, and usually composed of dense collagenous tissue (e.g. trabecula of the spleen.) Trabecula can be composed of other materials. For example, in the heart, trabeculae consist of muscles that form the trabeculae carneae of the ventricles.

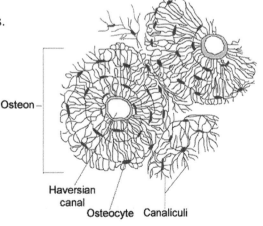

Diagram of compact bone from a transverse section of a long bone's cortex

27. D is correct.

28. E is correct. Bones are covered and lined by a protective tissue called periosteum that consists of dense irregular connective tissue. It is divided into an outer fibrous layer and inner osteogenic layer (or cambium layer). The fibrous layer contains fibroblasts, while the osteogenic layer contains progenitor cells that develop into osteoclasts and osteoblasts. These osteoblasts are responsible for increasing the width of a long bone and the overall size of the other bone types.

29. A is correct. Yellow bone marrow is usually found in the medullary cavity of long bones. Red blood cells, platelets, and most white blood cells arise in red marrow. Both red and yellow bone marrow contain numerous blood vessels and capillaries. At birth, all bone marrow is red. With age, more and more of it is converted to the yellow type; only around half of adult bone marrow is red. Red marrow is found mainly in the flat bones, such as the pelvis, scapulae, sternum, cranium, ribs and vertebrae, and in the spongy material at the epiphyseal ends of long bones, such as the femur and humerus.

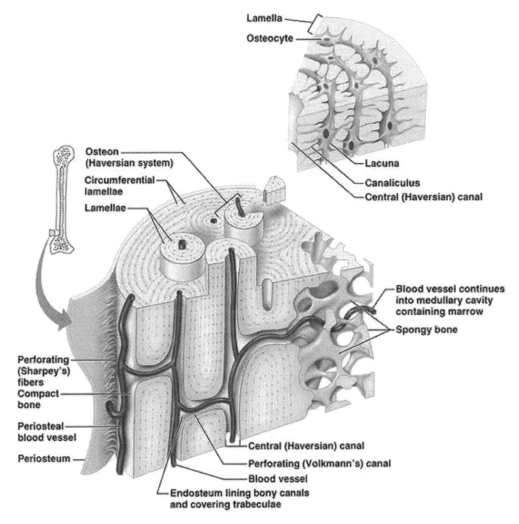

C: Haversian canals are a series of tubes around narrow channels formed by lamellae. This is the region of bone called compact bone. Osteons are arranged in parallel to the long axis of the bone. The Haversian canals surround blood vessels and nerve cells throughout the bone and communicate with osteocytes in lacunae (i.e. spaces within the dense bone matrix that contain the living bone cells) through canaliculi (i.e. canals between the lacunae of ossified bone).

D: Volkmann's canals are microscopic structures found in compact bone. They run within the osteons perpendicular to the Haversian canals, interconnecting the latter with each other and the periosteum (i.e. membrane that covers the surface of all bones). They usually run at obtuse angles to the Haversian canals and contain connecting vessels between Haversian capillaries. The Volkmann canals also carry small arteries throughout the bone.

30. B is correct.

31. D is correct. Endochondral ossification, with the presence of cartilage, is one of the two essential processes during fetal development of the mammalian skeletal system by which bone tissue is created. Endochondral ossification is also an essential process during the rudimentary formation of long bones, the growth of the length of long bones, and the natural healing of bone fractures.

Appositional growth occurs when the cartilage model grows in thickness due to the addition of more extracellular matrix on the peripheral cartilage surface, which is accompanied by new chondroblasts that develop from the perichondrium.

B: intramembranous ossification is another essential of rudimentary bone tissue creation during fetal development. It is also an essential process during the natural healing of bone fractures and the rudimentary formation of bones of the head. Unlike endochondral ossification, cartilage is not present during intramembranous ossification.

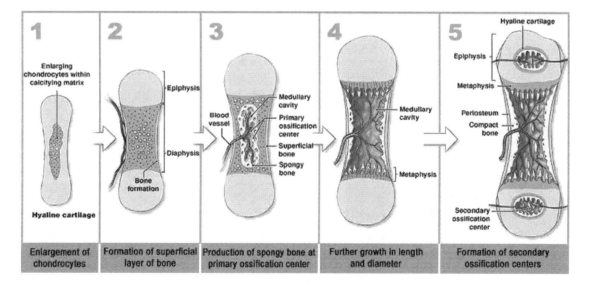

32. C is correct. Ligaments are fibrous connective tissue that joins bone to bone.

A: synovium is the smooth lining of synovial joints that produces synovial fluid.

B: osteoprogenitor cells become osteo*blasts* (not osteo*clasts*).

D: sockets are types of joints, such as the ball and socket joint of the hip.

E: muscles do connect bones via their origin and insertion points, but they are not connective tissue (i.e. contractile tissue).

33. B is correct.

34. E is correct.

35. D is correct. Up to 50% of bone by weight is a modified form of hydroxylapatite that consists of calcium and phosphate in a compound that includes hydroxyl (i.e. ~OH) groups. Bone stores calcium and phosphate.

36. B is correct.

37. E is correct. Osseous tissue, or bone tissue, is the major structural and supportive connective tissue that forms the rigid part of the bones of the skeletal system.

38. D is correct.

39. B is correct.

40. D is correct. Osteoclasts are the progenitor cells for white blood cells. Osteoclasts are derived from monocytes and macrophages. Osteoclasts and macrophages are involved in nonspecific phagocytosis and osteoclasts breakdown their surrounding matrix. PTH promotes activity, while calcitonin inhibits the activity of osteoclasts.

A: erythrocytes (red blood cells) are terminally differentiated

B: chondrocytes are terminally differentiated and are the only cells found in healthy cartilage. They produce and maintain the cartilaginous matrix, which consists mainly of collagen and proteoglycans. Mesenchymal stem cells are the progenitor of chondrocytes, but can also differentiate into several cell types including osteoblasts.

41. B is correct.

42. D is correct. The diaphysis is the main shaft of a long bone made of cortical bone and usually contains bone marrow and adipose tissue. It's a middle tubular part composed of compact bone that surrounds a central marrow cavity containing red or yellow marrow. Primary ossification occurs in diaphysis.

43. E is correct. The parathyroid glands are four small pea-shaped structures embedded in the posterior surface of the thyroid that synthesize and secrete PTH that (together with calcitonin and vitamin D) regulates plasma $[Ca^{2+}]$ via negative feedback.

PTH raises plasma $[Ca^{2+}]$ by increasing bone resorption and decreasing Ca^{2+} excretion in the kidneys. Also, PTH converts vitamin D to its active form, which then stimulates intestinal calcium absorption.

A: insulin lowers plasma glucose levels and increases glycogen storage in the liver.

B: glucagon stimulates the conversion of glycogen into glucose in the liver and therefore increases plasma glucose levels.

C: anti-diuretic hormone (ADH) stimulates the reabsorption of H_2O in the kidney by increasing the permeability of the nephron to H_2O.

D: aldosterone regulates plasma $[Na^+]$ and $[K^+]$ and therefore the total extracellular H_2O volume. Aldosterone stimulates active reabsorption of Na^+ and passive reabsorption of H_2O in the nephron.

44. C is correct.

45. E is correct. Spongy bone (i.e. cancellous bone) is typically found at the ends of long bones, proximal to joints and within the interior of vertebrae. Cancellous bone is highly vascular and frequently contains red bone marrow where haematopoiesis (i.e. production of blood cells) occurs. Spongy bone contains blood stem cells for differentiation into mature blood cells.

A: long bones are involved in fat storage.

D: erythrocyte cell storage occurs in the liver and spleen.

46. A is correct.

47. C is correct.

48. B is correct. Osteoporosis is a progressive bone disease characterized by a decrease in bone mass and density, which can lead to an increased risk of fracture. In osteoporosis, the bone mineral density is reduced, bone microarchitecture deteriorates and the amount and variety of proteins in bone is altered.

Trabecular bone (i.e. spongy bone in the ends of long bones and vertebrae) is more prone to turnover due to higher concentration of osteoclasts (breakdown bone) and osteoblasts (deposit bone) and remodeling of calcium/phosphate within the bone. Calcitonin stimulates the osteoblasts to build bone mass. Menopause contributes to osteoporosis by reducing estrogen levels and thereby leading to reduced osteoblast activity.

49. C is correct.

50. B is correct. The epiphyseal plate is a hyaline cartilage plate in the metaphysis at each end of a long bone. The plate is present in children and adolescents. In adults, it is replaced by an epiphyseal line.

51. A is correct. Hyaline cartilage is a type of cartilage (i.e. flexible connective tissue) found on many joint surfaces. It is pearly bluish in color with firm consistency, has a considerable amount of collagen and contains no nerves or blood vessels.

C: areolar connective tissue (also known as loose connective tissue) is a common type of connective tissue found beneath the dermis layer and underneath the epithelial tissue of all the body systems that have external openings. Areolar connective tissue holds organ in place and attaches epithelial tissue to other underlying tissues. It also serves as a reservoir of H_2O and salts for surrounding tissues.

D: fibrocartilage is a mixture of white fibrous tissue (i.e. toughness) and cartilaginous tissue (i.e. elasticity) in various proportions. Fibrocartilage is found in the pubic symphysis, the annulus fibrosus of intervertebral discs, menisci, and the TMJ (temporomandibular joint of the jaw).

52. D is correct.

53. B is correct.

54. C is correct. A tendon sheath is a layer of membrane around a tendon that permits the tendon to move.

55. E is correct.

56. A is correct. The epiphyseal plates are regions of cartilaginous cells separating the shaft of the long bone (i.e. diaphysis) from its two ends (i.e. epiphyses) and are the location of growth in long bones. Since growth occurs only at the epiphyseal plates, the strontium would be incorporated at the ends near the epiphyseal plates.

57. B is correct. Hyaluronic acid is an anionic, non-sulfated glycosaminoglycan distributed widely throughout epithelial, connective and neural tissues.

58. D is correct. Hydroxyapatite composes up to 50% of bone by weight. Carbonated calcium-deficient hydroxylapatite is the main mineral that dental enamel and dentin are composed of.

Chapter 4.10: Respiratory System

1. B is correct. Transport of gases to and from tissues is the function of the circulatory system. Respiratory system conducts the gas exchange in the alveoli of the lungs, while circulatory system carries the gases to and from tissues and cells.

2. A is correct.

3. D is correct. The autonomic nervous system (i.e. sympathetic and parasympathetic) regulates the internal environment (i.e. maintains homeostasis via parasympathetic division) by controlling involuntary processes. The ANS innervates the heart, smooth muscle around blood vessels and the digestive tract, as well as the endocrine, reproductive, excretory and respiratory systems. The sympathetic division of the ANS controls the vasoconstriction of arterioles in response to changes in temperature.

B: somatic nervous system innervates skeletal muscles in response to external stimuli.

C: sensory nervous system innervates the skin, taste and olfactory receptors.

4. A is correct. The walls of the alveoli are composed of two types of cells, type I and type II. Type I cells form the structure of the alveolar wall.

C: type II cells secrete surfactant.

5. E is correct.

6. B is correct. When pressure in the lungs exceeds atmospheric pressure, air moves out of the lungs by flowing down the pressure gradient.

7. C is correct. Unlike inspiration, expiration is a passive act because no muscular contractions are involved. Air is expelled from the lungs when the diaphragm relaxes and pushes up on the lungs.

8. E is correct. The diaphragm is a sheet of internal skeletal muscle that extends across the bottom of the rib cage. The diaphragm separates the thoracic cavity (heart, lungs and ribs) from the abdominal cavity. It performs the important function in respiration – as the diaphragm contracts, the volume of the thoracic cavity increases and air enters the lungs.

9. A is correct. Many animals do not have sweat glands and therefore sweating is not a method of heat loss. The panting mechanism serves as a substitute for sweating. When a large amount of air comes into the upper respiratory passages, it permits water evaporation from the mucosal surfaces to aid in heat loss. The preoptic area of the anterior hypothalamus is the thermoregulatory center that monitors blood temperature.

B: rapid breaths during panting are not deep enough to affect the rate of CO_2 expiration.

C: surface moisture of the respiratory mucosa is dehydrated via evaporation.

D: the respiratory muscles move faster and are engaged during panting.

10. B is correct. C-shaped rings made up of cartilage form the foundation of the trachea, which allows it to maintain its openness.

11. B is correct. Hyaline cartilage rings help support the trachea while simultaneously allowing it to be flexible during breathing.

12. D is correct. Intrapulmonary pressure is pressure within the lungs, which varies between inspiration and expiration. Intrapulmonary pressure must fall below atmospheric pressure to initiate inspiration: increase in lung volume during inspiration decreases intrapulmonary pressure to subatmospheric levels and air enters the lungs. Decrease in lung volume raises intrapulmonary pressure above atmospheric pressure expelling air from the lungs.

13. E is correct. The fenestrated (i.e. openings between cells) capillaries are endothelial cells of a single layer. The blood (with the erythrocytes) moves under hydrostatic pressure for oxygen exchange.

14. B is correct.

15. E is correct. [CO_2] in the blood affects Hb's affinity for O_2 and therefore affects the location of the curve on the O_2-dissociation graph, but [CO_2] is not the reason for the sigmoidal shape. A high [CO_2] in the blood decreases Hb's affinity for O_2 and shifts the curve to the right (known as the Bohr effect). The Bohr effect is a physiological observation where Hb's O_2 binding affinity is inversely related to both the [CO_2] and acidity of the blood. An increase in blood [CO_2] or decrease in blood pH (increase in H^+) results in Hb releasing its O_2 at the tissue.

Conversely, a decrease in CO_2 or an increase in pH results in hemoglobin binding O_2 and loading more O_2. CO_2 reacts with water to form carbonic acid, which causes a decrease in blood pH. Carbonic anhydrase (present in erythrocytes) accelerates the formation of bicarbonate and protons, which decreases pH at the tissue and promotes the dissociation of O_2 to the tissue. In the lungs, where PO_2 is high, binding of O_2 causes Hb to release H^+ which combines with bicarbonate to release CO_2 via exhalation. Since these two reactions are closely matched, homeostasis of the blood pH is maintained at about 7.35.

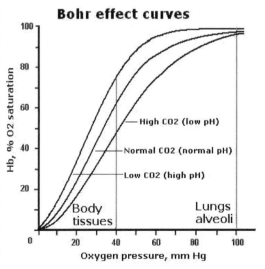

Bohr effect curves

16. D is correct. The increased concentration of carbon dioxide in blood are detected by respiratory centers in pons and medulla, which signal the body to breathe more deeply and frequently. Conversely, decreased levels of carbon dioxide lead to a decrease in frequency and depth of breathing.

17. A is correct.

18. C is correct. The rate of alveolar ventilation is directly proportional to the partial pressure of O_2 in the blood. The greater the ventilation rate, the greater is the amount of O_2 available in the alveoli to diffuse across the respiratory membrane and enter into the pulmonary capillaries. Alveolar ventilation increases during exercise (e.g. deeper and more rapid breathing) because metabolism is accelerated and there is a greater need for O_2.

From the graph, alveolar ventilation increases proportionally with increased O_2 consumption. The arterial partial pressure of O_2 remains (more or less) steady because the increased supply of O_2 is offset by the increased demand during exercise.

A: P_{O_2} does not decrease because of increased alveolar ventilation.

B: true only if alveolar ventilation did not increase with metabolism.

D: ventilation increases as metabolism increases.

19. D is correct. If ductus arteriosus fails to close, less O_2 is in the blood of the systemic circulation because some O_2 in blood is shunted from the aorta to the lower pressure pulmonary arteries. The pulmonary circulation carries blood that is more oxygenated than normal because highly oxygenated blood from the aorta is mixing with deoxygenated blood in the pulmonary circulation. The entire heart pumps more forcefully to compensate for the reduced O_2 delivery by bringing more blood to the tissues.

20. C is correct. In respiration, inhalation stretches the lungs, and *elastic recoil* refers to the ease with which the lungs rebound. During inhalation, the interpleural pressure (i.e. pressure within the pleural cavity) decreases. The diaphragm relaxes during expiration, the lungs recoil and the interpleural pressure restores. Lung compliance (i.e. measure of the lung's ability to stretch and expand) is inversely related to elastic recoil.

Elastic recoil of the lungs occurs because of 1) the elastic fibers in the connective tissue of the lungs and 2) the surface tension of the fluid that lines the alveoli. Due to cohesion, water molecules bond together (via hydrogen bonding), which pulls on the alveolar walls and causes the alveoli to recoil and become smaller.

The two factors that prevent the lungs from collapsing are 1) surfactant (i.e. surface-active lipoprotein complex formed by type II alveolar cells) and 2) the interpleural pressure.

21. B is correct. Chemoreceptors, located in the medulla oblongata, on the carotid arteries and on the aorta, detect and signal the respiratory centers in the medulla to modify the breathing rate when the partial pressure of the respiratory gases is too low or high. The chemoreceptors are most responsive to changes in the $[CO_2]$ and $[H^+]$ while only extreme changes in $[O_2]$ are relayed to the medulla oblongata. The $[H^+]$ of the blood is directly proportional to the partial pressure of $[CO_2]$. In erythrocytes, CO_2 combines with H_2O to form H_2CO_3, which then dissociates into HCO_3^- and H^+. An increase in $[H^+]$ decreases the pH of blood. When chemoreceptors detect an increase in the partial pressure of CO_2 or an increase in the $[H^+]$, the rate of breathing increases.

C: high partial pressure of O_2 in blood triggers a decrease, not an increase, in the rate of breathing.

22. A is correct. Lung compliance measures the lung's ability to stretch and expand. Surface tension of alveolar fluid is the force created by films of molecules that can reduce surface area, thus affecting lung compliance.

23. D is correct. Bronchodilation is the expansion of bronchial air passages and is caused by sympathetic stimulation, which increases muscle blood flow.

24. E is correct. *Tidal volume* is the amount of air that enters the lungs during normal inhalation at rest; the same amount leaves the lungs during exhalation. The average tidal volume is 500ml.

Inspiratory reserve volume is the amount of air inhaled above tidal volume during a deep breath, while *expiratory reserve rolume* is the amount of air exhaled above tidal volume during a forceful breath out.

Residual volume is the amount of air remaining in the lungs after a maximal exhalation; there is always some air left to prevent the lungs from collapsing.

Vital capacity is the most air that can be exhaled after taking the deepest possible breath; this amount can be up to ten times more than one would normally exhale.

Total lung capacity is the vital lung capacity plus the residual volume and equals the total amount of air the lungs can hold. The average total lung capacity is 6000ml, however this value varies with age, height, gender and health status.

25. B is correct. The exhalation of CO_2 is increased because of the low pH. The body removes excess plasma H^+ to raise the pH. Therefore, the reaction is shifted to the left to produce CO_2 that is expired by the lungs.

26. C is correct. Respiration performs gas exchange which relies on the laws of partial pressure. Partial pressure is the respective pressure of each gas in two regions that are connected. Gases tend to equalize their pressure, which affects the direction of movement. A gas moves from a higher partial pressure area to a lower partial pressure area. The greater the difference in pressure, the more rapidly the gases move.

27. D is correct. Fetal circulation differs from adult circulation whereby in fetal circulation, blood is oxygenated by the placenta because fetal lungs are nonfunctional prior to birth. Additionally, the umbilical vein carries oxygenated blood from the placenta to the fetus and fetal circulation contains three shunts that divert blood away from the fetal liver and lungs.

The three fetal shunts are: 1) ductus venosus diverts blood from the fetal liver before converging with the inferior vena cava that returns deoxygenated blood to the right atrium. Since the oxygenated blood from the umbilical vein mixes with the deoxygenated blood of the venae cavae, the blood entering the fetal right atrium is only partially oxygenated;

2) foramen ovale diverts blood away from the right ventricle and pulmonary artery as most blood bypasses the pulmonary circulation and enters the left atrium directly from the right atrium. The remaining blood in the right atrium empties into the right ventricle and is pumped to the lungs via the pulmonary artery;

3) ductus arteriosus reduces blood to the lungs by diverting it from the pulmonary artery to the aorta.

In the fetus, the pulmonary arteries carry partially oxygenated blood to the lungs where it is further deoxygenated because blood unloads its oxygen to the fetal lungs. Gas exchange occurs in the placenta (not in the fetal lungs). The deoxygenated blood then returns to the left atrium via pulmonary veins. Even though blood mixes with the partially oxygenated blood that crossed over from the right atrium (via the foramen ovale) before being pumped into the systemic circulation by the left ventricle, the blood delivered (via the aorta) has an even lower partial pressure of O_2 than the blood that was delivered to the lungs. Deoxygenated blood is returned to the placenta via the umbilical arteries.

Obstruction of the ductus arteriosus causes an increase in blood supply to the fetal lungs because all blood pumped into the pulmonary arteries by the right ventricle flows through the lungs.

A: ductus venosus obstruction greatly increases the blood supply to the fetal liver because oxygenated blood from the umbilical vein would pass through the liver before traveling to the heart.

B: pulmonary artery obstruction would result in a decrease in the blood supply to the fetal lungs.

C: aorta obstruction would greatly diminish the blood supply to the tissue, but would have no effect on the volume of blood delivered to the fetal lungs.

28. C is correct. Hypoxia is a pathological condition whereby tissue is deprived of an adequate oxygen supply.

29. E is correct. The lower respiratory tract begins at the trachea and extends into the primary bronchi and the lungs.

30. C is correct. Sensory organs of the brain, as well as those in the aorta and carotid arteries, monitor oxygen and carbon dioxide levels in the blood. To maintain the body's acid-base balance, carbon dioxide must be eliminated from the blood continuously. Chemoreceptors located near the respiratory center in the brain control acid-base balance by sensing the changes in the pH of cerobrospinal fluid. Normally, an increased concentration of CO_2 is the strongest stimulus to breathe more deeply and more frequently to remove excess CO_2.

In cases of chronic hypoventilation (e.g., patients with chronic obstructive pulmonary disease), when chemoreceptors lose their sensitivity and inadequately respond to increases in CO_2, peripheral chemoreceptors attempt to regulate respiratory function to restore acid-base balance. These peripheral chemoreceptors sensitive the concentration of oxygen in peripheral blood. Therefore, the stimulus to breathe is now low oxygen levels, rather than an increase in carbon dioxide levels. If the oxygen level is significantly increased by administering supplemental oxygen, the peripheral chemoreceptors will not stimulate breathing, resulting in apnea. This is the reason why supplemental oxygen must be given at very low levels to patients with chronic obstructive pulmonary disease.

Elevated blood pressure does not encourage breathing or cause any external symptoms.

31. C is correct. The esophagus is part of the digestive system and is not a part of the respiratory tract. The respiratory tract begins with the trachea, which divides into the two main bronchi. The main bronchi further subdivide into branching bronchioles, which lead to alveoli (site of gas exchange in the lungs).

32. D is correct. Respiratory centers are found in the medulla oblongata and pons, which are structures of the brain stem. The respiratory centers have receptors that receive neural, chemical and hormonal signals to control the depth and rate of breathing via movements of the diaphragm and other respiratory muscles.

33. E is correct. The exhalation process begins in the alveoli of the lungs, then air moves through the bronchioles to the bronchi, then passes through the trachea towards the larynx and finally the pharynx.

34. A is correct.

CO_2 is carried in blood in three ways. The percent concentration varies depending on whether it is venous or arterial blood.

70-80% of CO_2 is converted to bicarbonate ions HCO_3^- by the carbonic anhydrase in the red blood cells via reaction $CO_2 + H_2O \rightarrow H_2CO_3 \rightarrow H^+ + HCO_3^-$

5-10% is dissolved in blood plasma.

5-10% is bound to hemoglobin as carbamino compounds.

Hemoglobin is the main oxygen-carrying molecule in the RCB and also carries carbon dioxide. Carbon dioxide however does not bind to the same site as oxygen, it combines with the N-terminal groups on the four chains of globin. However, due to allosteric effect on the hemoglobin, the binding of CO_2 lowers the amount of oxygen bound for a given partial pressure of oxygen. Deoxygenation of the blood increases its ability to carry carbon dioxide – this is known as the Haldane effect. Conversely, oxygenated blood has a reduced capacity for carbon dioxide and this affects blood's ability to transport carbon dioxide from the tissues to the lungs. A rise in the partial pressure of CO_2 causes offloading of oxygen from hemoglobin (Bohr Effect).

35. B is correct. Nasal conchae (concha *sing.*) are long, spongy curled shelves of bone that protrude into the breathing passage of the nose.

36. C is correct. Diffusion is how O_2 and CO_2 are exchanged in the lungs and through all cell membranes.

37. B is correct. The capillaries are the most extensive capillary network of blood vessels, having the greatest cross sectional area and resistance to blood flow. Blood pressure is highest in the aorta and drops until the blood returns to the heart via the inferior and superior venae cavae.

38. C is correct.

39. E is correct. I: gas exchange is a passive process whereby gases diffuse down their partial pressure gradients.

II: inhalation is an active process requiring contraction of the diaphragm and the external intercostal muscles.

III: exhalation is a passive process from the elastic recoil of the lungs and relaxation of both the diaphragm and the external intercostal muscles. However, during vigorous exercise, active muscle contraction assists in exhalation.

40. A is correct.

41. C is correct. The carina is a ridge of cartilage in the lower part of the trachea that divides the two main bronchi.

42. B is correct.

43. E is correct. Hypovolemic shock is a state of decreased volume of blood plasma from such conditions as hemorrhaging and dehydration. The loss of blood plasma would be more rapid during arterial bleeding (i.e. high hydrostatic pressure) than during venous bleeding (i.e. low hydrostatic pressure).

44. A is correct.

45. A is correct. When the pressure of air within the lungs is less than the atmospheric pressure, air rushes into the lungs.

B: thoracic cavity enlargement causes the pressure of air within the lungs to fall.

C: low pressure inside the thoracic cavity is due to the expansion of the thoracic volume when the diaphragm contracts.

D: when the pressure drops, air rushes in and the ciliated membranes warm, moisten, and filter the inspired air. Air then travels through the bronchi, into the bronchioles and then into the alveoli where diffusion occurs to oxygenate the blood and release CO_2.

46. D is correct. Mucus in the nose traps dust and microbes, which are then carried away by cilia (tiny hairs that line the inside of nasal passages).

47. E is correct. The human pharynx is divided into three sections: nasopharynx, oropharynx and laryngopharynx. The pharynx is part of both the digestive and the respiratory systems, and also plays an important role in vocalization.

A: laryngopharynx (hypopharynx) is the part of the throat that connects to the esophagus. It lies inferior to (below) the epiglottis and extends to the location where pharynx diverges into the respiratory (larynx) and digestive (esophagus) pathways. Like the oropharynx above it, the laryngopharynx serves as a passageway for food and air.

B and D: oropharynx (mesopharynx) lies behind the oral cavity, extending from the uvula to the level of the hyoid bone. Both food and air pass through the oropharynx, and a flap of connective tissue (i.e. the epiglottis) closes over the glottis when food is swallowed to prevent aspiration.

C: nasopharynx (epipharynx) is the higher portion of the pharynx that extends from the base of the skull to the upper surface of the soft palate and lies above the oral cavity.

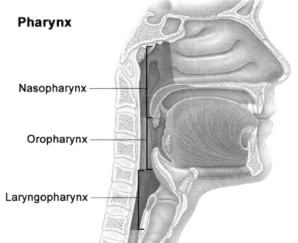

The pharyngeal tonsils (adenoids) are lymphoid tissue structures located in the posterior wall of the nasopharynx.

48. B is correct.

49. A is correct. Oxygenated blood returns from the lungs (i.e. pulmonary circulation) and feeds into the left atrium of the heart. From the left atrium, the oxygenated blood flows into the left ventricle that pumps it through the systemic circulation. The right atrium receives deoxygenated blood via the inferior and superior vena cava. The blood is then pumped to the right ventricle on its journey to the lungs.

50. B is correct.

51. D is correct. During normal inhalation, air is warmed and humidified by the extensive surface of the nasal passageway and particles are filtered from the air by nasal mechanisms.

When a patient undergoes a tracheotomy, the air entering the respiratory system bypasses the nasal cavities and is not warmed or moistened.

A: air does bypass the larynx but this does not account for the effects observed.

B: filtering the air as it enters the lungs is a separate process from making it warm and moist.

C: air does bypass the mouth and tongue, but this does not account for the effects observed.

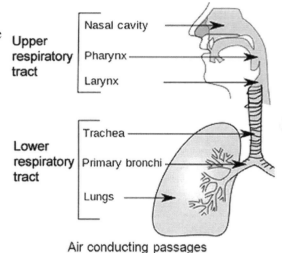

Air conducting passages

52. B is correct.

53. E is correct. The upper respiratory tract consists of the passage from the nose into the larynx and trachea.

54. A is correct.

55. D is correct. Respiratory bronchioles are the smallest bronchioles that connect terminal bronchioles to alveolar ducts. Ciliated epithelium lines the passageway from the terminal bronchiole to the alveoli.

56. E is correct. Voluntary cortical control allows breathing to be controlled consciously and unconsciously. It allows the rate and depth of breathing to be influenced by the cerebral cortex.

57. C is correct. The epiglottis is an elastic flap at the entrance of the larynx that is open during breathing and closed during swallowing directing food to go down the esophagus.

Chapter 4.11: Skin System

1. B is correct. Melanin is a natural pigment found in most organisms. Melanogenesis is the production of melanin in the skin after exposure to UV radiation, causing the skin to appear tan. Melanin is an effective absorber of light, whereby the pigment dissipates over 99.9% of absorbed UV radiation and protects skin cells from UVB radiation damage, thereby reducing the risk of cancer.

2. A is correct. The subcutaneous tissue (also known as hypodermis, subcutis or superficial fascia) is the lowermost layer of the integumentary system. However, it is not part of the skin, and it lies below the dermis. Its purpose is to attach the skin to underlying bone and muscle as well as to supply it with blood vessels and nerves. Hypodermis consists of loose connective tissue and elastin.

3. C is correct.

4. E is correct. A: nociception (i.e. pain receptors) detect mechanical, thermal or chemical changes above a set threshold.

5. B is correct. Apocrine glands, which begin to function at puberty under hormonal influence, are not important for thermoregulation. The odor from sweat is due to bacterial activity on the oily compound, which acts as a pheromone from the apocrine sweat glands (i.e. axillary, parts of the external genitalia, etc).The armpits are referred to as axillary.

6. A is correct.

7. E is correct.

8. C is correct.

9. B is correct. Sebaceous glands are glands in the skin that secrete an oily/waxy matter (sebum) to lubricate and waterproof the skin and hair. They are found in greatest abundance on the face and scalp, though they are distributed throughout the skin (except for the palms and soles). In the eyelids, meibomian sebaceous glands secrete a special type of sebum into tears.

10. D is correct.

11. D is correct.

12. A is correct. Sweat glands (also known as sudoriferous glands) are small tubular structures of the skin that produce sweat. There are two main types of sweat glands:

Eccrine sweat glands are distributed almost all over the body, though their density varies from region to region. Humans utilize eccrine sweat glands as a primary form of cooling.

Apocrine sweat glands are larger, have a different secretion mechanism, and are mostly limited to the axilla (armpits) and perianal areas in humans. Apocrine glands contribute little to cooling in humans but are the only effective sweat glands in hoofed animals such as cattle, camels, horses and donkeys.

The term eccrine is specifically used to designate merocrine secretions from sweat glands (eccrine sweat glands). Merocrine is used to classify exocrine glands and their secretions for histology. A cell is classified as merocrine if the secretions of that cell are excreted via exocytosis from secretory cells into an epithelial-walled duct or ducts and then onto a bodily surface (or into the lumen). Merocrine is the most common manner of secretion. The gland releases its product and no part of the gland is lost or damaged (compare holocrine and apocrine).

B: apocrine cells release their secretions via exocytosis of the plasma membrane. Apocrine secretion is less damaging to the gland than holocrine secretion (which destroys a cell) but more damaging than merocrine secretion (exocytosis). Mammary glands are apocrine glands responsible for secreting breast milk.

C: see explanation for question **47**.

D: see explanation for question **9**.

E: see explanation for question **23**.

13. D is correct.

14. C is correct.

Burn Type	Layers affected	Sensation	Healing time	Prognosis
Superficial (1°)	Epidermis	Painful	5–10 days	Heals well; Repeated sunburns increase the risk of skin cancer later in life
Superficial partial thickness (2°)	Extends into superficial (papillary) dermis	Very painful	less than 2–3 weeks	Local infection/cellulitis but no scarring typically
Deep partial thickness (2°)	Extends into deep (reticular) dermis	Pressure and discomfort	3–8 weeks	Scarring, contractures (may require excision and skin grafting)

Full thickness (3°)	Extends through entire dermis	Painless	Prolonged (months) and incomplete	Scarring, contractures, amputation (early excision recommended)
Fourth degree (4°)	Extends - entire skin, and into underlying fat, muscle and bone	Painless	Requires excision	Amputation, significant functional impairment and, in some cases, death

15. C is correct. The epidermis contains no blood vessels, and cells in the deepest layers are nourished by diffusion from blood capillaries that extend to the upper layers of the dermis.

16. B is correct.

17. E is correct. The dermis is a layer of skin between the epidermis (i.e. cutis) and subcutaneous tissues that consists of connective tissue and cushions the body from stress and strain. It is divided into two layers, the superficial area adjacent to the epidermis (i.e. papillary region) and a deep thicker area known as the reticular dermis. The dermis is tightly connected to the epidermis through a basement membrane.

Structural components of the dermis are collagen, elastic fibers, and extrafibrillar matrix. The dermis also contains mechanoreceptors that provide the sense of touch and heat, sweat glands, hair follicles, sebaceous glands, apocrine glands, blood vessels and lymphatic vessels. The blood vessels provide nourishment and waste removal for both dermal and epidermal cells.

18. A is correct.

19. E is correct.

20. A is correct.

21. D is correct.

22. A is correct. See explanation for question **53**.

23. E is correct. Holocrine secretions are produced in the cytoplasm of the cell and released by the rupture of the plasma membrane, which destroys the cell and results in the secretion of the product into the lumen. The sebaceous gland is an example of a holocrine gland because its product of secretion (sebum) is released with remnants of dead cells.

24. D is correct. Melanin refers to a group of natural pigments found in most organisms. Melanin is a derivative of the amino acid tyrosine, but is not itself made of amino acids and is not a protein. The pigment is produced in a specialized group of cells known as melanocytes.

25. D is correct.

26. A is correct.

27. B is correct.

28. C is correct.

29. E is correct.

30. A is correct.

31. B is correct. The lunula (i.e. lunulae) is the crescent-shaped whitish area of the bed of a fingernail or toenail and is the visible part of the root of the nail.

32. E is correct.

33. A is correct.

34. C is correct.

35. B is correct.

36. A is correct.

37. B is correct.

38. E is correct. See explanation for question **53**.

39. D is correct. See explanation for question **53**.

40. B is correct.

41. C is correct. A mechanoreceptor is a sensory receptor that responds to mechanical pressure (or distortion). The four main types are: Pacinian corpuscles, Meissner's corpuscles, Merkel's discs, and Ruffini endings.

Meissner's (i.e. tactile) corpuscles are mechanoreceptors and a type of nerve endings in the skin responsible for sensitivity to light touch. Merkel discs are mechanoreceptors found in the skin and mucosa (i.e. lining of endodermal organs) that detect pressure and texture and send the information to the brain.

A: Ruffinian endings are located in the deep layers of the skin, and register mechanical deformation within joints.

B: bulboid corpuscles (end-bulbs of Krause) are thermoreceptors that sense cold temperatures.

D: Pacinian corpuscles are nerve endings in the skin that sense vibrations and pressure.

42. D is correct.

43. E is correct.

44. D is correct.

45. C is correct.

46. D is correct. See explanation for question **9**.

47. A is correct. Ceruminous glands are specialized sudoriferous (sweat) glands located subcutaneously in the external auditory canal. They produce cerumen (earwax) by mixing their secretion with sebum and dead epidermal cells. Cerumen waterproofs the canal, kills bacteria, keeps the eardrum pliable, lubricates and cleans the external auditory canal and traps foreign particles (e.g. dust, fungal spores) by coating the guard hairs of the ear, making them sticky.

48. E is correct.

49. B is correct. Adipose tissue serves as an effective shock absorber.

C: basement membrane anchors the epithelium to its loose connective tissue (i.e. dermis) underneath by using cell-matrix adhesions via substrate adhesion molecules (SAMs).

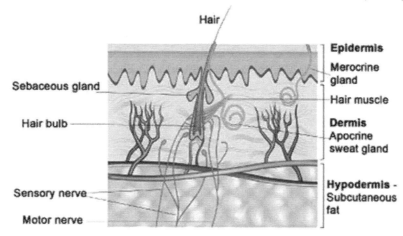

50. C is correct. Non-specific immunity refers to the basic resistance to disease. There are four types of barriers: 1) anatomic (e.g. skin and tears); 2) physiologic (e.g. temperature and pH); 3) phagocytic (e.g. macrophages and neutrophils) and 4) inflammatory (e.g. vasodilation). T cells are part of the specific immune response and recognize virally infected cells and stimulate the rest of the immune system.

51. D is correct. Keratinocytes accumulate melanin on their surface to form a UV-blocking pigment layer.

52. E is correct. A physiological mechanism used to dissipate excess heat is dilating arterioles leading to the skin (i.e. vasodilation). Vasodilation allows more blood to move closer to the surface of the skin where excess heat radiates away from the body.

53. A is correct. The epidermis consists of five layers of cells, whereby each layer has a distinct role in the health, well-being and functioning of the skin.

Stratum basale is the deepest layer of the five layers of the epidermis and is a continuous layer of cells. It is often described as one cell thick (though it may be two to three cells thick in hairless skin and hyperproliferative epidermis). Stratum basale is primarily made up of basal keratinocyte cells that divide to form the keratinocytes of the stratum spinosum, which migrate superficially. Other types of cells found within the stratum basale are melanocytes (pigment-producing cells), Langerhans cells (immune cells), and Merkel cells (touch receptors).

B: the stratum spinosum is where keratinization begins. This layer is composed of polyhedral keratinocytes active in synthesizing fibrilar proteins (cytokeratin), which build up within the cells by aggregating together to form tonofibrils. The tonofibrils form the desmosomes allowing strong connections to form between adjacent keratinocytes.

C: the stratum granulosum is a thin layer of cells where Keratinocytes that migrated from the underlying stratum spinosum, become granular cells containing keratohyalin granules. These granules are filled with proteins and promote hydration and crosslinking of keratin. At the transition between this layer and the stratum corneum, cells secrete lamellar bodies (containing lipids and proteins) into the extracellular space. This results in the formation of the hydrophobic lipid envelope responsible for the skin's barrier properties. Simultaneously, cells lose their nuclei and organelles, causing the granular cells to become non-viable corneocytes in the stratum corneum.

D: the stratum corneum is the outermost layer of the epidermis, consisting of dead cells (corneocytes). Its purpose is to form a barrier to protect underlying tissue from infection, dehydration, chemicals and mechanical stress.

E: the stratum lucidum is a thin, clear layer of dead skin cells in the epidermis named for its translucent appearance. It is composed of three to five layers of dead, flattened keratinocytes, which do not feature distinct boundaries and are filled with an intermediate form of keratin (eleidin).

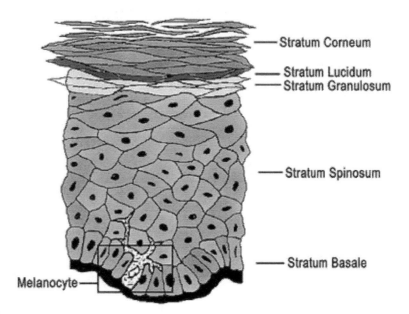

- Stratum Corneum
- Stratum Lucidum
- Stratum Granulosum
- Stratum Spinosum
- Stratum Basale

Melanocyte

54. D is correct.

55. C is correct.

56. E is correct. Eccrine glands (i.e. merocrine glands) are the major sweat glands found in almost all skin. They produce a clear, odorless substance, consisting primarily of water and NaCl.

Chapter 4.12: Reproductive System

1. D is correct. The menstrual cycle is regulated by several glands: the hypothalamus, anterior pituitary and ovaries. FSH is released by the anterior pituitary, which stimulates the maturation of the ova. Progesterone, synthesized by the ovary and the corpus luteum, regulates the development and shedding of the endometrial lining of the uterus.

The adrenal medulla synthesizes epinephrine (adrenaline) and norepinephrine, which assist the sympathetic nervous system in stimulating the *fight or flight* response, but are not involved in the menstrual cycle.

2. B is correct. Gametes are formed via meiosis and are haploid. Therefore, a single chromosome has only one strand because the sister chromatids were separated during anaphase.

3. A is correct. Oogenesis produces only one viable egg and (up to) three polar bodies that result from unequal distribution of the cytoplasm during meiosis. Gametes (e.g. egg and sperm) become haploid (1N) through reductive division (i.e. meiosis), in which a diploid cell (2N) gives rise to four haploid sperm or one haploid egg and (up to) three polar bodies.

B: interstitial cells are stimulated by LH to produce testosterone which, along with FSH, then stimulates the development of sperm within the seminiferous tubules.

C: FSH stimulates the eggs to develop in follicles within the ovaries.

D: FSH is involved in gamete production for both males and females.

4. C is correct. Let C designate wild type and c designate the color bind allele.

Mother is Cc and father is CY (normal allele with single copy of X gene). From mating, mother's gamete (as a carrier due to her dad) can be either C or c with 50% probability of the gamete inheriting either the C or c allele.

Assuming the child is a boy (i.e. father transmits Y and not X), so the father's allele of Y (boy) is 100% and the probability of the mother passing a c (colorblind) is 50%. Probability of the son being color blind is 50% or ½.

If the question had asked "what is the probability that they will have a color blind child?", the analysis changes to determine the probability among all children (not just among boys in the original question). Now, the same gametes are produced by the mother: C and c with 50% probability each. The only affected child is a boy. What is the probability that the father passes the X or Y gene to the offspring? Here, the probability is 50%.

The probabilities are multiplied to determine the overall probability: ½ × ½ = ¼

5. E is correct. Primary oocytes are arrested in meiotic prophase I within the ovaries from birth until ovulation.

6. C is correct. Seminiferous tubules are located in the testes and are the site of sperm production. Spermatozoa are the mature male gametes in many sexually reproducing organisms.

Spermatogenesis is the process by which spermatozoa are produced from male primordial germ cells through mitosis and meiosis. The initial cells in this pathway are called spermatogonia, which yield 1° spermatocytes by mitosis. The 1° spermatocyte divides via meiosis into two 2° spermatocytes.

Each 2° spermatocyte then completes meiosis as it divides into two spermatids. These develop into mature spermatozoa (known as sperm cells). Thus, the 1° spermatocyte gives rise to two 2° spermatocytes. The two 2° spermatocytes, by further meiosis, produce four spermatozoa. Meiosis converts 1° spermatocyte (2N) into four spermatids (1N).

Seminiferous tubules are located in the testes, and are the specific location of meiosis and the subsequent creation of gametes (e.g. spermatozoa for males or ova for females). The epithelium of the tubule consists of Sertoli cells whose main function is to nourish the developing sperm cells through the stages of spermatogenesis. Sertoli cells also act as phagocytes, consuming the residual cytoplasm during spermatogenesis. In between the Sertoli cells are spermatogenic cells, which differentiate through meiosis to sperm cells.

7. C is correct.

8. E is correct. FSH and LH are secreted by the anterior pituitary and influence the maturation of the follicle. FSH stimulates the production of estrogen that aids in the maturation of the primary follicle. LH stimulates ovulation and development of the corpus luteum; the mature corpus luteum secretes progesterone that causes the uterine lining to thicken and becomes more vascular in preparation for implantation of the fertilized egg (i.e. zygote).

9. A is correct. Sperm synthesis begins within the seminiferous tubules of the testes. Sperm undergo maturation and storage in the epididymis and, during ejaculation, are released from the vas deferens. Seminal vesicles are a pair of male accessory glands that produce about 50-60% of the liquid of the semen.

Mnemonic: SEVEN UP — the path of sperm during ejaculation:

Seminiferous tubules > **E**pididymis > **V**as deferens > **E**jaculatory duct > **N**othing > **U**rethra > **P**enis

10. D is correct. Genetic recombination is when two DNA strand molecules exchange genetic information (i.e. base composition within the nucleotides), which results in new combinations of alleles (i.e. alternative forms of genes). In eukaryotes, the natural process of genetic recombination during meiosis (i.e. formation of gametes – eggs and sperm) results in genetic information passed to progeny. Genetic recombination in eukaryotes involves the pairing of homologous chromosomes (i.e. a set of maternal and paternal chromosomes) which may involve nucleotide exchange between the chromosomes. The information exchange may occur without physical exchange (a section of genetic material is copied – duplicated – without a change in the donating chromosome) or by the breaking and rejoining of the DNA strands (i.e. forming new molecules of DNA).

Mitosis may involve recombination where two sister chromosomes are formed after DNA replication. In general, new combinations of alleles are not produced because the sister chromosomes are usually identical.

For both meiosis and mitosis, recombination occurs between similar DNA molecules (homologous chromosomes or sister chromatids, respectively). In meiosis, non-sister (i.e. same parent) homologous chromosomes pair with each other and recombination often occurs between non-sister homologues. For both, somatic cells (i.e. undergo mitosis) and gametes (i.e. undergo meiosis), recombination between homologous chromosomes or sister chromatids is a common DNA repair mechanism.

11. C is correct. Primary oocytes (2N cells) are arrested in prophase I of meiosis I until ovulation.

12. C is correct. The hypothalamus secretes releasing factors (i.e. tropic hormones) that cause the anterior pituitary to release LH and FSH. FSH stimulates the ovary to produce mature ovarian follicles. During the follicular stage, the ovary also produces estrogen. As estrogen is released, FSH levels drop and LH levels increase, which trigger the follicle to release the ovum (i.e. ovulation). LH continues to affect the corpus luteum (i.e. formerly the follicle) that secretes progesterone.

Estrogen is first secreted by the ovary (under the influence of FSH) during the follicular stage (day 1 to 14) that decreases FSH secretions. The decrease in FSH, along with an increase in LH, causes the rupture of the follicle (i.e. ovulation) and formation of the corpus luteum. As the corpus luteum matures, progesterone levels increase.

13. B is correct. The pseudoautosomal regions get their name because any genes located within them are inherited just like any autosomal genes. The function of these pseudoautosomal regions is that they allow the X and Y chromosomes to pair and properly segregate during meiosis in males.

Males have two copies of these genes: one in the pseudoautosomal region of their Y chromosome, the other in the corresponding portion of their X chromosome. Normal females also possess two copies of pseudoautosomal genes, as each of their two X chromosomes contains a pseudoautosomal region.

Crossing over (during prophase I) between the X and Y chromosomes is normally restricted to the pseudoautosomal regions. Thus, pseudoautosomal genes exhibit an autosomal, rather than sex-linked, pattern of inheritance. Females can inherit an allele originally present on the Y chromosome of their father, and males can inherit an allele originally present on the X chromosome of their father.

14. E is correct. A primary spermatocyte has completed the synthesis (S) phase of interphase, but has not completed the first meiotic division and is still diploid (2N) with 46 chromosomes (i.e. 23 pairs).

15. B is correct.

16. D is correct. At about day 14, LH surges, which causes the mature follicle to burst and release the ovum from the ovary (i.e. ovulation). Following ovulation, LH induces the ruptured follicle to develop into the corpus luteum, which then secretes progesterone and estrogen.

A: ovary secretes estrogen and progesterone, while the anterior pituitary secretes LH.

B: progesterone and estrogen inhibit GnRH release, which inhibits the release of FSH and LH and prevents additional follicles from maturing.

C: progesterone stimulates the development and maintenance of the endometrium in preparation for implantation of the embryo.

E: prolactin stimulates milk production after birth.

17. B is correct. Luteinizing hormone (LH) stimulates the release of testosterone in males by Leydig cells. Testosterone is necessary for the proper development of testes, the penis and seminal vesicles. Testosterone surges between the first and fourth month of life and testosterone inadequacy attributes as a primary cause for cryptorchidism.

During puberty, testosterone is necessary for secondary male sex characteristics (e.g. growth of body hair, broadening of the shoulders, enlarging of the larynx, deepening of the voice and increased secretions of oil and sweat glands)

A: cortisol is a stress hormone that elevates blood glucose levels.

E: FSH stimulates the maturation of the gametes (e.g. ova and spermatids).

18. C is correct. Leydig cells are located adjacent to the seminiferous tubules in the testicle. Leydig cells are androgens (i.e. hormones as 19-carbon steroids) that secrete testosterone, androstenedione and dehydroepiandrosterone (DHEA) when stimulated by luteinizing hormone (LH) released by the pituitary gland. LH increases the conversion of cholesterol to pregnenolone leading to testosterone synthesis and secretion by Leydig cells.

Prolactin (PRL) increases the response of Leydig cells to LH by increasing the number of LH receptors expressed on Leydig cells.

19. E is correct. In females, secondary oocytes are haploid (i.e. single copy of 23 chromosomes each with a pair of chromatids) and do not complete meiosis II (i.e. haploid with a single chromatid) until fertilized by a sperm.

Each month during puberty, one primary oocyte (i.e. diploid each with a pair of chromatids) completes meiosis I to produce a secondary oocyte (1N) and a polar body (1N). The 1N secondary oocyte (i.e. 23 chromosomes each with a pair of chromatids) is expelled as an ovum from the follicle during ovulation, but meiosis II does not occur until fertilization.

A: menarche is marked by the onset of menstruation and signals the possibility of fertility.

B: menstruation is the sloughing off of the endometrial lining of the uterus during the monthly hormonal cycle.

D: menopause is when menstruation ceases.

20. D is correct. Klinefelter syndrome describes the set of symptoms resulting from additional X genetic material in males.

A: Turner syndrome describes the condition in females resulting from a single X (monosomy X) chromosome.

B: XYY syndrome is a genetic condition in which a human male has an extra male (Y) chromosome, giving him a total of 47 chromosomes instead of 46. 47,XYY is not inherited but usually occurs during the formation of sperm cells. A nondisjunction error during anaphase II (of meiosis II) results in sperm cells with an extra copy of the Y chromosome. If one of these atypical sperm cells contributes to the genetic makeup of a child, the child will have an extra Y chromosome in each somatic cell.

C: Triple X syndrome is not inherited, but usually occurs during the formation of gametes (e.g. ovum and sperm) because of nondisjunction in cell division that results in reproductive cells with additional chromosomes. An error in cell division (non-disjunction) can result in gametes with additional chromosomes. An egg or a sperm may gain an extra X chromosome as a result of non-disjunction, and if one of these gametes contributes to the genetic makeup of the zygote, the child will have an extra X chromosome in each cell.

21. B is correct. Spermatozoon development produces motile, mature sperm, which fuses with an ovum to form the zygote. Eukaryotic cilia are structurally identical to eukaryotic flagella but distinctions are made based on function and length. Microtubules form the sperm's flagella, which is necessary for movement through the cervix, uterus and along the fallopian tubes (i.e. oviducts).

A: acrosomal enzymes digest the outer zona pellucida (a glycoprotein membrane surrounding the plasma membrane of an oocyte) to permit fusion of sperm and egg.

C: sperm's midpiece has many mitochondria that are used for ATP production for the sperm's movement (via the flagellum) through the female cervix, uterus and Fallopian tubes.

D: testosterone is necessary for spermatogenesis.

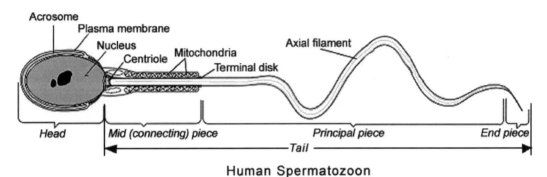

Human Spermatozoon

22. A is correct. Seminiferous tubules are located in the testes where meiosis occurs for the creation of gametes (e.g. spermatozoa). The epithelium of the tubule consists of Sertoli (i.e. *nurse cells*) cells that nourish the developing sperm and act as phagocytes by consuming the residual cytoplasm during spermatogenesis. In between the Sertoli cells are spermatogenic cells, which differentiate through meiosis to sperm cells.

23. C is correct. In males, spermatogonia are (2N) cells that undergo mitosis to produce (2N) cells called 1^0 spermatocytes, which undergo the first round of meiosis to yield $2°$ spermatocytes (1N). The 2^0 spermatocytes undergo another round of meiosis to produce four spermatids (1N) that mature into sperm (i.e. male gamete).

In females, the 1^0 oocyte is (2N) whereby the 2^0 oocyte undergoes the second meiotic division to produce two (1N) cells – a mature oocyte (i.e. ovum – female gamete) and another polar body.

A: primary oocyte is a diploid cell. During fertilization, a (1N) ovum and a (1N) sperm fuse to produce a (2N) zygote.

B: spermatogonium is a diploid (2N) cell.

24. D is correct. The probability that a child will be male (or female) is ½. Each event is independent; the fact that the first (or second or third) child is a particular gender does not affect future events. Therefore, the probability for 4 children is calculated as: ½ × ½ × ½ × ½ = 1/16. The same probability would be for the question of what is the probability that the first and third child is female with the second and fourth as male (or another combination).

25. C is correct. Chlamydia and other infections may scar the reproductive tract, which prevents the ova from reaching the uterus, resulting in infertility.

26. E is correct. The seminal vesicles secrete a significant proportion of the fluid that ultimately becomes semen. About 50-70% of the seminal fluid originates from the seminal vesicles, but is not expelled in the first ejaculate fractions, which are dominated by spermatozoa and zinc-rich prostatic fluid.

27. D is correct. Unequal division of the cytoplasm occurs during the meiotic process of oogenesis (i.e. production of an egg cell). Meiotic divisions occur in two stages. The first stage produces the precursor (2N) cell, whereby one of the daughter cells (precursor egg) receives most of the cytoplasm, while the other is devoid of sufficient cytoplasm (but genetically identical) and becomes a nonfunctioning polar body.

In the second division of oogenesis, the large (2N) daughter cell divides again, and again one of the (1N) daughter cells receives almost all of the cytoplasm (i.e. putative ovum) and the other becomes a small nonfunctional (1N) polar body. The polar body (from meiosis I) may also divide (via meiosis II) to form two nonfunctional (1N) polar bodies. The result is a potential of four (1N) cells, but only one—the one with a greater amount of cytoplasm during each meiotic division—becomes a functional egg (i.e. ovum) cell along with (up to) three (1N) polar bodies.

A: bacterial cells divide for reproduction, and cytoplasmic division is about equal.

B: during mitosis of kidney cells, the cytoplasm is distributed equally.

C: during spermatogenesis, one (2N) precursor cell forms four functional (1N) sperm cells because both meiosis divisions are equal and contain an equal amount of cytoplasm.

28. B is correct. Dynein is a motor protein (also molecular motor) that uses ATP to perform mechanical movement. Dynein transports various cellular contents by *walking* along cytoskeletal microtubules towards the minus-end of the microtubule, which is usually oriented towards the cell center (i.e. *minus-end directed motors.*) This is known as retrograde transport. In contrast, kinesins are motor proteins that move toward the plus end of the microtubules (i.e. *plus-end directed motors*).

Ovulation is the phase of the female menstrual cycle in which a partially mature ovum that has yet to complete meiosis II is released from the ovarian follicles into the Fallopian tube (i.e. oviduct). After ovulation, during the luteal phase, the egg can be fertilized by sperm. Ovulation is determined by levels of circulating hormones and would not be affected by a defect in the motor protein dynein.

A: lungs require cilia to remove bacteria and other particulate.

C: a male with Kartagener's syndrome would be infertile due to sperm immobility.

D: in females, ova would not enter the Fallopian tubes (i.e. oviduct) normally because of the lack of cilia, and this would cause an increased risk of ectopic pregnancy.

29. A is correct. The estrous cycle (i.e. sexual desire) comprises the recurring physiologic changes that are induced by reproductive hormones in most mammalian females. Estrous cycles start after sexual maturity in females and are interrupted by anestrous phases or pregnancies.

The menstrual cycle is the cycle of changes that occurs in the uterus and ovary for the purpose of sexual reproduction. The menstrual cycle is essential for the production of eggs and for the preparation of the uterus for pregnancy. In humans, the length of a menstrual cycle varies greatly among women (ranging from 21 to 35 days), with 28 days designated as the average length. Each cycle can be divided into three phases based on events in the ovary (ovarian cycle) or in the uterus (uterine cycle).

The ovarian cycle consists of the *follicular phase*, *ovulation*, and *luteal phase*, whereas the uterine cycle is divided into menstruation, proliferative phase, and secretory phase. Both cycles are controlled by the endocrine system, and the normal hormonal changes that occur can be interfered with using hormonal contraception to prevent reproduction.

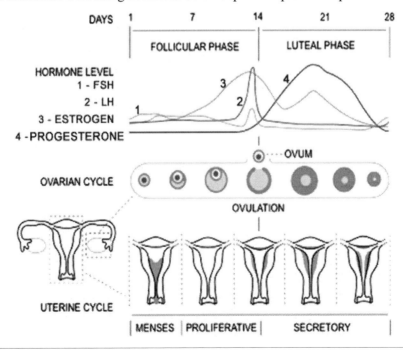

Copyright © 2018 Sterling Test Prep. Any duplication is illegal.

30. E is correct. DNA replication occurs during the synthesis (S) phase of interphase to form sister chromatids joined by the centromere.

31. C is correct. Progesterone is a steroid hormone involved in the female menstrual cycle, pregnancy (supports *gestation*) and embryogenesis. In women, progesterone levels are relatively low during the pre-ovulatory phase of the menstrual cycle, rise after ovulation, and are elevated during the luteal phase.

If pregnancy occurs, human chorionic gonadotropin is released, which maintains the corpus luteum and allows it to maintain levels of progesterone. At around 12 weeks, the placenta begins to produce progesterone in place of the corpus luteum – this process is called the luteal-placental shift. After the luteal-placental shift, progesterone levels start to rise further. After delivery of the placenta and during lactation, progesterone levels are very low.

Progesterone levels are relatively low in children and postmenopausal women. Adult males have levels similar to those in women during the follicular phase of the menstrual cycle.

32. A is correct. During meiosis, the gamete (e.g. ovum and sperm) reduces its genetic component from diploid (2N) to haploid (1N) with half the normal chromosome number for somatic (i.e. body cells). When a haploid egg and sperm unite, they form a diploid zygote. All ova contain an X chromosome, while sperm contain either an X or a Y chromosome. These gametes are formed during meiosis (i.e. two reduction divisions) but without an intervening chromosome replication – S phase. During prophase I of meiosis I, tetrads form and sister chromatids (i.e. chromosome that has replicated in a prior S phase) undergo homologous recombination known as crossing over. Crossing over increases genetic variance within the progeny and is a driving force in evolution of a species.

33. B is correct. The fertilized egg is the zygote that undergoes rapid cell divisions without significant growth to produce a cluster of cells of the same total size as the original zygote. The cells derived from cleavage are blastomeres and form a solid mass known as a morula. Cleavage ends with the formation of the blastula.

In different species, depending mostly on the amount of yolk in the egg, the cleavage can be holoblastic (i.e. total or entire cleavage) or meroblastic (i.e. partial cleavage). The pole of the egg with the highest concentration of yolk is referred to as the vegetal pole, while the opposite is referred to as the animal pole. Humans undergo holoblastic cleavage.

34. C is correct. Estrogen is a steroid hormone necessary for normal female maturation. Estrogen stimulates the development of the female reproductive tract and contributes to the development of secondary sexual characteristics and libido. Estrogen is also secreted by the follicle during the menstrual cycle and is responsible for the thickening of the endometrium in preparation for implantation of the fertilized egg.

A: luteinizing hormone (LH) induces the ruptured follicle to develop into the corpus luteum.

B: LH also stimulates testosterone synthesis in males.

D: FSH is released by the anterior pituitary and promotes the development of the follicle, which matures and then begins secreting estrogen.

E: insulin is a peptide hormone secreted in response to high blood glucose concentrations and it stimulates the uptake of glucose by muscle and adipose cells and also the conversion of glucose into the storage molecule of glycogen.

35. E is correct. X-linked dominance is a mode of inheritance whereby a dominant gene is carried on the X chromosome. X-linked dominant is less common than X-linked recessive. For X-linked dominant inheritance, only one copy of the allele is sufficient to cause the disorder when inherited from a parent who has the disorder. X-linked dominant traits do not necessarily affect males more than females (unlike X-linked recessive traits). An affected father will give rise to all affected daughters, but no affected sons will be affected (unless the mother is also affected).

36. A is correct. Increased estrogen secretions initiate the luteal surge that increases LH secretion → ovulation.

37. E is correct. A: X-linked recessive inheritance is a mode of inheritance whereby a mutation in a gene on the X chromosome causes the phenotype to be expressed (1) in males who are hemizygous with a single copy of the mutation because they have only one X chromosome and (2) in females who are homozygous for the gene mutation (i.e. a copy of the gene mutation on each of their two X chromosomes). X-linked inheritance indicates that the gene is located on the X chromosome. Females have two X, while males have one X and one Y. Carrier females have only one copy and do not usually express the phenotype.

38. C is correct. During prophase I, the chromatin condenses into chromosomes, the centrioles migrate to the poles, the spindle fibers begin to form, and the nucleoli and nuclear membrane disappear. Homologous chromosomes physically pair and intertwine in a process called synapsis. During prophase I, chromatids (i.e. a strand of the replicated chromosome) of homologous chromosomes break and exchange equivalent pieces of DNA via crossing over.

A: in anaphase I, homologous pairs of chromosomes separate and are pulled by the spindle fibers to opposite poles of the cell. Disjunction (i.e. separation of the homologous chromosomes) is important for segregation, as described by Mendel.

B: in telophase I, a nuclear membrane forms around each new nucleus and each chromosome still consist of sister chromatids joined at the centromere. The cell divides into two daughter cells, each receiving a haploid nucleus (1N) of chromosomes.

D: in metaphase I, homologous pairs align at the equatorial plane and each pair attaches to a separate spindle fiber by its kinetochore (i.e. protein collar around the centromere of the chromosome).

E: interkinesis is a short period between the two reduction cell divisions of meiosis I and II and during which the chromosomes partially uncoil.

39. B is correct. Sperm mature and are stored until ejaculation in the epididymis. During ejaculation, sperm flow from the lower portion of the epididymis. They have not been activated by products from the prostate gland and they are unable to swim. The sperm are transported via the peristaltic action of muscle layers within the vas deferens and form semen as they are mixed with the diluting fluids of the seminal vesicles and other accessory glands prior to ejaculation.

A: testosterone is secreted by the seminiferous tubules. The primary functions of the testes are to produce sperm (i.e. spermatogenesis) and to produce androgens (e.g. testosterone). Leydig cells are localized between seminiferous tubules and produce and secrete testosterone and other androgens (e.g. DHT, DHEA) important for sexual development and puberty, secondary sexual characteristics (e.g. facial hair, sexual behavior and libido), supporting spermatogenesis and erectile function.

C: luteinizing hormone (LH) results in testosterone release. The presence of both testosterone and follicle-stimulating hormone (FSH) is needed to support spermatogenesis.

40. D is correct.

41. A is correct. In prophase I, tetrads form, genetic recombination occurs and the spindle apparatus forms. Chromosomes migrate to the poles of the cell during anaphase from the splitting of the centromere.

42. E is correct. Spermatogenesis and oogenesis are both gametogenesis, because the haploid gametes (e.g. ova and sperm) are produced through reduction divisions (i.e. meiosis) of diploid cells. Spermatogenesis occurs in the gonads whereby the cytoplasm is equally divided during meiosis with the production of four viable sperm 1N cells.

Oogenesis also occurs in the gonads, whereby the cytoplasm is divided unequally and only one 1N ovum (e.g. egg), which received the bulk of the cytoplasm, is produced plus (up to) three additional 1N polar bodies. The polar bodies contain a 1N genome (like the sperm and egg) but lack sufficient cytoplasm to produce a viable gamete.

43. B is correct. The inner lining of the Fallopian tubes is covered with cilia that help the egg move towards the uterus where it implants if it is fertilized. Fertilization usually happens when a sperm fuses with an egg in the Fallopian tube.

C: cilia lining the respiratory tract perform this function.

D: Fallopian tubes are isolated from the external environment so pH fluctuations are not an issue.

44. B is correct. A Barr body is the inactive X chromosome in a female somatic cell that is rendered inactive in a process called lyonization for those species in which sex is determined by the presence of the Y chromosomes (i.e. humans and some other species). The Lyon hypothesis states that in cells with multiple X chromosomes, all but one X are inactivated during mammalian embryogenesis, and this randomly occurs early in embryonic development. In humans with more than one X chromosome, the number of Barr bodies visible at interphase is always one less than the total number of X chromosomes. For example, a man with Klinefelter syndrome of 47,XXY karyotype has a single Barr body, whereas a woman with a 47,XXX karyotype has two Barr bodies.

45. A is correct. Polar bodies are formed in females as 1N, nonfunctional cells during meiosis. Meiosis is a two-stage process whereby a 2N cell undergoes a reduction division to form two 1N cells. In the second meiotic division, each 1N cell undergoes a second division to form two 1N cells each for a total of four 1N cells from an original 2N germ cell.

For sperm cells, four functional, unique 1N gamete sperm cells are formed.

For egg cells, the first meiotic division involves an unequal division of cytoplasm that results in the formation of one large cell and one small cell (i.e. first polar body). The large cell divides again (second meiotic division) to generate one functional egg (1N ovum containing most of the cytoplasm) and another polar body. The first polar body divides again to form two other polar bodies. Therefore, during a female meiotic division, one large ovum (i.e. functional egg cell) and three polar bodies are formed.

Polar bodies are not formed in mitosis, which is how somatic (i.e. body cells) cell divisions occur; equal cell divisions do not form gametes (i.e. germ line cells).

46. B is correct. In males, testosterone is primarily synthesized in Leydig cells in the testis. The synthesis of testosterone is regulated by the hypothalamic–pituitary–testicular axis. When testosterone levels are low, hypothalamus releases gonadotropin-releasing hormone (GnRH), which stimulates the pituitary gland to release FSH and LH. FSH and LH stimulate the testis to produce testosterone. When levels are high, testosterone acts on the hypothalamus and pituitary to inhibit the release of GnRH and FSH/LH through a negative feedback loop.

47. E is correct. Turner syndrome (i.e. 45,X) describes several conditions in females, of which monosomy X (i.e. absence of an entire sex (X) chromosome, the Barr body) is most common. It is a chromosomal abnormality in which all or part of one of the sex chromosomes is absent or has other abnormalities.

48. B is correct. If fertilization of the ovum does not occur, corpus luteum stops secreting progesterone and degenerates.

A: menstruation phase does follow decreased progesterone secretion, but does not result from increased estrogen levels.

C: increased estrogen secretion (not LH) causes the luteal surge, which occurs earlier in the cycle.

D: thickening of the endometrial lining occurs while estrogen and progesterone levels are high.

49. A is correct.

Chapter 4.13: Development

1. A is correct. See explanations for question **21**.

2. D is correct. Indeterminate cleavage results in cells that maintain the ability to develop into a complete organism.

A: once the zygote is implanted into the uterus, cell migration transforms the blastula from a single cell layer into a three-layered (i.e. ectoderm, mesoderm and endoderm) gastrula.

B: blastulation begins when the morula develops the blastocoel as a fluid-filled cavity that (by the fourth day) becomes the blastula as a hollow sphere of cells.

C: determinate cleavage results in cells whose future differentiation potentials are determined at an early developmental stage.

3. B is correct. See explanations for questions **18**, **32** and **45**.

4. E is correct. The ectoderm gives rise to the hair, nails, skin, brain and nervous system.

A: connective tissue is derived from the mesoderm.

B: bone is derived from the mesoderm.

C: rib cartilage is derived from the mesoderm.

D: epithelium of the digestive system is derived from the endoderm.

5. D is correct. In humans and most mammals, the chorion is one of the membranes between the developing fetus and mother. The chorion consists of two layers: an outer layer formed by the trophoblast, and an inner layer formed by the somatic mesoderm (in contact with amnion).

6. A is correct.

7. C is correct. The mesoderm develops into the circulatory system, the musculoskeletal system, the excretory system, the outer coverings of internal organs, the gonads and various types of muscle tissue.

A: ectoderm develops into the brain and nervous system, hair and nails, the lens of the eye, the inner ear, sweat glands, the lining of the nose and mouth and the epidermis of the skin.

B: epidermis is not an embryonic germ layer but the layer of the skin that covers the dermis.

D: endoderm develops into the epithelial lining of the digestive tract, respiratory tracts and lining of the liver, the thyroid, bladder, pancreas and alveoli of the lungs.

8. B is correct. Gastrulation is a phase early in the embryonic development of animals, during which the single-layered blastula is reorganized into a trilaminar (*three-layered*) structure of the gastrula. These three germ layers are the *ectoderm, mesoderm* and *endoderm.*

Gastrulation occurs after cleavage, the formation of the blastula and primitive streak that is followed by organogenesis, when individual organs develop within the newly formed germ layers. Following gastrulation, cells in the body are either organized into sheets of connected cells (as in epithelia), or as a mesh of isolated cells (i.e. mesenchyme).

Each germ layer gives rise to specific tissues and organs in the developing embryo. The *ectoderm* gives rise to the epidermis and other tissues that will form the nervous system.

The *mesoderm* is found between the ectoderm and the endoderm and gives rise to somites that form muscle, cartilage of the ribs and vertebrae, dermis, notochord, blood and blood vessels, bone and connective tissue.

The *endoderm* gives rise to the epithelium of the respiratory system, digestive system and organs associated with the digestive system (e.g. liver and pancreas).

Gastrulation occurs when a blastula, made up of one layer, folds inward and enlarges to create the three primary germ layers (e.g. endoderm, mesoderm and ectoderm). Archenteron gives rise to the digestive tube.

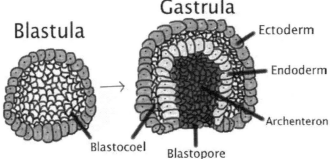

9. A is correct.

10. E is correct. The ectoderm germ layers give rise to the epidermis of the skin and also the nervous system.

The endoderm germ layer gives rise to the lining of the digestive system, its associated glands and organs (e.g. liver and pancreas) and the lungs.

The mesoderm gives rise to most of the other organs and systems of the body including the excretory system, the reproductive system, the muscular and skeletal systems and the circulatory system. If a particular tissue is not specifically endoderm or ectoderm, there is a high probability that it is one of the many mesoderm-derived tissues.

For example, all the incorrect choices above are tissues or structures of mesoderm origin.

11. C is correct. Fate mapping is a method of tracing the embryonic origin of various tissues in the adult organism by establishing the correspondence between individual cells (or groups of cells) at one stage of development, and their progeny at later stages of development. When carried out at single-cell resolution, this process is termed cell lineage tracing.

A: phylogenetic tree (i.e. evolutionary tree) is a branching diagram showing the inferred evolutionary relationships (phylogeny) among various biological species or other entities based upon similarities and differences in their physical and/or genetic characteristics. The taxa joined together in the tree are implied to have descended from a common ancestor.

B: pedigree diagram shows the occurrence and appearance (i.e. phenotypes) of a particular gene and its ancestors among generations. A pedigree results in the presentation of family information in the form of an easily readable chart. Pedigrees use a standardized set of symbols: squares represent males and circles represent females. Pedigree construction is a family history, and details about an earlier generation may be uncertain as memories fade. If the sex of the person is unknown, a diamond is used. Someone with the phenotype in question is represented by a filled-in (darker) symbol. Heterozygotes are indicated by a shade dot inside a symbol or a half-filled symbol.

D: linkage map is a genetic map of an experimental population that shows the position of its known genes or genetic markers relative to each other based on recombination frequency, rather than a specific physical distance along each chromosome. Genetic linkage is the tendency of genes that are located proximal to each other on a chromosome to be inherited together during meiosis. Genes whose loci (i.e. position) are closer to each other are less likely to be separated onto different chromatids during chromosomal crossover (i.e. during prophase I) and are therefore said to be genetically *linked*.

12. A is correct.

13. D is correct. Gene expression changes as development proceeds, and different proteins are encoded by expressed genes. During development, cells change in their ability to respond to signals from other tissues and also in their ability to induce changes in other cells. These changes during development are inherited by daughter cells.

A: microtubules are involved in mitosis but do not change during development.

C: somatic cells have the same genes (exceptions include gametes and B and T cells).

14. C is correct. See explanation for question **23**.

E: A pronucleus is the nucleus of a sperm or an egg cell after the sperm enters the ovum but before they fuse. The male and female pronuclei don't fuse but their genetic material do. Instead, their membranes dissolve, leaving no barriers between the male and female chromosomes.

15. E is correct. The three primary germ layers are formed during gastrulation of embryological development: *ectoderm*, *mesoderm* and *endoderm*.

Ectoderm is the external germ layer and gives rise to the skin, fingernails and the nervous system (including the eye).

Mesoderm is the middle germ layer and gives rise to most organ systems, including the musculoskeletal, cardiovascular, reproductive and excretory systems.

Endoderm is the innermost germ layer and gives rise to the gallbladder, liver, pancreas and epithelial lining of luminal structures and accessory digestive organs.

16. D is correct. In the developing fetus, the ductus arteriosus is a blood vessel connecting the pulmonary artery to the proximal descending aorta. It allows most of the blood from the right ventricle to bypass the fetus's fluid-filled non-functioning lungs. Upon closure at birth, it becomes the ligamentum arteriosum. Ligamentum teres (i.e. round ligament) refers to several structures: ligamentum teres uteri (i.e. round ligament of uterus), ligamentum teres hepatis (round ligament of liver) and ligamentum teres femoris (i.e. ligament of head of femur).

17. C is correct. During gastrulation, invagination of the blastula forms the mesoderm as the third primary germ layer.

18. E is correct. Homeotic genes encode the related homeodomain protein involved in developmental patterns and sequences. Homeodomain is a protein structural domain that binds DNA or RNA and is often found in transcription factors. The homeodomain fold consists of a 60-amino acid helix-turn-helix structure in which three alpha helices are connected by short loop regions. The N-terminus of the two helices is antiparallel and the longer C-terminal helix is roughly perpendicular to the axes established by the first two. It is the third helix that functions as a transcription factor and interacts directly with DNA.

19. D is correct. Neurulation and organogenesis follow gastrulation.

A: mitosis continues throughout development, but cleavage is a specific term reserved for the first few cell divisions when the zygote becomes the morula. During cleavage, no growth occurs and the morula is the same approximate size as the zygote.

B: blastula formation precedes gastrulation.

C: blastocoel is the fluid-filled central region of a blastula and forms early after fertilization when the zygote divides into many cells.

E: once a cell is differentiated, it does not reverse this process of differentiation unless it is a cancerous cell.

20. A is correct. Trophoblast cells give rise to the outer layer of a blastocyst, provide nutrients to the embryo and develop into a large part of the placenta. Trophoblasts are formed during the first stage of pregnancy and are the first cells to differentiate from the fertilized egg.

21. B is correct. The placenta connects the developing fetus to the uterine wall and allows nutrient uptake, waste elimination and gas exchange via the mother's blood supply.

The umbilical cord is a conduit between the developing embryo (or fetus) and the placenta. During prenatal development, the umbilical cord is physiologically and genetically part of the fetus and contains two arteries (the umbilical arteries) and one vein (the umbilical vein). The umbilical vein supplies the fetus with oxygenated, nutrient-rich blood from the placenta. Conversely, the fetal heart pumps deoxygenated, nutrient-depleted blood through the umbilical arteries back to the placenta.

22. E is correct. For vertebrates, induction is defined as the process by which a group of cells causes differentiation in another group of cells. For example, cells that form the notochord induce the formation of the neural tube. Other examples of induction in vertebrate development include the formation of the eye, where the optic vesicles induce the ectoderm to thicken and form the lens placode (i.e. thickened portion of the ectoderm that becomes the lens), which then induces the optic vesicle to form the optic cup, which then induces the lens placode to form the cornea.

A: the neural tube does develop into the nervous system but is not induced by another group of cells or tissue.

B: TSH (thyroxin stimulating hormone) does stimulate the secretion of thyroxine but is not induction.

C: ectoderm does develop into the nervous system but does not answer the question.

D: neurons synapse with other neurons via neurotransmitters (i.e. chemical messenger), but they don't induce changes in other tissues as does induction.

23. D is correct. The acrosome is at the tip of the sperm and contains specialized secretory molecules. The acrosomal reaction is due to a signaling cascade involving the glycoproteins on the egg's surface. Acrosin digests the zona pellucida and membrane of the oocyte. Part of the sperm's cell membrane then fuses with the egg cell's membrane, and the contents of the head enter the egg that is fused with the plasma membrane. This allows the sperm to release its degradation enzymes to penetrate the egg's tough coating and allow the sperm to bind to and fuse with the egg.

24. C is correct.

A: endoderm is the innermost germ layer and gives rise to the inner lining of the respiratory and digestive tracts, and associated organs.

B: blastula refers to the hollow ball of embryonic cells that arises from the morula. The blastula is not a germ layer.

D: ectoderm is the outermost germ layer and gives rise to the hair, nails, eyes, skin and central nervous system.

25. E is correct. The inner cell mass are cells inside the primordial embryo that give rise to the definitive structures of the fetus. Before implantation, the inner cell mass forms the endometrium of the uterus. The primitive streak is a structure that forms in the blastula during the early stages of embryonic development. The presence of the primitive streak establishes bilateral symmetry, determines the site of gastrulation and initiates germ layer formation.

26. D is correct. At birth, the fetal circulatory system changes as the newborn begins using its lungs. Resistance in the pulmonary blood vessels decreases, which causes an increase in blood flow to the lungs.

At birth, umbilical blood flow ceases and blood pressure in the inferior vena cava decreases, which causes a decrease in pressure in the right atrium. In contrast, the left atrial pressure increases due to increased blood flow from the lungs. Increased left atrial pressure, coupled with decreased right atrial pressure, causes the closure of the foramen ovale.

Additionally, the ductus arteriosus (i.e. connects pulmonary artery to aorta) constricts and subsequently is sealed closed. The ductus venosus, which shunts blood from the left umbilical vein directly to the inferior vena cava to allow oxygenated blood from the placenta to bypass the liver, completely closes within three months after birth. The infant begins to produce adult hemoglobin a few weeks before birth (2 α and 2 β chains; alpha and beta) though lower amounts of fetal hemoglobin continue to be produced until the production completely stops. After the first year, only low levels of fetal hemoglobin (2 α and 2 γ chains; alpha and gamma) are present in the blood.

27. B is correct. The homeobox is a stretch of DNA about 180 nucleotides long that encodes a homeodomain (i.e. protein) in both vertebrates and invertebrates. Only exons are retained during RNA processing of the primary transcript into mRNA that is used for translation into proteins.

28. A is correct. All somatic cells have the same genome, but different cells express different genes. This difference in expression can be spatial (i.e. cell type) or temporal (i.e. different stages of development: fetal *vs.* adult).

I: ectoderm tissue arises after gastrulation as one of the primary germ layers. Each of the three germ layers retains its determination.

II: somatic cells have identical genomes compared to their parents. Only gametes (via segregation and recombination during prophase I) have unique genomes compared to their parents.

29. C is correct.

30. A is correct. See explanations for questions **5** and **20**.

31. C is correct. *Spina bifida* is a bone abnormality that arises from the embryonic mesoderm germ layer. A lesion to the mesoderm would likely also affect the development of other structures based on different types of connective tissue (e.g. blood, blood vessels, muscles and connective tissue of various organs). Thus, this lesion affects the development of both blood vessels and muscles.

I: intestinal epithelium develops from the endoderm.

II: skin, hair and the nervous system develop from the ectoderm.

32. B is correct. Homeotic genes are genes involved in developmental patterns and sequences. For example, homeotic genes are involved in determining where, when and how body segments develop. Alterations in these genes cause changes in patterns of body parts and structures, sometimes resulting in dramatic effects.

33. D is correct. The mesoderm forms muscles, blood, bone, reproductive organs and kidneys.

34. E is correct.

35. A is correct.

36. D is correct. The ectoderm cells are determined but not differentiated, because they have the potential to develop into more than one type of tissue, but not any type. In the experiment, development of the ectoderm cells was influenced (i.e. induced) by changing their location within the developing embryo.

The underlying mesoderm has already differentiated because it released the molecular inducers as very specific signals to the overlying ectoderm (which is still undifferentiated).

A: the location of ectoderm cells is important because their development (i.e. differentiation into specific tissue/structures) is induced by the underlying mesoderm cells that send chemical substances (i.e. inducers) that are specific to the position of the mesoderm. Therefore, the mesoderm, not the ectoderm itself, determines which types of differentiated cells are produced at this stage of development.

B: differentiated fate of the cells has not yet been determined because the transplanted cells would develop into wing feathers instead of claws.

C: ectoderm cells cannot develop into any type of tissue because location influences the development of the transplanted ectoderm cells.

37. D is correct. Recognition of the sperm by the vitelline envelope (i.e. membrane between the outer zona pellucid and the inner plasma membrane) triggers the cortical reaction, which transforms it into the hard *fertilization membrane* that serves as a physical barrier to other spermatozoa (i.e. *slow block to polyspermy*). The cortical reaction within the egg is analogous to the acrosomal reaction within the sperm.

38. C is correct.

39. E is correct. The mesoderm germ layer gives rise to many of the human body tissues.

A: intestinal mucosa is derived from the endoderm

B and C: nerve and skin are derived from the ectoderm.

D: lung epithelium is derived from the endoderm.

40. C is correct. Genomic imprinting is an epigenetic (i.e. heritable change) phenomenon by which certain genes can be expressed in a parent-of-origin-specific manner. Genomic imprinting is an epigenetic process that can involve DNA methylation and histone modulation to achieve single allelic gene expression without altering the original genetic sequence.

41. C is correct. An embryo requires a greater rate of translation using the ribosomes to create proteins by reading mRNA transcripts.

42. E is correct.

43. A is correct. Capacitation is the final step in the maturation of spermatozoa and is required for them to be competent to fertilize an oocyte. Capacitation involves the destabilization of the acrosomal sperm head membrane, allowing greater binding between sperm and oocyte by removing steroids (e.g. cholesterol) and non-covalently bound glycoproteins, which increases membrane fluidity and Ca^{2+} permeability. An influx of Ca^{2+} produces increased intracellular cAMP levels and an increase in sperm motility.

44. C is correct. The endoderm develops into the epithelial linings of the digestive and respiratory tracts, parts of the liver, thyroid, pancreas and lining of the bladder.

45. E is correct. Homeotic genes influence the development of specific structures in plants and animals, such as the Hox and ParaHox genes, which are important for segmentation. For example, homeotic genes are involved in determining where, when and how body segments develop in flies. Alterations in these genes cause changes in patterns of body parts, sometimes causing dramatic effects, such as legs growing in place of antennae or an extra set of wings.

B: loss-of-function mutations result in the gene product having less or no function. When the allele has a complete loss of function (i.e. null allele), it is called an amorphic (i.e. complete loss of gene function) mutation. Phenotypes associated with such mutations are most often recessive. Exceptions are when the organism is haploid, or in haploinsufficiency (i.e. diploid organism has only a single copy of the functional gene) when the reduced dosage of a normal gene product is not enough for a normal phenotype.

46. A is correct. The cells can assume several different fates and are not yet terminally differentiated.

B: gastrula's cells that can be influenced by their surroundings are competent.

C: ectoderm layer gives rise to the eye, among other structures.

D: cells can become other ectoderm tissue (e.g. gills).

47. B is correct. See explanation for question **23**.

48. A is correct.

49. B is correct. The ectoderm develops into the nervous system, the epidermis, the lens of the eye and the inner ear.

I: endoderm develops into the lining of the digestive tract, lungs, liver and pancreas.

III: mesoderm develops into the connective tissue, muscles, skeleton, circulatory system, gonads and kidneys.

50. A is correct. See explanation for question **26**.

51. C is correct. Genomic imprinting is an epigenetic (i.e. heritable changes in gene activity) process that involves DNA methylation and histone remodeling to achieve monoallelic (i.e. only one of the two alleles) gene expression without altering the genetic sequence within the genome. These epigenetic marks are established in the germline (i.e. cells that give rise to gametes of egg/sperm) and can be maintained through mitotic divisions.

52. E is correct. The amniotic fluid, surrounded by the amnion, is a liquid environment that surrounds the egg and protects it from shock. The embryonic membranes include: 1) chorion, which lines the inside of the shell and permits gas exchange; 2) allantois, which is a saclike structure developed from the digestive tract and functions in respiration, excretion and gas exchange with external environment; 3) amnion, which encloses the amniotic fluid that provides a watery environment for development and protection against shock; 4) yolk sac, which encloses the yolk and transfers food to the developing embryo.

53. A is correct. Mesoderm gives rise to the entire circulatory system, muscle and most of the other tissue between the gut and the skin (excluding the nervous system).

B: ectoderm gives rise to skin, the nervous system, retina, lens, etc.

C: differentiated endoderm has a restricted fate that does not include heart tissue.

D: endoderm gives rise to the inner lining of the gut.

54. C is correct.

55. E is correct.

56. B is correct. Blastulation begins when the morula develops a fluid-filled cavity (i.e. blastocoel) that, by the fourth day, becomes a hollow sphere of cells (i.e. blastula).

A: gastrula is the embryonic stage characterized by the presence of the three primary germ layers (e.g. endoderm, ectoderm and mesoderm), the blastocoel and the archenteron. The early gastrula is two-layered (i.e. ectoderm and endoderm) and shortly afterwards, a third layer (i.e. mesoderm) develops. Gastrulation is followed by organogenesis, when individual organs develop within the newly formed germ layers.

C: morula is the solid ball of cells from the early stages of cleavage in the zygote.

D: zygote is the (2N) cell formed by the fusion of two (1N) gametes (e.g. ovum and sperm).

57. B is correct. See explanation for question **23**.

58. C is correct.

Chapter 4.14: Animal Behavior

1. A is correct.

2. E is correct. Behavior is the internally coordinated responses (actions or inactions) of living organisms (individuals or groups) to internal and/or external stimuli. Behaviors can be either innate (instinct) or learned. Behavior can be regarded as any action of an organism that changes its relationship to its environment. Behavior provides outputs from the organism to the environment.

3. B is correct.

4. C is correct. Operant conditioning is a type of learning in which an individual's behavior is modified by its antecedents (i.e. before trained behavior) and consequences. It is distinguished from *classical conditioning* (or *respondent conditioning*) because operant conditioning deals with the reinforcement and punishment to change behavior. Operant behavior operates on the environment and is maintained by its antecedents and consequences, while classical conditioning deals with the conditioning of reflexive (reflex) behaviors which are also elicited by prior conditions.

A: see explanation for question **10**.

D: see explanation for question **8**.

5. A is correct.

6. D is correct. Innate behavior, also called instinct or inborn behavior, is the inherent inclination of a living organism towards a particular complex behavior. The simplest example of an instinctive behavior is a fixed action pattern, in which a very short to medium length sequence of actions, without variation, is carried out in response to a clearly defined stimulus. Any behavior is instinctive if it is performed without being based upon prior experience (that is, in the absence of learning), and is therefore an expression of innate biological factors.

7. B is correct.

8. D is correct. Classical conditioning is a learning that occurs when a conditioned stimulus is paired with an unconditioned stimulus. Classical conditioning differs from *operant conditioning* because behavior is strengthened or weakened, depending on its consequences (i.e. reward or punishment). Behaviors conditioned through a classical conditioning procedure are not maintained by consequences.

9. C is correct.

10. E is correct. Imprinting is a phase-sensitive learning (learning occurring at a particular age or a particular life stage) that is rapid and apparently independent of the consequences of behavior. Imprinting is believed to have a critical period.

11. A is correct.

12. D is correct. See explanation for question **8**.

13. C is correct.

14. B is correct. See explanation for question **10**.

15. E is correct.

16. A is correct. Circadian rhythms are any biological processes that tend to reoccur every 24 hours. These rhythms are driven by a circadian clock and have been widely observed in plants, animals, fungi and cyanobacteria.

17. D is correct.

18. C is correct. Kin selection is the evolutionary strategy that favors the reproductive success of an organism's relatives, even at a cost to the organism's own survival and reproduction. Kin altruism is altruistic behavior whose evolution is driven by kin selection. Kin selection is an instance of inclusive fitness, which combines the number of offspring produced with the number an individual can produce by supporting others, such as siblings. Examples include honey bees that don't reproduce but leave that function to their relatives, adoption of orphans within animal populations, caring for young by other related females.

19. E is correct.

20. B is correct. See explanation for question **4**.

21. E is correct.

22. A is correct. Many animal species have mate-selection rituals also referred to as "courtship". It can involve complicated dances or touching, vocalizations, or displays of beauty or fighting prowess. Most animal courtship occurs out of sight of humans, so it is often the least documented of animal behaviors.

23. D is correct.

24. B is correct. Insight learning is also known as faculty of reason or rationality. Insight learning is a kind of learning or problem solving that occurs suddenly through understanding the relationships of the different parts of a problem rather than through test and error. It involves a sudden realization distinct from cause-and-effect problem solving. Insight learning manifests as a spontaneous occurrence, and is a noteworthy phenomenon in the learning process.

25. C is correct.

26. D is correct. See explanation for question **6**.

27. E is correct.

28. C is correct. Nocturnality is an animal behavior characterized by activity during the night and sleeping during the day. Nocturnal creatures generally have highly developed senses of hearing. Such trait helps these animals avoid predators and communicate.

29. A is correct.

30. B is correct. See explanation for question **4**.

31. A is correct.

32. C is correct.

UNIT 5. EVOLUTION AND DIVERSITY

Chapter 5.1: Evolution, Natural Selection, Classification, Diversity

1. D is correct.

2. D is correct. Use the mnemonic: **D**arn **K**ing **P**hillip **C**ame **O**ver **F**or **G**ood **S**oup:

Domain → kingdom → phylum → class → order → family → genus → species.

The term *Canis lupus* indicates that the wolf's genus is *Canis* and species is *lupus*.

Family is a more inclusive category than either genus or species so any member of the species *lupus* must also be in the genus *Canis* of the Canidae family.

3. A is correct. Organisms are classified according to evolutionary relationships:

Kingdom → Phylum → Class → Order → Family → Genus → Species

The largest group (i.e. kingdom) is divided into smaller and smaller subdivisions. Each smaller group has more common specific characteristics. Of the answers, genus is the smallest subdivision, and organisms in the same genus would be more similar than organisms classified as the same family, order, class or kingdom.

4. E is correct.

5. A is correct. Chloroplast is one of the plant cell organelles. Chloroplasts are considered to have originated from cyanobacteria through endosymbiosis. The theory is that around a billion years ago, a eukaryotic cell engulfed a photosynthesizing cyanobacterium and it became a permanent resident in the cell because it escaped the phagocytic vacuole it was contained in. Chloroplasts contain their own DNA and ribosomes (similar to prokaryotic ribosomes), and undergo autosomal replication (i.e. replicate independently of the cell cycle) providing further evidence to support this theory.

Cyanobacteria obtain their energy through photosynthesis and are named after the color of the bacteria (i.e. blue). Although often called blue-green algae, that name is a misnomer because cyanobacteria are prokaryotes, while algae are eukaryotes. By producing oxygen as a by-product gas of photosynthesis, cyanobacteria are thought to have converted Earth's early reducing atmosphere into an oxidizing one, which dramatically changed the composition of life forms by stimulating biodiversity and leading to the near-extinction of oxygen-intolerant organisms. The endosymbiotic theory proposes that chloroplasts in plants, mitochondria in eukaryotes and eukaryotic algae evolved from cyanobacteria ancestors.

C: mitochondria evolved from free living prokaryotic heterotrophs that, similarly to chloroplasts, entered eukaryotic cells and established a symbiotic relationship with the host (theory known as endosymbiotic). Mitochondria contain circular DNA (similar in size and composition to prokaryotes), ribosomes (similar to prokaryotic ribosomes), unique proteins in the organelle membrane (similar in composition to prokaryotes), and replicate independently, providing further evidence to support this theory.

6. D is correct. Evolution is the set of long-term changes in a population gene pool caused by environmental selection pressures.

I: random mutation creates new alleles which can be selected for (or against) as phenotypic variation for natural selection (i.e. survival of the fittest).

II: reproductive isolation of a population is a key component for speciation. Geographically isolated populations often diverge to yield reproductive isolation leading to speciation.

III: speciation might also happen in a population with no specific extrinsic barrier to gene flow. For example, a population extends over a broad geographic range, and mating throughout the population is not random. Individuals in the far west would have zero chance of mating with individuals in the far east of the range. This results in reduced gene flow, but not total isolation. Such situation may or may not be sufficient to cause speciation. Speciation would also be promoted by different selective pressures at opposite ends of the range, which would alter gene frequencies in groups at different ends of the range so much that they would not be able to mate if they were reunited.

7. C is correct.

8. A is correct. A single class encompasses several orders.

Use the mnemonic: **D**arn **K**ing **P**hillip **C**ame **O**ver **F**or **G**ood **S**oup:

Domain → kingdom → phylum → class → **order** → family → genus → species.

9. E is correct. Prokaryotes are mostly unicellular organisms although a few such as mycobacterium have multicellular stages in their life cycles or create large colonies like cyanobacteria.

Eukaryotes are often multicellular and are typically much larger than prokaryotes. They have internal membranes and structures (i.e. organelles) and a cytoskeleton composed of microtubules, microfilaments and intermediate filaments, which play an important role in defining the cell's organization and shape.

10. D is correct.

11. A is correct. Urey and Miller demonstrated in 1953 that organic molecules may be created from inorganic molecules under the primordial earth conditions. They conducted an experiment that simulated the conditions thought at the time to be present on the early Earth, and tested for the occurrence of chemical origins of life. Urey-Miller did not prove the existence of life on earth, but the original experiments were able to show that there were actually well over 20 different amino acids produced in Miller's original experiments.

12. B is correct. Nature uses many methods for fertilization, development and care of offspring. One method is internal fertilization, internal development, and then much care given to the offspring. These organisms (e.g. humans and many mammals) produce few offspring, but a large percentage of the offspring reach adulthood.

At the other end of the spectrum would be external fertilization, external development and little or no care for offspring. These organisms (e.g. many species of fish) must produce large numbers of both sperm and eggs because relatively few sperm and eggs interact to produce a zygote. Without any physical protection from predators, few zygotes survive and millions of eggs and sperm must be released in order to perpetuate the species.

A: protective coloring might be useful for initial survival against predators, but it is not related to whether the young are cared for.

C: laying eggs in itself is not relevant to not caring for young (most birds care for their young).

D: is not relevant to the survival rate of the offspring.

13. E is correct. See explanation for question **14**.

14. A is correct. Homology is a relationship defined between structures or DNA derived from a common ancestor. Homologous traits of organisms are therefore explained by descent from a common ancestor. The opposite of homologous organs are analogous organs, which have similar functions in two taxa that were not present in the last common ancestor but rather evolved separately.

Homologous sequences are *orthologous* if they descended from the same ancestral sequence separated by a speciation event: when a species diverges into two separate species, the copies of a single gene in the two resulting species are said to be orthologous. Orthologs (i.e. orthologous genes) are genes in different species that originated by vertical descent from a single gene of the last common ancestor.

Orthology is strictly defined in terms of ancestry. Because the exact ancestry of genes in different organisms is difficult to ascertain due to gene duplication and genome rearrangement events, the strongest evidence that two similar genes are orthologous is usually found through phylogenetic analysis of the gene lineage. Orthologous genes often, but not always, have the same function.

15. C is correct. In mutualism, two species live in close association from which both species benefit.

A: parasitism involves a relationship between two organisms where one organism benefits while the other is harmed.

B: commensalism involves a relationship between two organisms where one organism benefits without affecting the other.

D: mimicry, in evolutionary biology, is the similarity of one species to another which protects one or both, and occurs when a group of organisms, the mimics, evolve to share common perceived characteristics with another group, the models. This similarity can be in appearance, behavior, sound, scent and location, with the mimics found in similar places to their models.

E: amensalism is the type of relationship where one species is inhibited or completely obliterated while the other is unaffected. An example is a sapling growing under the shadow of a mature tree. The mature tree can deprive the sapling of sunlight, rainwater and soil nutrients. Throughout the process the mature tree is unaffected. Additionally, if the sapling dies, the mature tree would gain nutrients from the decaying sapling. Since the nutrients become available due to the dead sapling's decomposition (as opposed to the living sapling), this would not be a case of parasitism.

16. C is correct. See explanation for question **14**.

D: these are analogous structures.

17. B is correct. Memorize the taxonomy of humans:

Domain: Eukaryota → Kingdom: Animalia → Phylum: Chordata → Subphylum: Vertebrata → Class: Mammalia → Order: Primates → Family: Hominidae → Genus: Homo → Species: *H. sapiens* → Subspecies: *H. s. sapiens*

18. D is correct. Disruptive selection is the shift in allele frequencies towards variants of either extreme, which leads to the creation of two subpopulations. These subpopulations do not favor the intermediate variants of the original population and grow more disparate over time, likely leading to speciation.

A: sexual selection gives an individual an advantage in finding a mate. Sexual selection often acts opposite to the effects of natural selection (e.g., a brighter plume may make a bird more vulnerable to predators, but gives it an advantage during reproduction). It has been hypothesized that sexual selection is the factor leading to sexual dimorphism (i.e. phenotypic differences between males and females of the same species).

B: stabilizing selection is a shift in phenotypes towards an intermediate by disfavoring variants at either extreme. Stabilizing selection reduces variation within the species and perpetuates similarities between generations.

C: directional selection is the population shift in allele frequencies towards variants of one extreme.

19. D is correct.

20. E is correct. Mutualism (i.e. symbiotic) is a class of relationship between two organisms where both organisms benefit. Soybeans depend on humans for their survival while humans are provided with food. Both species benefit from the relationship.

Commensalism is a class of relationship between two organisms where one organism benefits without affecting the other.

Mutualism (i.e. symbiotic) is a class of relationship between two organisms where both organisms benefit.

Parasitism is a class of relationship between two organisms where one benefits while the other is harmed.

21. D is correct. Ecological succession in a rocky barren area to a final climax community is: lichen → mosses → annual grasses → perennial grasses → shrubs → deciduous trees (e.g. thick shade trees such as oak and hemlock).

22. B is correct.

23. A is correct.

24. E is correct.

25. D is correct.

26. B is correct. Ants are not chordates.

C: tunicates are marine invertebrates and members of the subphylum Tunicata within Chordata (i.e. phylum includes all animals with dorsal nerve cords and notochords). Some tunicates live as solitary individuals, but others replicate by budding and become colonies (each unit known as a zooid).

Tunicates are marine filter feeders with a water-filled, sac-like body structure and two tubular openings, known as siphons, through which they draw in and expel water. During their respiration and feeding, they take in water through the incurrent siphon and expel the filtered water through the excurrent siphon. Most adult tunicates are sessile and are permanently attached to rocks or other hard surfaces on the ocean floor; others swim in the pelagic zone (i.e. open sea) as adults.

27. C is correct. Genetic drift is the change in allele frequency within a population due to random sampling. Offspring have the same alleles as the parents and chance determines whether a given individual survives and reproduces. The allele frequency of a population is the fraction of the same alleles. Genetic drift may cause allele variants to disappear and thereby reduce genetic variation.

28. E is correct. Echinoderms are the invertebrate predecessors of the chordates and include starfish, sea urchin and the sea cucumber. They are characterized by having a primitive vascular system known as the water vascular system. Adult echinoderms have radial symmetry, while the larvae are bilaterally symmetrical. They move with structures known as tube feet, have no backbone and are heterotrophic (i.e. do not synthesize their own food).

Crayfish are in the phylum Arthropoda which includes insects. Arthropoda are characterized by segmented bodies covered in a chitin exoskeleton and jointed appendages.

29. A is correct.

30. C is correct. The oldest known fossilized prokaryotes were laid down approximately 3.6 billion years ago, only about 1 billion years after the formation of the Earth's crust. Eukaryotes appear in the fossil record later, and may have formed from the aggregation of multiple prokaryote ancestors. The oldest known fossilized eukaryotes are about 1.7 billion years old. However, some genetic evidence suggests eukaryotes appeared as early as 3 billion years ago.

D: protists comprise a large and diverse group of eukaryotic microorganisms, which belong to the kingdom Protista. There have been attempts to remove the kingdom from the taxonomy, but it is still commonly used. The name Protista is also preferred by various organizations and institutions.

Besides their relatively simple levels of organization, protists do not have much in common. They are unicellular, or they are multicellular without specialized tissues. This simple cellular organization distinguishes protists from other eukaryotes (e.g. fungi, animals and plants). Protists live in almost any environment that contains liquid water.

Many protists, such as algae, are photosynthetic and are vital primary producers in ecosystems, particularly in the ocean as part of the plankton. Other protists include pathogenic species, such as the kinetoplastid *Trypanosoma brucei* which causes sleeping sickness, and species of the apicomplexan *Plasmodium* that causes malaria.

31. D is correct. The definition of a species involves organisms that can mate and produce viable offspring that are fertile (offspring that can give rise to additional offspring). Morphological or physical similarity between organisms is not sufficient for classification as the same species. Likewise, organisms that may appear quite different from each other (e.g., different breeds of dog) can be the same species because they can mate and produce fertile offspring.

A: organisms can be the same species and yet have different varieties of gene (i.e. alleles) at the same gene locus. For example, blue-eyed and green-eyed indviduals can mate and produce fertile offspring.

B: there is no relationship between the ability to produce viable, fertile offspring and blood type.

C: somatic cells are body cells and are not reproductive cells (i.e. gametes) nor are they haploid (i.e. 1N or monoploid). Somatic cells are diploid (i.e. 2N) and undergo mitosis for cell division. Gametes undergo meiosis for cell division.

32. C is correct.

33. E is correct.

34. C is correct.

35. B is correct. A single kingdom encompasses many phyla.

Use the mnemonic: **D**arn **K**ing **P**hillip **C**ame **O**ver **F**or **G**ood **S**oup:

Domain → kingdom → **phylum** → class → order → family → genus → species.

36. E is correct. Natural selection is the process by which mutations are selected for (or against) in the environment. If the resulting phenotype offers some degree of fitness, the genes are passed to the next generation.

A: natural selection includes selection pressures, but also needs a population with genetic variation (random mutations) to select the most fit organisms.

C: Darwin's theory depends on more than merely mutations. It is also based on over-reproduction, and these offspring are then selected for (or against) whether their genetic makeup is the most fit for the environment where the organism is located. These organisms that are most fit (for their environment) pass on their genes (i.e. gametes) to the next generation. The result is survival (greatest reproduction) of the fittest organisms within the environment.

D: Lamarck proposed that if some traits were used (e.g. stretching of a giraffe's neck), these acquired traits are passed to the next generation. Acquired characteristics (phenotypic changes) do not affect the genes and therefore are not passed to the next generation via the gametes (e.g. sperm and egg).

37. B is correct. Fewer copies of an allele magnify the effect of genetic drift, while the effects are smaller when there are many copies of an allele.

38. D is correct. See explanation for question **17**.

39. B is correct. Organisms of the phylum Chordata have a dorsal notochord, while members of the subphylum Vertebrata have a backbone. Nonvertebrate chordates include the amphioxus (lancelet) or the unrelated tunicate worm.

Lancelet (also known as amphioxus) is the modern representative of the subphylum Cephalochordata and is important in the study of zoology – it provides indications about the origins of the vertebrates. Lancelets serve as a comparison point for tracing how vertebrates have evolved and adapted. Although lancelets split from vertebrates more than 520 million years ago, their genomes hold clues about evolution, specifically how vertebrates have employed old genes for new functions. They are regarded as similar to the archetypal vertebrate form.

Vertebrates include the majority of the phylum Chordata with about 64,000 species. Vertebrates include the amphibians, reptiles (e.g. lizard – A), mammals and birds, as well as the jawless fish, bony fish, sharks and rays.

C and D: shark and lamprey eel are both vertebrates with cartilaginous skeletons.

40. D is correct.

41. A is correct.

42. B is correct.

43. B is correct. Life originated in an atmosphere with little or no oxygen. Free oxygen did not exist in the atmosphere until about 2.4 billion years ago during the Great Oxygenation Event (GOE). Cyanobacteria, which appeared about 200 million years before the GOE, began producing oxygen by photosynthesis. Before the GOE, any free oxygen they produced was chemically captured by dissolved iron or organic matter. The GOE was the point when these oxygen sinks became saturated and could not capture all of the oxygen that was produced by cyanobacterial photosynthesis. After the GOE, the excess free oxygen started to accumulate in the atmosphere.

Free oxygen is toxic to obligate anaerobic organisms, and the rising concentrations are believed to have wiped out most of the Earth's anaerobic inhabitants at the time. Cyanobacteria were therefore responsible for one of the most significant extinction events in Earth's history. Additionally the free oxygen reacted with the atmospheric methane (a greenhouse gas) reducing its concentration and thereby possibly triggering the longest snowball Earth episode. Free oxygen has been an important constituent of the atmosphere ever since. Periods with much oxygen in the atmosphere are associated with rapid development of animals. Today's atmosphere contains about 21% oxygen, which is high enough for the rapid development of animals.

44. A is correct. The notochord is a flexible rod-shaped structure found in embryos of all chordates. It is composed of cells derived from the mesoderm and defines the primitive axis of the embryo. In lower chordates, this chord remains throughout the life of the animal. In higher chordates (e.g. humans), the notochord exists only during embryonic development then disappears. If it persists throughout life, it functions as the main axial support of the body, while in most vertebrates it becomes the nucleus pulposus of the intervetebral disc. The notochord is found ventral to the neural tube. It forms during gastrulation and then induces the formation of the neural plate (during neurulation) to synchronize the development of the neural tube (precursor to the central nervous system).

B: notochord remains in the lower chordates, such as the amphioxus and the tunicate worm. It is not vestigial (not a genetically determined structure that lost its ancestral function) because it disappears.

D: echinoderms are the invertebrate predecessors of the chordates and include starfish, sea urchin, and the sea cucumber. Echinoderms do not possess a notochord.

E: the notochord is not part of the nervous system.

45. B is correct.

46. E is correct. Chordates are animals that, for at least some period of their life cycles, possess all of the following: a notochord, a dorsal neural tube, pharyngeal slits, an endostyle, and a post-anal tail. Taxonomically, the phylum includes the subphyla Vertebrata (e.g. mammals, fish, amphibians, reptiles, birds), Tunicata (e.g. salps and sea squirts) and Cephalochordata (e.g. lancelets). Members of the Vertebrata subphylum have the backbone, but it is not a shared characteristic of all chordates.

Chordates make up a phylum of creatures that share a bilateral body plan, and are defined by having at some stage in their lives all of the following:

- A notochord is a fairly stiff rod of cartilage that extends along the inside of the body. Among the vertebrate sub-group of chordates, the notochord develops into the spine, and in wholly aquatic species this helps the animal to swim by flexing its tail.

- A dorsal neural tube (in vertebrates, including fish) develops into the spinal cord, the main communications trunk of the nervous system.

- Pharyngeal slits, whereby the pharynx is the part of the throat immediately behind the mouth, in fish are modified to form gills, but in some other chordates they are part of a filter-feeding system that extracts particles of food from the water in which the animals live.

- An endostyle is a groove in the ventral wall of the pharynx. In filter-feeding species it produces mucus to gather food particles, which helps in transporting food to the esophagus. It also stores iodine and may be a precursor of the vertebrate thyroid gland.

- Post-anal tail is a muscular tail that extends backwards behind the anus.

47. D is correct. Mutualism is the way two organisms of different species exist in a relationship where each individual benefits (+/+ relationship). Mutualism is a type of symbiosis. Symbiosis is a broad category, defined to include relationships that are mutualistic, parasitic or commensal. Mutualism is only one *type*.

The body of most lichens is different from those of either the fungus or alga growing separately. The fungus surrounds the algal cells, often enclosing them within complex fungal tissues. The algal cells are photosynthetic and reduce atmospheric carbon dioxide into organic carbon sugars to feed both symbionts. Both partners gain water and mineral nutrients mainly from the atmosphere, through rain and dust. The fungal partner protects the alga by retaining water, serving as a larger capture area for mineral nutrients and, in some cases, provides minerals obtained from the substrate. If a cyanobacterium is present as another symbiont in addition to green alga (tripartite lichens), they can fix atmospheric nitrogen, complementing the activities of the green alga.

A: nematodes (i.e. roundworms) are unlike cnidarians or flatworms because nematodes have tubular digestive systems with openings at both ends. Nematodes are obligate aquatic organisms consisting of cylindrical unsegmented worms. They are either free-living saprophytes or parasites.

B: bread mold is a saprophyte because it is an organism that feeds on dead and decaying material.

C: tapeworms are parasites because they extract nutrition (they benefit) from the host while the host is depleted (host is harmed) of the nutrients. This is a +/– relationship.

E: epiphytes are plants that live on the branches of other plants and receive greater exposure to sunlight than they would normally be able to. This is an example of commensalism and a +/0 relationship.

48. E is correct.

49. C is correct. D: the first living organism had nothing else to eat. Even when there were millions of living organisms, one organism would have to eat many others and that would have exhausted the supply.

50. A is correct. A climax community is a historic term that describes a biological community of plants, animals and fungi which, through the process of ecological succession (i.e. the development of vegetation in an area over time), had reached a steady state. This equilibrium was thought to occur because the climax community is composed of species best adapted to average conditions in that area. In the Midwest, the climax community is the grasslands, while in the northeastern part of the United States, it is the deciduous forest.

B: the climax community is especially dependent on the environment. Factors such as temperature, type of soil and amount of rainfall determine which organisms survive and thrive.

C: biomes are geographically and climatically defined as contiguous areas with similar conditions on the Earth, such as communities of plants, animals, and soil organisms, and are often referred to as ecosystems. Many species live in a biome.

D: pioneer species are hardy species that are the first to colonize previously disrupted or damaged ecosystems, beginning a chain of ecological succession that ultimately leads to a more biodiverse steady-state ecosystem. Pioneer species are therefore the earliest species in a biome and initially colonize it (e.g. lichen on rocks).

E: dead and decaying matter is a part of all communities but not the only aspect of any of them.

51. B is correct.

52. B is correct. Migratory birds nesting on different islands do not represent a type of behavioral, temporal or geographical isolation that leads to speciation. The birds are migratory and are not geographically isolated.

53. A is correct. Lamarck's theory of evolution proposed that new structures or changes in existing structures (e.g. giraffe's neck) arise because of the needs of the organism. The amount of change was thought to be based on the use (or disuse) of the structure. This theory was based on a false understanding of genetics because useful characteristics acquired in one generation were thought to be transmitted to the next generation. A classic example was that giraffes stretched their necks to reach leaves on higher branches of trees. It was proposed that offspring inherited the valuable trait of longer necks as a result of this prior use. This is false because only changes in the DNA of gametes (e.g. egg and sperm) can be inherited.

B: de Vries confirmed Mendel's observations with different plant species.

C: Mendel defined classical genetics through experiments with inheritable traits (e.g. seed color, wrinkled vs. round seeds, etc.) in pea plants and he described the principles of dominance, segregation, and independent assortment. At Mendel's time, the unit of inheritance (i.e. genes) was unknown.

D: Darwin's Theory of Natural Selection states that pressures in the environment select the organism most fit to survive and reproduce. His theory is based on six principles.

1) Overpopulation: more offspring are produced than can survive, and there is insufficient food, air, light and space to support the entire population;

2) Variations: offspring have differences (i.e. variations) in their characteristics as compared to the population. Darwin did not know the reason for their differences, but de Vries later suggested genetic mutations as the cause of variations. Some mutations are beneficial, but most are detrimental;

3) Competition: the developing population must compete for the necessities of life (e.g. food, air, light). Many young must die, while the number of adults remains about constant within generations;

4) Natural selection: some organisms have phenotypes (i.e. variations) that confer an advantage over other members;

5) Inheritance of variations: the individuals that survive and reproduce, and thus transmit these favorable phenotypes to their offspring. These favored alleles (i.e. variations of genes) eventually dominate the gene pool;

6) Evolution of a new species: over many generations of natural selection, the favorable genes/phenotypes are perpetuated in the species.

The accumulation of these favorable changes eventually results in significant changes of the gene pool for the evolution of a new species (i.e. organisms that reproduce and yield fertile offspring).

E: Morgan induced mutation in *Drosophila* and studied the inheritance of these mutations and also described sex-linked inheritance (i.e. Morgan unit for frequency of recombination within an organism).

54. D is correct.

55. A is correct. Imprinting is a process in which environmental patterns (or objects) presented to a developing organism during a *critical period* in early life becomes accepted as permanent aspects of their behavior. For example, a duckling passes through a critical period in which it learns that its mother is the first large moving object it sees.

B: instrumental conditioning involves reward or reinforcement to stimuli to establish a conditioning response.

C: discrimination involves the ability of the learning organism to respond differently to slightly different stimuli.

D: many animals secrete substances called pheromones (e.g. sex attractants) that influence the behavior of other members of the species.

56. E is correct.

57. C is correct. Asexual reproduction is more efficient than sexual reproduction in the number of offspring produced per reproduction, in the amount of energy invested in this process, and in the amount of time invested in the development of the young (both before and after birth). But asexual reproduction relies on genetic mutation for phenotypic variability to be passed to future generations, since it produces genetic clones of the parent. Sexual reproduction, on the other hand, involves the process of meiosis – two rounds of cell division and likelihood of cross-over (during prophase I).

A: new phenotype may be either disadvantageous *or* advantageous.

B: there is a much greater risk of mutation occurring with sexual reproduction (i.e. at the chromosomal level), which is known as chromosomal aberrations.

D: sexual reproduction requires more energy and time per progeny (i.e. offspring).

Species that reproduce sexually have a selective advantage because gene recombination occurs during fertilization (the fusion of two genetically unique nuclei – haploid sperm nucleus and haploid egg nucleus) to yield a genetically unique (2N) zygote. Fertilization of two unique gametes introduces phenotypic variability into a population – this variability may neither be a benefit nor detriment to the individual. If the new phenotype is advantageous, this individual is more likely to survive and pass his genes on to future generations consistent with natural selection (i.e. survival of the fittest).

58. D is correct.

59. E is correct. Nucleotides are made up of three components: sugar (either ribose for RNA or deoxyribose for DNA), a phosphate group and a base. The nitrogenous bases are guanine, adenine, cytosine, thymine (for DNA) or uracil (for RNA). Adenine and guanine are purines, while thymine, cytosine, and uracil are pyrimidines.

A: lipids (i.e. fats) are composed of a glycerol (i.e. 3-carbon chain) and (up to) three fatty acids.

B: monosaccharides are the monomers of carbohydrates.

C: nucleoside does not contain the phosphate group, only nitrogenous base and sugar, while nucleotides contain a phosphate group, nitrogenous base and sugar.

D: amino acids are the monomers for proteins that do not contain sugar, phosphate or bases.

60. A is correct.

61. D is correct. Sponges are animals of the Porifera (pore-bearing) phylum. They are multicellular, heterotrophic and lack cell walls, like other animals. However, sponges lack tissues and organs and have no body symmetry, making them different from other animals. Heterotrophy refers to the ability to obtain food and energy by consuming other organic substances, rather than through sunlight or inorganic compounds.

62. A is correct. Heterotrophic organisms obtain nutrients and energy through the consumption of other organic substances, whereas autotrophic organisms synthesize food from inorganic substances in the surrounding environment, such as light or chemical energy.

63. B is correct. See explanation for question **114**.

64. C is correct. Arthropods are invertebrate animals with an exoskeleton, segmented body and jointed appendages. These include insects, arachnids, myriapods and crustaceans.

65. A is correct. The presence of embryos indicates that the organisms these fossils belong to reproduce sexually, because embryos develop from zygotes, the single cell that results from fertilizing the female egg cell with a male sperm cell. The development of an embryo involves two distinct reproductive cells fusing together, characteristic of sexual reproduction.

66. D is correct. The Cambrian Explosion, which occurred around 542 million years ago, was the rapid appearance of most major animal phyla. This was accompanied by major diversification of other organisms as shown by the fossil record. Before about 580 million years ago, most organisms were simple, composed of individual cells occasionally organized into colonies. Over the following 70 or 80 million years, the rate of evolution accelerated by an order of magnitude, and diversity of life began to become similar to today's.

67. E is correct. Animal cells do not contain cell walls, making them distinct from plant cells. Another distinct feature of plant cells is the presence of chloroplasts in the structure.

68. D is correct. Extracellular digestion allows for the digestion of food outside of the cell, which permits the ingestion of large pieces of food. Intracellular digestion is limited to particles small enough to be taken into the individual cells. More complex animals typically have a digestive tract with enzymes that act on the food material.

69. C is correct. Annelids are a large phylum of segmented worms with each segment having the same set of organs.

A coelom is a cavity lined by a mesoderm-derived epithelium. Organs formed inside a coelom can grow, freely move and develop independently of the body wall while protected by fluid cushions.

Organisms can be placed into one of three groups based upon their type of body cavity:

Coelomates are animals that have a fluid filled body cavity (i.e. coelom) with a complete lining (peritoneum derived from the mesoderm) that allows organs to be attached to each other and suspended in a particular order while still being able to move freely within the cavity. Coelomates include most bilateral animals, including all the vertebrates.

Pseudocoelomates are animals that have a pseudocoelom that is a fully functional body cavity (e.g. roundworm). Mesoderm-derived tissue only partly lines the fluid filled body cavity. Although organs are held in place loosely, they are not as well organized as in coelomates. All pseudocoelomates are protostomes, but not all protostomes are pseudocoelomates.

Acoelomates are animals that have no body cavity at all (i.e. flatworms). Semi-solid mesodermal tissues between the gut and body wall hold their organs in place.

70. B is correct. According to embryological studies, the most ancient chordates were closely related to the ancestors of echinoderms, which are marine animals such as sea stars and sea urchins. Sea anemones are a group of water-dwelling, predatory animals.

71. C is correct. Echinoderms are marine animals such as sea stars, sea urchins and sand dollars, which are characterized by radial (usually five-point) symmetry. Most echinoderms can regenerate tissue, organs and limbs, and reproduce asexually.

72. A is correct. Cnidaria is a phylum of animals that contains over 10,000 species found exclusively in aquatic and mostly marine environments. Their distinguishing feature is cnidocytes, specialized cells that they use mainly for capturing prey. Their bodies consist of a non-living jelly-like substance, sandwiched between two layers of epithelium that are mostly one cell thick. Cnidaria have two basic body forms: swimming medusae and sessile polyps, both of which are radially symmetrical with mouths surrounded by tentacles that bear cnidocytes. Both forms have a single orifice and body cavity that are used for digestion and respiration. Many cnidarian species produce colonies that are single organisms composed of medusa-like or polyp-like zooids, or both.

B: echinoderms are a phylum of marine animals, and adults have radial (usually five-point) symmetry (e.g. sea urchins, sand dollars, starfish and sea cucumbers).

73. C is correct. Chordates are animals that have a notochord (stiff rod of cartilage that extends inside the body), dorsal neural tube, and pharyngeal slits, which are openings of the pharynx that allow aquatic organisms to extract oxygen from water and excrete carbon dioxide (i.e., gills).

74. E is correct. A hollow nerve cord is a single hollow tract of nervous tissue that constitutes the central nervous system of chordates.

75. B is correct. Cephalization is an evolutionary trend, in which nervous tissue becomes concentrated toward one end of an organism over many generations. This process eventually produces a head region with sensory organs. The process of cephalization is intrinsically connected with a change in symmetry. It accompanied the shift to bilateral symmetry in flatworms. In addition to a concentration of sense organs, all animals from annelids on also place the mouth in the head region. This process is also tied to the development of an anterior brain in the chordates from the notochord. A notable exception to the trend of cephalization throughout evolutionary advancement is phylum Echinodermata, which, despite having a bilateral ancestor, developed into a pentaradial animal (5 point symmetry) with no concentrated neural ganglia or sensory head region. However, some echinoderms have developed bilateral symmetry secondarily.

76. E is correct. Deuterostomes differ from protostomes in their embryonic development; in deuterostomes, the first opening (i.e., blastopore) becomes the anus, whereas in protostomes it becomes the mouth. Chordates, echinoderms and hemichordates are all deuterostomes.

77. A is correct. Pharyngeal pouches are filter-feeding organs found in fish that develop into gills, used for respiration.

78. D is correct. Corals are marine invertebrates (phylum Cnidaria) typically living in compact colonies of many identical individual *polyps*. The group includes the reef builders of tropical oceans that secrete calcium carbonate to form a hard skeleton.

A coral "head" is a colony of myriad genetically identical polyps, whereby each polyp is a spineless animal typically only a few millimeters in diameter and a few centimeters in length. Tentacles surround a central mouth opening and an exoskeleton is excreted near the base. Over many generations, the colony creates a large skeleton structure. Individual heads grow by asexual reproduction of polyps. Corals also breed sexually by spawning: polyps of the same species release gametes simultaneously over a period of one to several nights cycling around a full moon.

79. B is correct. The first groups of amphibians developed from lobe-finned fish around 370 million years ago during the Devonian period. Lobe-finned fish had multi-jointed leg-like fins that allowed them to crawl along the bottom of the sea. Their bony fins eventually evolved into limbs, becoming the ancestors to all tetrapods, including amphibians, reptiles, birds and mammals.

80. E is correct. The notochord is a flexible rod that has an essential role in vertebrate development as a major skeletal element of developing embryo. It is most closely related to cartilage, serving as the axial skeleton of the embryo until other elements form.

81. D is correct. See explanations for questions **44** and **46**.

82. C is correct. Earthworms are part of the Annelid phylum. The pseudocoelom is a fluid-filled body cavity between the body wall and intestine of certain invertebrates such as roundworms and hookworms. All other statements are true.

83. C is correct. Notochords are similar to cartilage in their consistency, made of type II collagen. This makes them both soft and flexible. The notochord is a flexible rod between the nerve cord and the digestive track.

84. B is correct. See explanation for question **72**.

85. A is correct. Chordates share several key features, such as the presence of a notochord, a dorsal hollow nerve cord, pharyngeal slits and a post-anal tail. Of the characteristics visible on the outside of a cat, a tail that extends beyond the anus makes it a chordate.

86. B is correct. The vertebral column is the backbone or spine, a segmented series of bones separated by intervertebral discs.

87. E is correct. Ovoviviparous fish (e.g. guppies and angel sharks) have eggs that develop inside the mother's body after internal fertilization, whereby each embryo develops within its own egg. The yolk is used for nourishment and the embryo receives little or no nourishment directly from the mother.

Viviparous fish are the species where the mother retains the eggs and nourishes the embryos. Viviparous fish have a structure analogous to the placenta (i.e. as in mammals), connecting the mother's blood supply with the embryo.

88. C is correct. One of the defining characteristics of a chordate is the presence of a notochord. Not all chordates are vertebrates, have paired appendages or have backbones. Chordates can only be deuterostomes.

89. A is correct.

90. B is correct. See explanation for question **99**.

91. D is correct. The evolution of fish began with the appearance of jawless fish, followed by bony fish with a jaw, and lastly fish with leg-like fins, which allowed for movement along the sea bed, eventually evolving into limbs of tetrapods.

92. E is correct.

93. A is correct. See explanation for question **52** of chapter *4.13 Development*.

94. C is correct. During the course of evolution, primates needed to be able to walk on two legs to free up the hands. Hands were then put to use in creating weapons and other tools to gather food and protect the young.

95. E is correct. External fertilization refers to the process of male sperm fertilizing a female egg outside of the female's body, seen in amphibians and fish. Mammals, which are further down in the evolution of vertebrate groups, use internal fertilization for reproduction.

96. A is correct. Prehensile refers to an appendage or organ that has adapted for holding or grasping.

97. B is correct. Eyes that face forward allow two fields of vision to overlap, which enables the perception of depth. As primates moved into trees to escape predators, they needed to navigate tree branches, meaning evolution favored good depth perception. Animals with eyes that do not face forward most likely lacked the ability to judge the location of tree branches due to a flawed perception of depth.

98. D is correct. All mammals are endothermic (warm-blooded) vertebrates with hair on their bodies and the ability to feed babies with milk. In the embryonic development of vertebrates (e.g., mammals), pharyngeal pouches form that develop into essential structures such as the ear drum and thymus gland.

99. E is correct. Hominoids are also called *apes*, but the term "ape" is used in several different meanings. It has been used as a synonym for "monkey" or for any tailless primate with a humanlike appearance.

Hominines are a subfamily of Hominidae that includes humans, gorillas, chimpanzees, bonobos and includes all hominids that arose after the split from orangutans. The Homininae cladogram has three main branches, which lead to gorillas, chimpanzees and bonobos, and humans.

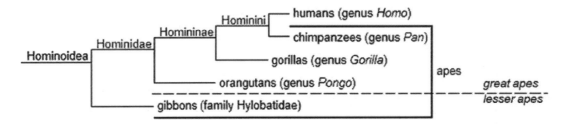

100. D is correct. Sexual reproduction increases fitness and allows for modifications in traits, which can help adapt to new living conditions in a changing environment. Sexual reproduction adds to the diversity of the offspring. All other answer choices characterize asexual reproduction.

101. C is correct. Cartilaginous fishes have a skeleton made of cartilage, rather than bone. Sharks, skates, rays and sturgeons have cartilaginous skeletons, while hagfish only have cartilaginous skulls.

102. A is correct. Old World monkeys are a group of primates in the clade of Catarrhin. Old World monkeys are native to Africa and Asia today, inhabiting environments that range from tropical rain forest to savanna, shrubland and mountainous terrain. Old World monkeys include many of the familiar species of nonhuman primates (e.g. baboons and macaques). Old World monkeys are medium to large in size, and range from arboreal (i.e. locomotion in trees) to fully terrestrial (e.g. baboons).

New World monkeys are the five families of primates that are found in Central and South America and portions of Mexico. New World monkeys are small to mid-sized primates. New World monkeys differ slightly from Old World monkeys in several aspects. The most prominent phenotypic distinction is the nose, which is the feature used most commonly to distinguish between the two groups. New World monkeys have noses that are flatter than the narrow noses of the Old World monkeys and have side-facing nostrils. New World monkeys are the only monkeys with prehensile tails, compared with the shorter, non-grasping tails of the Old World monkeys.

103. C is correct. A reptile's body temperature is determined by the temperature of its surrounding environment, meaning it could fluctuate more often than endothermic (warm-blooded) animals. Birds are endothermic, so their body temperature remains constant and is warmer than that of reptiles.

104. D is correct. Earthworms are soft-bodied segmented worms with nephridia, which are paired organs in each segment of the worm. Nephridia remove nitrogenous waste.

105. E is correct. Monotremes are mammals that lay eggs instead of giving birth to live young like marsupials and placental mammals. The only surviving species of monotremes are all indigenous to Australia and New Guinea, though evidence suggests that they were once more widespread. The existing monotreme species are the platypus and four species of echidnas (i.e. spiny anteaters).

106. B is correct.

107. D is correct. Lemurs are a clade of strepsirrhine primates, characterized by a moist nose tip. Lemurs are primarily nocturnal animals and are found on the island of Madagascar.

108. E is correct. Birds have one of the most complex respiratory systems of all animal groups. Upon inhalation, 75% of the fresh air bypasses the lungs and flows directly into a posterior air sac that extends from the lungs and connects with air spaces in the bones and fills them with air. The other 25% of the air goes directly into the lungs.

When the bird exhales, the used air flows out of the lung and the stored fresh air from the posterior air sac is simultaneously forced into the lungs. Therefore, a bird's lungs receive a constant supply of fresh air during both inhalation and exhalation.

Sound production is achieved using the syrinx, a muscular chamber incorporating multiple tympanic membranes which diverges from the lower end of the trachea. The trachea being elongated in some species increases the volume of vocalizations and the perception of the bird's size.

109. D is correct. The grasping hands of primates (i.e., having opposable thumbs) are an adaptation to life in the trees. The thumbs help primates grasp branches (and other objects) firmly.

110. B is correct. Chordates share several key features, such as the presence of a notochord, a dorsal hollow nerve cord, pharyngeal slits and a post-anal tail. Chordates can be both vertebrates and invertebrates. In vertebrate chordates, the notochord is replaced by a vertebral column.

111. D is correct. Arboreal locomotion is the movement of animals in trees. In habitats where trees are present, some animals have evolved to move in them. Animals may only scale trees occasionally or become exclusively arboreal.

112. B is correct. Radial symmetry is present in organisms with symmetry around a central axis. This is present in starfish or sea urchins, whose body parts extend equally outward from a center point.

113. A is correct. Australopithecus afarensis is an extinct species that shows hip, knee and foot morphology distinctive to bipedalism. Footprints belonging to this species shows that Australopithecus afarensis walked with an upright posture and a strong heel strike, one of the earliest evidence of bipedalism.

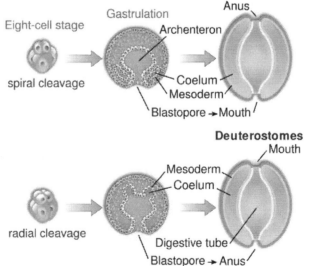

114. C is correct. Deuterostome are distinguished by their embryonic development, whereby the first opening (i.e. blastopore) becomes the anus, while in protostomes it becomes the mouth. The deuterostome mouth develops at the opposite end of the embryo from the blastopore and a digestive tract develops in the middle connecting the two.

Humans are deuterostomes, while all invertebrates are protostomes.

115. E is correct. Amniotic eggs are laid on land by certain reptiles, birds and mammals. They are different from anamniotic eggs, which are typically laid in water (by fishes and amphibians).

116. A is correct.

117. E is correct. *Homo habilis* (also known as *Australopithecus habilis*) is a species of the *Hominini* tribe, which lived from approximately 2.33 to 1.44 million years ago. While there has been controversy regarding its placement in the genus *Homo* rather than the genus *Australopithecus*, its brain size has been shown to be in the range 550-687 cm^3, rather than 363-600 cm^3 as formerly thought. These more recent findings about the brain size support its traditional placement in the genus *Homo* and therefore the theory that *Homo habilis* is indeed the common ancestor.

118. A is correct. The gluteus muscles of the pelvis are important to propulsion and stability while walking. Bipedal humans are different than quadrupedal apes in the lateral orientation of the ilium, which constitutes the bowl-shaped pelvis.

119. C is correct. Homo sapiens evolved after the Homo erectus and are what is now known as the modern man.

120. B is correct. Neanderthals are very closely related to modern humans with differences based upon DNA by only 0.3%, but twice the greatest DNA difference among contemporary humans. Genetic evidence suggests that Neanderthals contributed to the DNA of anatomically modern humans, probably through interbreeding between 80,000 and 30,000 years ago. Remains left by Neanderthals include bones and stone tools. Recently, evidence suggests that Neanderthals practiced burial behavior and intentionally buried their dead.

121. B is correct. Birds are the only chordates that have feathers. Other chordates such as mammals have fur or hair. Feathers are the single most important characteristic that separates birds from other living animals.

122. B is correct. A double-loop circulatory system is one in which blood flows through the heart twice. Pulmonary circulation involves blood flowing between the heart and lungs, and systemic circulation is blood movement from the heart to the rest of the body and back. Lungs are essential for a double-loop circulatory system.

123. A is correct. B: cephalization is an evolutionary trend, whereby nervous tissue (over many generations) becomes concentrated toward one end of an organism to produce a head region with sensory organs.

To access the online SAT tests at a special pricing visit:
http://SAT.Sterling-Prep.com/bookowner.htm

We want to hear from you

Your feedback is important to us because we strive to provide the highest quality prep materials. If you have any questions, comments or suggestions, email us, so we can incorporate your feedback into future editions.

Customer Satisfaction Guarantee

If you have any concerns about this book, including printing issues, contact us and we will resolve any issues to your satisfaction.

info@sterling-prep.com

81873839R10315

Made in the USA
San Bernardino, CA
12 July 2018